HOW THE WORLD WORKS

NOAM CHOMSKY

INTERVIEWED BY DAVID BARSAMIAN

EDITED BY ARTHUR NAIMAN

HAMISH HAMILTON
an imprint of
PENGUIN BOOKS

HAMISH HAMILTON

Published by the Penguin Group
Penguin Books Ltd, 80 Strand, London wc2r 0rl, England
Penguin Group (USA) Inc., 375 Hudson Street, New York, New York 10014, USA
Penguin Group (Canada), 90 Eglinton Avenue East, Suite 700, Toronto, Ontario, Canada m4p 2y3
(a division of Pearson Penguin Canada Inc.)
Penguin Ireland, 25 St Stephen's Green, Dublin 2, Ireland (a division of Penguin Books Ltd)
Penguin Group (Australia), 250 Camberwell Road, Camberwell, Victoria 3124, Australia
(a division of Pearson Australia Group Pty Ltd)
Penguin Books India Pvt Ltd, 11 Community Centre,
Panchsheel Park, New Delhi – 110 017, India
Penguin Group (NZ), 67 Apollo Drive, Rosedale, Auckland 0632, New Zealand
(a division of Pearson New Zealand Ltd)
Penguin Books (South Africa) (Pty) Ltd, Block D, Rosebank Office Park,
181 Jan Smuts Avenue, Parktown North, Gauteng 2193, South Africa

Penguin Books Ltd, Registered Offices: 80 Strand, London wc2r 0rl, England

www.penguin.com

First published in the United States of America by Soft Skull Press 2011
First published in Great Britain by Hamish Hamilton 2012
002

Printed in Great Britain by Clays Ltd, St Ives plc

A CIP catalogue record for this book is available from the British Library

isbn: 978-0-241-14538-8

www.greenpenguin.co.uk

Penguin Books is committed to a sustainable
future for our business, our readers and our planet.
This book is made from Forest Stewardship
Council™ certified paper.

ALWAYS LEARNING   PEARSON

TABLE OF CONTENTS

ABOUT THE AUTHOR

Noam Chomsky has long been the most cited living author; on the all-time list, he's eighth (after Marx, Lenin, Shakespeare, Aristotle, the Bible, Plato and Freud). Lionized abroad, he's by far the most important social critic in the world, but his political ideas are marginalized here in the United States. The modern-day equivalent of an Old Testament prophet, he's truly a prophet without honor in his own land.

The *New York Times* may grudgingly admit that he's "arguably the most important intellectual alive," but they do it in the context of deploring his politics. He's a media star in other countries and attracts standing-room-only audiences wherever he speaks here, but his appearances on American television are few and far between. The acceptable range of opinion stops long before it gets to him.

Yet the accuracy of his insights and his analyses is uncanny. In one of the classic books collected here, originally published in 1994, he warned: "In 1970, about 90% of international capital was used for trade and long-term investment—more or less productive things—and 10% for speculation. By 1990, those figures had reversed."

We know where things went from there; it probably got to 99.9% speculation before it all came crashing down. We're paying now for not heeding him then (not that you and I had much control over it).

Here's what he said back in the 1990s about money lent to Third World goons, long before Western nations and international lenders like the World Bank and the IMF [the International Monetary Fund] began forgiving such loans:

"As happened almost everywhere in the Third World, Brazil's generals, their cronies and the super-rich borrowed huge amounts of money and sent much of it abroad. The need to pay off that debt is a stranglehold that prevents Brazil from doing anything to solve its problems; it's what limits social spending and equitable, sustainable development.

"But if I borrow money and send it to a Swiss bank, and then can't pay my creditors, is that your problem or mine? The people in the slums didn't borrow the money, nor did the landless workers.

"In my view, it's no more the debt of 90% of the people of Brazil than it is the man in the moon's. Let the people who *borrowed* the money pay it back."

Fortunately, Brazil has advanced quite a bit from the sorry state it was in back then—thanks in no small part to Chomsky's efforts on its behalf.

Avram Noam Chomsky was born December 7, 1928 in Philadelphia. His father William was a famous Hebrew scholar and Noam spent time on a kibbutz in his early twenties. The father of three children, he lost his wife Carol in 2008, after almost sixty years of marriage.

Since 1955 he's taught philosophy and linguistics—a field his theories have revolutionized—at MIT, where he became a full professor at the age of 32. In addition to his paradigm-shifting linguistic theories, he's written many books on political issues and has received countless honors and awards (including 37 honorary degrees). A nonstop activist with a relentless lecture schedule, he does more than any three normal people, but feels he's never doing enough,

Chomsky is an electrifying speaker, and that's due solely to what he says, not to the unpretentious, straightforward way in which he says it (he consciously avoids rhetorical flourishes). Sharp as a razor in debate but warm and amiable in conversation, he's both the most moral and the most knowledgable person I've ever met.

I hope he lives to be 100. You should too. The world will be an emptier, lonelier and less just place without him.

Arthur Naiman

EDITOR'S NOTE

Made up of intensively edited speeches and interviews, this book offers something it's not easy to find—pure Chomsky, with every dazzling idea and penetrating insight intact, delivered in clear, accessible, reader-friendly prose.

The idea for what you have here—and for the Real Story Series in general—began when I heard a talk by Chomsky broadcast on KPFA radio in Berkeley. I was struck by how much more accessible I found his ideas when he spoke than when I read them, so I sent him a letter suggesting that I edit some of his talks together into a short, informal book.

He agreed and put me in touch with David Barsamian, who had been recording his speeches—and conducting recorded interviews with him—since 1986. (He's still at it).*

Working from transcripts of seven talks and interviews that David provided, I spent several months grouping together all the different things Chomsky said at various times on a wide range of topics. I then picked what I thought were the best turns of phrase, removed the repetition that's inevitable when discussing the same subjects on widely separated dates, put everything back together so that it sounded coherent, and sent the result to Chomsky for final corrections. He supplemented my compilation of what he said with new, written material that amplified and clarified it.

We produced four books using this method: *What Uncle Sam Really Wants; The Prosperous Few and the Restless*

---

* Barsamian's Alternative Radio series is heard on 150 stations worldwide. Alternative Radio offers mp3 audio downloads, CDs and transcripts of hundreds of Chomsky interviews and talks, as well as ones by many other progressive speakers.

- web: http://www.alternativeradio.org
- email: info@alternativeradio.org
- phone: 800 444 1977 • fax: 303 245 8292
- mail: P.O. Box 551, Boulder CO 80306 USA

Many; Secrets, Lies and Democracy; and *The Common Good.* There was apparently a lot of demand for this new, conversational Chomsky, because the four books sold a total of 593,000 copies.

At the beginning of our three-way collaboration, I wasn't sure of the best way to present this material, so in the first book, I eliminated David's questions entirely. They are, however, included in the other three books, where they appear in this typeface (as do phoned-in questions from radio listeners).

Each of the original books is presented here in the order in which it was published, with its own title page and table of contents. The index, however, was newly prepared for this volume and covers everything in it. (And I do mean *everything*—it isn't the sorry excuse for an index you find in most books.)

I've tried to define terms and identify people you may be unfamiliar with the first time they're mentioned. These explanatory notes appear in this typeface [inside square brackets]. If you run across a term or name you don't recognize, check its index entry for an italicized page number (which will usually be the first page on which the term appears).

Some of the original books contained supplemental material: notes, titles of other books by Chomsky, lists of organizations worth supporting, and so on. While Chomsky's ideas haven't gone out of date, those sections of the books mostly have, so they aren't included in this volume.

Although the talks and interviews compiled in this book originally took place in the 1990s (and some even in the late 1980s), I think you'll find Chomsky's take on things more insightful than virtually anything you hear on the airwaves or read in the papers today. His analyses are so deep and farsighted that they only seem to get more timely—and startling—with age. Read a few pages and see if you don't agree.

Arthur Naiman

WHAT UNCLE SAM REALLY WANTS

FIRST PUBLISHED 1992

15 US PRINTINGS

11 FOREIGN EDITIONS

244,000 COPIES IN PRINT

TABLE OF CONTENTS

THE MAIN GOALS OF US FOREIGN POLICY

Protecting our turf

Relations between the United States and other countries obviously go back to the origins of American history, but World War II was a real watershed, so let's begin there.

While most of our industrial rivals were either severely weakened or totally destroyed by the war, the United States benefited enormously from it. Our national territory was never under attack and American production more than tripled.

Even before the war, the US had been by far the leading industrial nation in the world—as it had been since the turn of the century. Now, however, we had literally 50% of the world's wealth and controlled both sides of both oceans. There'd never been a time in history when one power had had such overwhelming control of the world, or such overwhelming security.

The people who determine American policy were well aware that the US would emerge from WWII as the first global power in history, and during and after the war they were carefully planning how to shape the postwar world. Since this *is* an open society, we can read their plans, which were very frank and clear.

American planners—from those in the State Department to those on the Council on Foreign Relations (one major channel by which business leaders influence foreign policy)—agreed that the dominance of the United States had to be maintained. But there was a spectrum of opinion about how to do it.

At the hard-line extreme, you have documents like National Security Council Memorandum 68 (1950). NSC 68 developed the views of Secretary of State Dean Acheson and was written by Paul Nitze, who's still around (he was one of Reagan's arms-control negotiators). It called for a "roll-back strategy" that would "foster the seeds of destruction within the Soviet system" so that we could then negotiate a settlement on our terms "with the Soviet Union (or a successor state or states)."

The policies recommended by NSC 68 would require "sacrifice and discipline" in the United States—in other words, huge military expenditures and cutbacks on social services. It would also be necessary to overcome the "excess of tolerance" that allows too much domestic dissent.

These policies were, in fact, already being implemented. In 1949, US espionage in Eastern Europe had been turned over to a network run by Reinhard Gehlen, who had headed Nazi military intelligence on the Eastern Front. This network was one part of the US-Nazi alliance that quickly absorbed many of the worst criminals, extending to operations in Latin America and elsewhere.

These operations included a "secret army" under US-Nazi auspices that sought to provide agents and military supplies to armies that had been established by Hitler and which were still operating inside the Soviet Union and Eastern Europe through the early 1950s. (This is known in the US but considered insignificant—although it might raise a few eyebrows if the tables were turned and we discovered that, say, the Soviet Union had dropped agents and supplies to armies established by Hitler that were operating in the Rockies.)

The liberal extreme

NSC 68 is the hard-line extreme, and remember: the policies weren't just theoretical—many of them were actually being implemented. Now let's turn to the other extreme, to the doves. The leading dove was undoubtedly George Kennan, who headed the State Department planning staff until 1950, when he was replaced by Nitze—Kennan's office, incidentally, was responsible for the Gehlen network.

Kennan was one of the most intelligent and lucid of US planners, and a major figure in shaping the postwar world. His writings are an extremely interesting illustration of the dovish position. One document to look at if you want to understand your country is Policy Planning Study 23, written by Kennan for the State Department planning staff in 1948. Here's some of what it says:

We have about 50% of the world's wealth but only 6.3% of its population....In this situation, we cannot fail to be the object of envy and resentment. Our real task in the coming period is to devise a pattern of relationships which will permit us to maintain this position of disparity....To do so, we will have to dispense with all sentimentality and daydreaming; and our attention will have to be concentrated everywhere on our immediate national objectives....We should cease to talk about vague and...unreal objectives such as human rights, the raising of the living standards, and democratization. The day is not far off when we are going to have to deal in straight power concepts. The less we are then hampered by idealistic slogans, the better.

PPS 23 was, of course, a top-secret document. To pacify the public, it was necessary to trumpet the "idealistic slogans" (as is still being done constantly), but here planners were talking to one another.

Along the same lines, in a briefing for US ambassadors to Latin American countries in 1950, Kennan observed that a major concern of US foreign policy must be "the protection of our [i.e. Latin America's] raw materials." We must therefore combat a dangerous heresy which, US intelligence reported, was spreading through Latin America: "the idea that the government has direct responsibility for the welfare of the people."

US planners call that idea *Communism,* whatever the actual political views of the people advocating it. They can be church-based self-help groups or whatever, but if they support this heresy, they're Communists.

This point is also made clear in the public record. For example, a high-level study group in 1955 stated that the essential threat of the Communist powers (the real meaning of the term *Communism* in practice) is their refusal to fulfill their service role—that is, "to complement the industrial economies of the West."

Kennan went on to explain the means we have to use against our enemies who fall prey to this heresy:

The final answer might be an unpleasant one, but...we should not hesitate before police repression by the local government. This is not shameful since the Communists are essentially traitors....It is better to have a strong regime in power than a liberal government if it is indulgent and relaxed and penetrated by Communists.

12

Policies like these didn't begin with postwar liberals like Kennan. As Woodrow Wilson's Secretary of State had already pointed out 30 years earlier, the operative meaning of the Monroe Doctrine is that "the United States considers its own interests. The integrity of other American nations is an incident, not an end." Wilson, the great apostle of self-determination, agreed that the argument was "unanswerable," though it would be "impolitic" to present it publicly.

Wilson also acted on this thinking by, among other things, invading Haiti and the Dominican Republic, where his warriors murdered and destroyed, demolished the political system, left US corporations firmly in control, and set the stage for brutal and corrupt dictatorships.

The "Grand Area"

During World War II, study groups of the State Department and Council on Foreign Relations developed plans for the postwar world in terms of what they called the "Grand Area," which was to be subordinated to the needs of the American economy.

The Grand Area was to include the Western Hemisphere, Western Europe, the Far East, the former British Empire (which was being dismantled), the incomparable energy resources of the Middle East (which were then passing into American hands as we pushed out our rivals France and Britain), the rest of the Third World and, if possible, the entire globe. These plans were implemented, as opportunities allowed.

Every part of the new world order was assigned a specific function. The industrial countries were to be guided by the "great workshops"—Germany and Japan—who had demonstrated their prowess during the war and now would be working under US supervision.

The Third World was to "fulfill its major function as a source of raw materials and a market" for the industrial capitalist societies, as a 1949 State Department memo put it. It was to be "exploited" (in Kennan's words) for the reconstruction of Europe and Japan. (The references are to Southeast Asia and Africa, but the points are general.)

Kennan even suggested that Europe might get a psychological lift from the project of "exploiting" Africa. Naturally, no one suggested that Africa should exploit Europe for its reconstruction, perhaps also improving its state of mind. These declassified documents are read only by scholars, who apparently find nothing odd or jarring in all this.

The Vietnam War emerged from the need to ensure this service role. Vietnamese nationalists didn't want to accept it, so they had to be smashed. The threat wasn't that they were going to conquer anyone, but that they might set a dangerous example of national independence that would inspire other nations in the region.

The US government had two major roles to play. The first was to secure the far-flung domains of the Grand Area. That requires a very intimidating posture, to ensure that no one interferes with this task—which is one reason why there's been such a drive for nuclear weapons. The government's second role was to organize a public subsidy for high-technology industry. For various reasons, the method adopted has been military spending, in large part.

Free trade is fine for economics departments and newspaper editorials, but nobody in the corporate world or the government takes the doctrines seriously. The parts of the US economy that are able to compete internationally are primarily the state-subsidized ones: capital-intensive agriculture (*agribusiness*, as it's called), high-tech industry, pharmaceuticals, biotechnology, etc.

The same is true of other industrial societies. The US government has the public pay for research and development and provides, largely through the military, a state-guaranteed market for waste production. If something is marketable, the private sector takes it over. That system of public subsidy and private profit is what is called *free enterprise.*

Restoring the traditional order

Postwar planners like Kennan realized right off that it was going to be vital for the health of US corporations that the other Western industrial societies reconstruct from wartime damage so they could import US manufactured goods and provide investment opportunities (I'm counting Japan as part of the West, following the South African convention of treating Japanese as "honorary whites"). But it was crucial that these societies reconstruct in a very specific way.

The traditional, right-wing order had to be restored, with business dominant, labor split and weakened, and the burden of reconstruction placed squarely on the shoulders of the working classes and the poor.

The major thing that stood in the way of this was the antifascist resistance, so we suppressed it all over the world, often installing fascists and Nazi collaborators in its place. Sometimes that required extreme violence, but other times it was done by softer measures, like subverting elections and withholding desperately needed food. (This ought to be Chapter 1 in any honest history of the postwar period, but in fact it's seldom even discussed.)

The pattern was set in 1942, when President Roosevelt installed a French Admiral, Jean Darlan, as Governor-General of all of French North Africa. Darlan was a leading Nazi collaborator and the author of the antisemitic laws promulgated by the Vichy government (the Nazis' puppet regime in France).

But far more important was the first area of Europe liberated—southern Italy, where the US, following Churchill's advice, imposed a right-wing dictatorship headed by Fascist war hero Field Marshall Badoglio and the King, Victor Emmanuel III, who was also a Fascist collaborator.

US planners recognized that the "threat" in Europe was not Soviet aggression (which serious analysts, like Dwight Eisenhower, did not anticipate) but rather the antifascist resistance with its radical democratic ideals, and the political power and appeal of the local Communist parties. To prevent an economic collapse that would enhance their influence, and to rebuild Western Europe's state-capitalist economies, the US instituted the Marshall Plan (under which Europe was provided with more than $12 billion in loans and grants between 1948 and 1951, funds used to purchase a third of US exports to Europe in the peak year of 1949).

In Italy, a worker- and peasant-based movement, led by the Communist party, had held down six German divisions during the war, and liberated northern Italy. As US forces advanced through Italy, they dispersed this antifascist resistance and restored the basic structure of the prewar Fascist regime.

Italy has been one of the main areas of CIA subversion ever since the agency was founded. The CIA was concerned about Communists winning power legally in the crucial Italian elections of 1948. A lot of techniques were used, including restoring the Fascist police, breaking the unions and withholding food. But it wasn't clear that the Communist party could be defeated.

The very first National Security Council memorandum, NSC 1 (1948), specified a number of actions the US would take if the Communists won these elections. One planned response was armed intervention, by means of military aid for underground operations in Italy.

Some people, particularly George Kennan, advocated military action *before* the elections—he didn't want to take a chance. But others convinced him we could carry it off by subversion, which turned out to be correct.

In Greece, British troops entered after the Nazis had withdrawn. They imposed a corrupt regime that evoked renewed resistance, and Britain, in its postwar decline, was unable to maintain control. In 1947, the United States moved in, supporting a murderous war that resulted in about 160,000 deaths.

This war was complete with torture, political exile for tens of thousands of Greeks, what we called "re-education camps" for tens of thousands of others, and the destruction of unions and of any possibility of independent politics.

It placed Greece firmly in the hands of US investors and local businessmen, while much of the population had to emigrate in order to survive. The beneficiaries included Nazi collaborators, while the primary victims were the workers and the peasants of the Communist-led, anti-Nazi resistance.

Our successful defense of Greece against its own population was the model for the Vietnam War—as Adlai Stevenson explained to the United Nations in 1964. Reagan's advisors used exactly the same model in talking about Central America, and the pattern was followed many other places.

In Japan, Washington initiated the so-called "reverse course" of 1947 that terminated early steps towards democratization taken by General MacArthur's military administration. The reverse course suppressed the unions and other democratic forces and placed the country firmly in the hands of corporate elements that had backed Japanese fascism—a system of state and private power that still endures.

When US forces entered Korea in 1945, they dispersed the local popular government, consisting primarily of antifascists who had resisted the Japanese, and inaugurated a brutal repression, using Japanese fascist police and Koreans who had collaborated with them during the Japanese occupation. About 100,000 people were murdered in South Korea prior to what we call the Korean War, including 30,000 to 40,000 killed during the suppression of a peasant revolt in one small region, Cheju Island.

A fascist coup in Colombia, inspired by Franco's Spain, brought little protest from the US government; neither did a military coup in Venezuela, nor the restoration of an admirer of fascism in Panama. But the first democratic government in the history of Guatemala, which modeled itself on Roosevelt's New Deal, elicited bitter US antagonism.

In 1954, the CIA engineered a coup that turned Guatemala into a hell on earth. It's been kept that way ever since, with regular US intervention and support, particularly under Kennedy and Johnson.

One aspect of suppressing the antifascist resistance was the recruitment of war criminals like Klaus Barbie, an SS officer who had been the Gestapo chief of Lyon, France. There he earned his nickname: the Butcher of Lyon. Although he was responsible for many hideous crimes, the US Army put him in charge of spying on the French.

When Barbie was finally brought back to France in 1982 to be tried as a war criminal, his use as an agent was explained by Colonel (ret.) Eugene Kolb of the US Army Counterintelligence Corps: Barbie's "skills were badly needed....His activities had been directed against the underground French Communist party and the resistance," who were now targeted for repression by the American liberators.

Since the United States was picking up where the Nazis left off, it made perfect sense to employ specialists in antiresistance activities. Later on, when it became difficult or impossible to protect these useful folks in Europe, many of them (including Barbie) were spirited off to the United States or to Latin America, often with the help of the Vatican and fascist priests.

There they became military advisers to US-supported police states that were modeled, often quite openly, on the Third Reich. They also became drug dealers, weapons merchants, terrorists and educators—teaching Latin American peasants torture techniques devised by the Gestapo. Some of the Nazis' students ended up in Central America, thus establishing a direct link between the death camps and the death squads—all thanks to the postwar alliance between the US and the SS.

Our commitment to democracy

In one high-level document after another, US planners stated their view that the primary threat to the new US-led world order was Third World nationalism—sometimes called *ultranationalism:* "nationalistic regimes" that are responsive to "popular demand for immediate improvement in the low living standards of the masses" and production for domestic needs.

The planners' basic goals, repeated over and over again, were to prevent such "ultranationalist" regimes from ever taking power—or if, by some fluke, they did take power, to remove them and to install governments that favor private investment of domestic and foreign capital, production for export and the right to bring profits out of the country. (These goals are never challenged in the secret documents. If you're a US policy planner, they're sort of like the air you breathe.)

Opposition to democracy and social reform is never popular in the victim country. You can't get many of the people living there excited about it, except a small group connected with US businesses who are going to profit from it.

The United States expects to rely on force, and makes alliances with the military—"the least anti-American of any political group in Latin America," as the Kennedy planners put it—so they can be relied on to crush any indigenous popular groups that get out of hand.

The US has been willing to tolerate social reform—as in Costa Rica, for example—*only* when the rights of labor are suppressed and the climate for foreign investment is preserved. Because the Costa Rican government has always respected these two crucial imperatives, it's been allowed to play around with its reforms.

Another problem that's pointed to over and over again in these secret documents is the excessive liberalism of Third World countries. (That was particularly a problem in Latin America, where the governments weren't sufficiently committed to thought control and restrictions on travel, and where the legal systems were so deficient that they required evidence for the prosecution of crimes.)

This is a constant lament right through the Kennedy period (after that, the documentary record hasn't yet been declassified). The Kennedy liberals were adamant about the need to overcome democratic excesses that permitted "subversion"—by which, of course, they meant people thinking the wrong ideas.

The United States was not, however, lacking in compassion for the poor. For example, in the mid-1950s, our ambassador to Costa Rica recommended that the United Fruit Company, which basically ran Costa Rica, introduce "a few relatively simple and superficial human-interest frills for the workers that may have a large psychological effect."

Secretary of State John Foster Dulles agreed, telling President Eisenhower that to keep Latin Americans in line, "you have to pat them a little bit and make them think that you are fond of them."

Given all that, US policies in the Third World are easy to understand. We've consistently opposed democracy if its results can't be controlled. The problem with real democracies is that they're likely to fall prey to the heresy that governments should respond to the needs of their own population, instead of those of US investors.

A study of the inter-American system published by the Royal Institute of International Affairs in London concluded that, while the US pays lip service to democracy, the real commitment is to "private, capitalist enterprise." When the rights of investors are threatened, democracy has to go; if these rights are safeguarded, killers and torturers will do just fine.

Parliamentary governments were barred or overthrown, with US support and sometimes direct intervention, in Iran in 1953, in Guatemala in 1954 (and in 1963, when Kennedy backed a military coup to prevent the threat of a return to democracy), in the Dominican Republic in 1963 and 1965, in Brazil in 1964, in Chile in 1973 and often elsewhere. Our policies have been very much the same in El Salvador and in many other places across the globe.

The methods are not very pretty. What the US-run contra forces did in Nicaragua, or what our terrorist proxies do in El Salvador or Guatemala, isn't only ordinary killing. A major element is brutal, sadistic torture—beating infants against rocks, hanging women by their feet with their breasts cut off and the skin of their face peeled back so that they'll bleed to death, chopping people's heads off and putting them on stakes. The point is to crush independent nationalism and popular forces that might bring about meaningful democracy.

The threat of a good example

No country is exempt from this treatment, no matter how unimportant. In fact, it's the weakest, poorest countries that often arouse the greatest hysteria.

Take Laos in the 1960s, probably the poorest country in the world. Most of the people who lived there didn't even know there was such a thing as Laos; they just knew they had a little village and there was another little village nearby.

But as soon as a very low-level social revolution began to develop there, Washington subjected Laos to a murderous "secret bombing," virtually wiping out large settled areas in operations that, it was conceded, had nothing to do with the war the US was waging in South Vietnam.

Grenada has a hundred thousand people who produce a little nutmeg, and you could hardly find it on a map. But when Grenada began to undergo a mild social revolution, Washington quickly moved to destroy the threat.

From the Bolshevik Revolution of 1917 till the collapse of the Communist governments in Eastern Europe in the late 1980s, it was possible to justify every US attack as a defense against the Soviet threat. So when the United States invaded Grenada in 1983, the chairman of the Joint Chiefs of Staff explained that, in the event of a Soviet attack on Western Europe, a hostile Grenada could interdict oil supplies from the Caribbean to Western Europe and we wouldn't be able to defend our beleaguered allies. Now this sounds comical, but that kind of story helps mobilize public support for aggression, terror and subversion.

The attack against Nicaragua was justified by the claim that if we don't stop "them" there, they'll be pouring across the border at Harlingen, Texas—just two days' drive away. (For educated people, there were more sophisticated variants, just about as plausible.)

As far as American business is concerned, Nicaragua could disappear and nobody would notice. The same is true of El Salvador. But both have been subjected to murderous assaults by the US, at a cost of hundreds of thousands of lives and many billions of dollars.

There's a reason for that. The weaker and poorer a country is, the more dangerous it is *as an example.* If a tiny, poor country like Grenada can succeed in bringing about a better life for its people, some other place that has more resources will ask, "Why not us?"

This was even true in Indochina, which is pretty big and has some significant resources. Although Eisenhower and his advisers ranted a lot about the rice and tin and rubber, the real fear was that if the people of Indochina achieved independence and justice, the people of Thailand would emulate it, and if that worked, they'd try it in Malaya, and pretty soon Indonesia would pursue an independent path, and by then a significant part of the Grand Area would have been lost.

If you want a global system that's subordinated to the needs of US investors, you can't let pieces of it wander off. It's striking how clearly this is stated in the documentary record—even in the public record at times. Take Chile under Allende.

Chile is a fairly big place, with a lot of natural resources, but again, the United States wasn't going to collapse if Chile became independent. Why were we so concerned about it? According to Kissinger, Chile was a "virus" that would "infect" the region with effects all the way to Italy.

Despite 40 years of CIA subversion, Italy still has a labor movement. Seeing a social democratic government succeed in Chile would send the wrong message to Italian voters. Suppose they got funny ideas about taking control of their own country and revived the workers' movements the CIA undermined in the 1940s?

US planners from Secretary of State Dean Acheson in the late 1940s to the present have warned that "one rotten apple can spoil the barrel." The danger is that the "rot"—social and economic development—may spread.

This "rotten apple theory" is called the *domino theory* for public consumption. The version used to frighten the public has Ho Chi Minh getting in a canoe and landing in California, and so on. Maybe some US leaders believe this nonsense—it's possible—but rational planners certainly don't. They understand that the real threat is the "good example."

Sometimes the point is explained with great clarity. When the US was planning to overthrow Guatemalan democracy in 1954, a State Department official pointed out that "Guatemala has become an increasing threat to the stability of Honduras and El Salvador. Its agrarian reform is a powerful propaganda weapon; its broad social program of aiding the workers and peasants in a victorious struggle against the upper classes and large foreign enterprises has a strong appeal to the populations of Central American neighbors where similar conditions prevail."

In other words, what the US wants is "stability," meaning security for the "upper classes and large foreign enterprises." If that can be achieved with formal democratic devices, OK. If not, the "threat to stability" posed by a good example has to be destroyed before the virus infects others.

That's why even the tiniest speck poses such a threat, and may have to be crushed.

The three-sided world

From the early 1970s, the world has been drifting into what's called *tripolarism* or *trilateralism*—three major economic blocs that compete with each other. The first is a yen-based bloc with Japan as its center and the former Japanese colonies on the periphery.

Back in the thirties and forties, Japan called that The Greater East Asia Co-Prosperity Sphere. The conflict with the US arose from Japan's attempt to exercise the same kind of control there that the Western powers exercised in their own spheres. But after the war, we reconstructed the region for them. We then had no problem with Japan exploiting it—they just had to do it under our overarching power.

There's a lot of nonsense written about how the fact that Japan became a major competitor proves how honorable we are and how we built up our enemies. The actual policy options, however, were narrower. One was to restore Japan's empire, but now all under our control (this was the policy that was followed). The other option was to keep out of the region and allow Japan and the rest of Asia to follow their independent paths, excluded from the Grand Area of US control. That was unthinkable.

Furthermore, after WWII, Japan was not regarded as a possible competitor, even in the remote future. It was assumed that maybe somewhere down the road Japan would be able to produce knickknacks, but nothing beyond that. (There was a strong element of racism in this.) Japan recovered in large part because of the Korean War and then the Vietnam War, which stimulated Japanese production and brought Japan huge profits.

A few of the early postwar planners were more far-sighted, George Kennan among them. He proposed that the United States encourage Japan to industrialize, but with one limit: the US would control Japanese oil imports. Kennan said this would allow us "veto power" over Japan in case it ever got out of line. The US followed this advice, keeping control over Japan's oil supplies and refineries. As late as the early 1970s, Japan still controlled only about 10% of its own oil supplies.

That's one of the main reasons the United States has been so interested in Middle Eastern oil. We didn't need the oil for ourselves; until 1968, North America led the world in oil production. But we do want to keep our hands on this lever of world power, and make sure that the profits flow primarily to the US and Britain. That's one reason why we have maintained military bases in the Philippines. They're part of a global intervention system aimed at the Middle East to make sure indigenous forces there don't succumb to "ultranationalism."

The second major competitive bloc is based in Europe and is dominated by Germany. It's taking a big step forward with the consolidation of the European Common Market. Europe has a larger economy than the United States, a larger population and a better educated one. If it ever gets its act together and becomes an integrated power, the United States could become a second-class power. This is even more likely as German-led Europe takes the lead in restoring Eastern Europe to its traditional role as an economic colony, basically part of the Third World.

The third bloc is the US-dominated, dollar-based one. It was recently extended to incorporate Canada, our major trading partner, and will soon include Mexico and other parts of the hemisphere through "free trade agreements" designed primarily for the interests of US investors and their associates.

We've always assumed that Latin America belongs to us by right. As Henry Stimson [Secretary of War under FDR and Taft, Secretary of State under Hoover], once put it, it's "our little region over here, which never has bothered anybody." Securing the dollar-based bloc means that the drive to thwart independent development in Central America and the Caribbean will continue.

Unless you understand our struggles against our industrial rivals and the Third World, US foreign policy appears to be a series of random errors, inconsistencies and confusions. Actually, our leaders have succeeded rather well at their assigned chores, within the limits of feasibility.

Our Good Neighbor policy

How well have the precepts put forth by George Kennan been followed? How thoroughly have we put aside all concern for "vague and unreal objectives such as human rights, the raising of the living standards, and democratization"? I've already discussed our "commitment to democracy," but what about the other two issues?

Let's focus on Latin America, and begin by looking at human rights. A study by Lars Schoultz, the leading academic specialist on human rights there, shows that "US aid has tended to flow disproportionately to Latin American governments which torture their citizens." It has nothing to do with how much a country *needs* aid, only with its willingness to serve the interests of wealth and privilege. Broader studies by economist Edward [S.] Herman reveal a close correlation worldwide between torture and US aid, and also provide the explanation: both correlate independently with improving the climate for business operations. In comparison with that guiding moral principle, such matters as torture and butchery pale into insignificance.

How about raising living standards? That was supposedly addressed by President Kennedy's Alliance for Progress, but the kind of development imposed was oriented mostly towards the needs of US investors. It entrenched and extended the existing system in which Latin Americans are made to produce crops for export and to cut back on subsistence crops like corn and beans grown for local consumption. Under Alliance programs, for example, beef production increased while beef consumption declined.

This agro-export model of development usually produces an "economic miracle" where GNP goes up while much of the population starves. When you pursue such policies, popular opposition inevitably develops, which you then suppress with terror and torture.

(The use of terror is deeply ingrained in our character. Back in 1818, John Quincy Adams hailed the "salutary efficacy" of terror in dealing with "mingled hordes of lawless Indians and negroes." He wrote that to justify Andrew Jackson's rampages in Florida which virtually annihilated the native population and left the Spanish province under US control, much impressing Thomas Jefferson and others with his wisdom.)

The first step is to use the police. They're critical because they can detect discontent early and eliminate it before "major surgery" (as the planning documents call it) is necessary. If major surgery does become necessary, we rely on the army. When we can no longer control the army of a Latin American country—particularly one in the Caribbean-Central American region—it's time to overthrow the government.

Countries that have attempted to reverse the pattern, such as Guatemala under the democratic capitalist governments of Arévalo and Arbenz, or the Dominican Republic under the democratic capitalist regime of Bosch, became the target of US hostility and violence.

The second step is to use the military. The US has always tried to establish relations with the military in foreign countries, because that's one of the ways to overthrow a government that has gotten out of hand. That's how the basis was laid for military coups in Indonesia in 1965 and in Chile in 1973. Before the coups, we were very hostile to the Chilean and Indonesian governments, but we continued to send them arms. Keep good relations with the right officers and they overthrow the government *for* you. The same reasoning motivated the flow of US arms to Iran via Israel from the early 1980s, according to the high Israeli officials involved, facts well-known by 1982, long before there were any hostages.

During the Kennedy administration, the mission of the US-dominated Latin American military was shifted from "hemispheric defense" to "internal security" (which basically means war against your own population). That fateful decision led to "direct [US] complicity" in "the methods of Heinrich Himmler's extermination squads," in the retrospective judgment of Charles Maechling, who was in charge of counterinsurgency planning from 1961 to 1966.

The Kennedy administration prepared the way for the 1964 military coup in Brazil, helping to destroy Brazilian democracy, which was becoming too independent. The US gave enthusiastic support to the coup, while its military leaders instituted a neo-Nazi-style national security state with torture, repression, etc. That inspired a rash of similar developments in Argentina, Chile and all over the hemisphere, from the mid-sixties to the eighties—an extremely bloody period.

(I think, legally speaking, there's a very solid case for impeaching every American president since the Second World War. They've all been either outright war criminals or involved in serious war crimes.)

The military typically proceeds to create an economic disaster, often following the prescriptions of US advisers, and then decides to hand the problem over to civilians to administer. Overt military control is no longer necessary as new devices become available—for example, controls exercised through the International Monetary Fund (which, like the World Bank, lends Third World nations funds largely provided by the industrial powers).

In return for its loans, the IMF imposes "liberalization": an economy open to foreign penetration and control, sharp cutbacks in services to the general population, etc. These measures place power even more firmly in the hands of the wealthy classes and foreign investors ("stability") and reinforce the classic two-tiered societies of the Third World—the super-rich (and a relatively well-off professional class that serves them) and an enormous mass of impoverished, suffering people.

The indebtedness and economic chaos left by the military pretty much ensures that the IMF rules will be followed—unless popular forces attempt to enter the political arena, in which case the military may have to reinstate "stability."

Brazil is an instructive case. It is so well endowed with natural resources that it ought to be one of the richest countries in the world, and it also has high industrial development. But, thanks in good measure to the 1964 coup and the highly praised "economic miracle" that followed (not to speak of the torture, murder and other devices of "population control"), the situation for many Brazilians is now probably on a par with Ethiopia—vastly worse than in Eastern Europe, for example.

The Ministry of Education reports that over a third of the education budget goes to school meals, because most of the students in public schools either eat at school or not at all.

According to *South* magazine (a business magazine that reports on the Third World), Brazil has a higher infant mortality rate than Sri Lanka. A third of the population lives below the poverty line and "seven million abandoned children beg, steal and sniff glue on the streets. For scores of millions, home is a shack in a slum...or increasingly, a patch of ground under a bridge." That's Brazil, one of the naturally richest countries in the world.

The situation is similar throughout Latin America. Just in Central America, the number of people murdered by US-backed forces since the late 1970s comes to something like 200,000, as popular movements that sought democracy and social reform were decimated. These achievements qualify the US as an "inspiration for the triumph of democracy in our time," in the admiring words of the liberal *New Republic.* Tom Wolfe tells us the 1980s were "one of the great golden moments that humanity has ever experienced." As Stalin used to say, we're "dizzy with success."

The crucifixion of El Salvador

For many years, repression, torture and murder were carried on in El Salvador by dictators installed and supported by our government, a matter of no interest here. The story was virtually never covered. By the late 1970s, however, the US government began to be concerned about a couple of things.

One was that Somoza, the dictator of Nicaragua, was losing control. The US was losing a major base for its exercise of force in the region. A second danger was even more threatening. In El Salvador in the 1970s, there was a growth of what were called "popular organizations"—peasant associations, cooperatives, unions, church-based Bible study groups that evolved into self-help groups and the like. That raised the threat of democracy.

In February 1980, the Archbishop of El Salvador, Oscar Romero, sent a letter to President Carter in which he begged him not to send military aid to the junta that ran the country. He said such aid would be used to "sharpen injustice and repression against the people's organizations" which were struggling "for respect for their most basic human rights" (hardly news to Washington, needless to say).

A few weeks later, Archbishop Romero was assassinated while saying a mass. The neo-Nazi Roberto d'Aubuisson is generally assumed to be responsible for this assassination (among countless other atrocities). D'Aubuisson was "leader-for-life" of the ARENA party, which now governs El Salvador; members of the party, like current Salvadoran president Alfredo Cristiani, had to take a blood oath of loyalty to him.

Thousands of peasants and urban poor took part in a commemorative mass a decade later, along with many foreign bishops, but the US was notable by its absence. The Salvadoran Church formally proposed Romero for sainthood.

All of this passed with scarcely a mention in the country that funded and trained Romero's assassins. The *New York Times*, the "newspaper of record," published no editorial on the assassination when it occurred or in the years that followed, and no editorial or news report on the commemoration.

On March 7, 1980, two weeks before the assassination, a state of siege had been instituted in El Salvador, and the war against the population began in force (with continued US support and involvement). The first major attack was a big massacre at the Rio Sumpul, a coordinated military operation of the Honduran and Salvadoran armies in which at least 600 people were butchered. Infants were cut to pieces with machetes, and women were tortured and drowned. Pieces of bodies were found in the river for days afterwards. There were church observers, so the information came out immediately, but the mainstream US media didn't think it was worth reporting.

Peasants were the main victims of this war, along with labor organizers, students, priests or anyone suspected of working for the interests of the people. In Carter's last year, 1980, the death toll reached about 10,000, rising to about 13,000 for 1981 as the Reaganites took command. In October 1980, the new archbishop condemned the "war of extermination and genocide against a defenseless civilian population" waged by the security forces. Two months later they were hailed for their "valiant service alongside the people against subversion" by the favorite US "moderate," José Napoleón Duarte, as he was appointed civilian president of the junta.

The role of the "moderate" Duarte was to provide a fig leaf for the military rulers and ensure them a continuing flow of US funding after the armed forces had raped and murdered four churchwomen from the US. That had aroused some protest here; slaughtering Salvadorans is one thing, but raping and killing American nuns is a definite PR mistake. The media evaded and downplayed the story, following the lead of the Carter administration and its investigative commission.

The incoming Reaganites—notably Secretary of State Alexander Haig and UN Ambassador Jeane Kirkpatrick—went much further, seeking to justify the atrocity. But it was still deemed worthwhile to have a show trial a few years later, while exculpating the murderous junta—and, of course, the paymaster.

The independent newspapers in El Salvador, which might have reported these atrocities, had been destroyed. Although they were mainstream and pro-business, they were still too undisciplined for the military's taste. The problem was taken care of in 1980 to 1981, when the editor of one was murdered by the security forces; the other fled into exile. As usual, these events were considered too insignificant to merit more than a few words in US newspapers.

In November 1989, six Jesuit priests, their cook and her daughter, were murdered by the army. That same week, at least 28 other Salvadoran civilians were murdered, including the head of a major union, the leader of the organization of university women, nine members of an Indian farming cooperative and ten university students.

The news wires carried a story by AP correspondent Douglas Grant Mine, reporting how soldiers had entered a working-class neighborhood in the capital city of San Salvador, captured six men, added a 14-year-old boy for good measure, then lined them all up against a wall and shot them. They "were not priests or human rights campaigners," Mine wrote, "so their deaths have gone largely unnoticed"—as did his story.

The Jesuits were murdered by the Atlacatl Battalion, an elite unit created, trained and equipped by the United States. It was formed in March 1981, when fifteen specialists in counterinsurgency were sent to El Salvador from the US Army School of Special Forces. From the start, the Battalion was engaged in mass murder. A US trainer described its soldiers as "particularly ferocious....We've always had a hard time getting [them] to take prisoners instead of ears."

In December 1981, the Battalion took part in an operation in which over a thousand civilians were killed in an orgy of murder, rape and burning. Later it was involved in the bombing of villages and murder of hundreds of civilians by shooting, drowning and other methods. The vast majority of victims were women, children and the elderly.

The Atlacatl Battalion was being trained by US Special Forces shortly before murdering the Jesuits. This has been a pattern throughout the battalion's existence—some of its worst massacres have occurred when it was fresh from US training.

In the "fledgling democracy" that was El Salvador, teenagers as young as 13 were scooped up in sweeps of slums and refugee camps and forced to become soldiers. They were indoctrinated with rituals adopted from the Nazi SS, including brutalization and rape, to prepare them for killings that often had sexual and satanic overtones.

The nature of Salvadoran army training was described by a deserter who received political asylum in Texas in 1990, despite the State Department's request that he be sent back to El Salvador. (His name was withheld by the court to protect him from Salvadoran death squads.)

According to this deserter, draftees were made to kill dogs and vultures by biting their throats and twisting off their heads, and had to watch as soldiers tortured and killed suspected dissidents—tearing out their fingernails, cutting off their heads, chopping their bodies to pieces and playing with the dismembered arms for fun.

In another case, an admitted member of a Salvadoran death squad associated with the Atlacatl Battalion, César Vielman Joya Martínez, detailed the involvement of US advisers and the Salvadoran government in death-squad activity. The Bush administration has made every effort to silence him and ship him back to probable death in El Salvador, despite the pleas of human rights organizations and requests from Congress that his testimony be heard. (The treatment of the main witness to the assassination of the Jesuits was similar.)

The results of Salvadoran military training are graphically described in the Jesuit journal *America* by Daniel Santiago, a Catholic priest working in El Salvador. He tells of a peasant woman who returned home one day to find her three children, her mother and her sister sitting around a table, each with its own decapitated head placed carefully on the table in front of the body, the hands arranged on top "as if each body was stroking its own head."

The assassins, from the Salvadoran National Guard, had found it hard to keep the head of an 18-month-old baby in place, so they nailed the hands onto it. A large plastic bowl filled with blood was tastefully displayed in the center of the table.

According to Rev. Santiago, macabre scenes of this kind aren't uncommon.

> *People are not just killed by death squads in El Salvador—they are decapitated and then their heads are placed on pikes and used to dot the landscape. Men are not just disemboweled by the Salvadoran Treasury Police; their severed genitalia are stuffed into their mouths. Salvadoran women are not just raped by the National Guard; their wombs are cut from their bodies and used to cover their faces. It is not enough to kill children; they are dragged over barbed wire until the flesh falls from their bones, while parents are forced to watch.*

Rev. Santiago goes on to point out that violence of this sort greatly increased when the Church began forming peasant associations and self-help groups in an attempt to organize the poor.

By and large, our approach in El Salvador has been successful. The popular organizations have been decimated, just as Archbishop Romero predicted. Tens of thousands have been slaughtered and more than a million have become refugees. This is one of the most sordid episodes in US history—and it's got a lot of competition.

Teaching Nicaragua a lesson

It wasn't just El Salvador that was ignored by the mainstream US media during the 1970s. In the ten years prior to the overthrow of the Nicaraguan dictator Anastasio Somoza in 1979, US television—all networks—devoted exactly *one hour* to Nicaragua, and that was entirely on the Managua earthquake of 1972.

From 1960 through 1978, the *New York Times* had three editorials on Nicaragua. It's not that nothing was happening there—it's just that whatever was happening was unremarkable. Nicaragua was of no concern at all, as long as Somoza's tyrannical rule wasn't challenged.

When his rule *was* challenged, by the Sandinistas in the late 1970s, the US first tried to institute what was called "Somocismo [Somoza-ism] without Somoza"—that is, the whole corrupt system intact, but with somebody else at the top. That didn't work, so President Carter tried to maintain Somoza's National Guard as a base for US power.

The National Guard had always been remarkably brutal and sadistic. By June 1979, it was carrying out massive atrocities in the war against the Sandinistas, bombing residential neighborhoods in Managua, killing tens of thousands of people. At that point, the US ambassador sent a cable to the White House saying it would be "ill-advised" to tell the Guard to call off the bombing, because that might interfere with the policy of keeping them in power and the Sandinistas out.

Our ambassador to the Organization of American States also spoke in favor of "Somocismo without Somoza," but the OAS rejected the suggestion flat out. A few days later, Somoza flew off to Miami with what was left of the Nicaraguan national treasury, and the Guard collapsed.

The Carter administration flew Guard commanders out of the country in planes with Red Cross markings (a war crime) and began to reconstitute the Guard on Nicaragua's borders. They also used Argentina as a proxy. (At that time, Argentina was under the rule of neo-Nazi generals, but they took a little time off from torturing and murdering their own population to help reestablish the Guard—soon to be renamed the Contras, or "freedom fighters.")

Reagan used them to launch a large-scale terrorist war against Nicaragua, combined with economic warfare that was even more lethal. We also intimidated other countries so they wouldn't send aid either.

And yet, despite astronomical levels of military support, the United States failed to create a viable military force in Nicaragua. That's quite remarkable, if you think about it. No real guerillas anywhere in the world have ever had resources even remotely like what the United States gave the Contras. You could probably start a guerilla insurgency in mountain regions of the US with comparable funding.

Why did the US go to such lengths in Nicaragua? The international development organization Oxfam explained the real reasons, stating that, from its experience of working in 76 developing countries, "Nicaragua was...exceptional in the strength of that government's commitment...to improving the condition of the people and encouraging their active participation in the development process." Of the four Central American countries where Oxfam had a significant presence (El Salvador, Guatemala, Honduras and Nicaragua), only in Nicaragua was there a substantial effort to address inequities in land ownership and to extend health, educational and agricultural services to poor peasant families.

Other agencies told a similar story. In the early 1980s, the World Bank called its projects "extraordinarily successful in Nicaragua in some sectors, better than anywhere else in the world." In 1983, the Inter-American Development Bank concluded that "Nicaragua has made noteworthy progress in the social sector, which is laying the basis for long-term socio-economic development."

The success of the Sandinista reforms terrified US planners. They were aware that—as José Figueres, the father of Costa Rican democracy, put it—"for the first time, Nicaragua has a government that cares for its people." (Although Figueres was the leading democratic figure in Central America for forty years, his unacceptable insights into the real world were completely censored from the US media.)

The hatred that was elicited by the Sandinistas for trying to direct resources to the poor (and even succeeding at it) was truly wondrous to behold. Virtually all US policymakers shared it, and it reached a virtual frenzy.

Back in 1981, a State Department insider boasted that we would "turn Nicaragua into the Albania of Central America"— that is, poor, isolated and politically radical—so that the Sandinistas' dream of creating a new, more exemplary political model for Latin America would be in ruins.

George Shultz called the Sandinistas a "cancer, right here on our land mass," that has to be destroyed. At the other end of the political spectrum, leading Senate liberal Alan Cranston said that if it turned out not to be possible to destroy the Sandinistas, then we'd just have to let them "fester in [their] own juices."

So the US launched a three-fold attack against Nicaragua. First, we exerted extreme pressure to compel the World Bank and Inter-American Development Bank to terminate all projects and assistance.

Second, we launched the contra war along with an illegal economic war to terminate what Oxfam rightly called "the threat of a good example." The Contras' vicious terrorist attacks against "soft targets" under US orders did help, along with the boycott, to end any hope of economic development and social reform. US terror ensured that Nicaragua couldn't demobilize its army and divert its pitifully poor and limited resources to reconstructing the ruins that were left by the US-backed dictators and Reaganite crimes.

One of the most respected Central America correspondents, Julia Preston (who was then working for the *Boston Globe*), reported that "administration officials said they are content to see the Contras debilitate the Sandinistas by forcing them to divert scarce resources toward the war and away from social programs." That's crucial, since the social programs were at the heart of the good example that might have infected other countries in the region and eroded the American system of exploitation and robbery.

We even refused to send disaster relief. After the 1972 earthquake, the US sent an enormous amount of aid to Nicaragua, most of which was stolen by our buddy Somoza. In October 1988, an even worse natural disaster struck Nicaragua—Hurricane Joan. We didn't send a penny for that, because if we had, it would probably have gotten to the people, not just into the pockets of some rich thug. We also pressured our allies to send very little aid.

This devastating hurricane, with its welcome prospects of mass starvation and long-term ecological damage, reinforced our efforts. We wanted Nicaraguans to starve so we could accuse the

Sandinistas of economic mismanagement. Because they weren't under our control, Nicaraguans had to suffer and die.

Third, we used diplomatic fakery to crush Nicaragua. As Tony Avirgan wrote in the Costa Rican journal *Mesoamerica,* "the Sandinistas fell for a scam perpetrated by Costa Rican president Oscar Arias and the other Central American presidents, which cost them the February [1990] elections." For Nicaragua, the peace plan of August 1987 was a good deal, Avirgan wrote: they would move the scheduled national elections forward by a few months and allow international observation, as they had in 1984, "in exchange for having the Contras demobilized and the war brought to an end...." The Nicaraguan government did what it was required to do under the peace plan, but no one else paid the slightest attention to it.

Arias, the White House and Congress never had the slightest intention of implementing any aspect of the plan. The US virtually tripled CIA supply flights to the Contras. Within a couple of months, the peace plan was totally dead.

As the election campaign opened, the US made it clear that the embargo that was strangling the country and the Contra terror would continue if the Sandinistas won the election. You have to be some kind of Nazi or unreconstructed Stalinist to regard an election conducted under such conditions as free and fair—and south of the border, few succumbed to such delusions.

If anything like that were ever done by our *enemies*...I leave the media reaction to your imagination. The amazing part of it was that the Sandinistas still got 40% of the vote, while *New York Times* headlines proclaimed that Americans were "United in Joy" over this "Victory for US Fair Play."

US achievements in Central America in the past fifteen years are a major tragedy, not just because of the appalling human cost, but because a decade ago there were prospects for real progress towards meaningful democracy and meeting human needs, with early successes in El Salvador, Guatemala and Nicaragua. These efforts might have worked and might have taught useful lessons to others plagued with similar problems—which, of course, was exactly what US planners feared. The threat has been successfully aborted, perhaps forever.

Making Guatemala a killing field

There was one place in Central America that did get some US media coverage before the Sandinista revolution, and that was Guatemala. In 1944, a revolution there overthrew a vicious tyrant, leading to the establishment of a democratic government that basically modeled itself on Roosevelt's New Deal. In the ten-year democratic interlude that followed, there were the beginnings of successful independent economic development.

That caused virtual hysteria in Washington. Eisenhower and Dulles warned that the "self-defense and self-preservation" of the United States was at stake unless the virus was exterminated. US intelligence reports were very candid about the dangers posed by capitalist democracy in Guatemala. A CIA memorandum of 1952 described the situation in Guatemala as "adverse to US interests" because of the "Communist influence...based on militant advocacy of social reforms and nationalistic policies." The memo warned that Guatemala "has recently stepped-up substantially its support of Communist and anti-American activities in other Central American countries." One prime example cited was an alleged gift of $300,000 to José Figueres.

As mentioned above, José Figueres was the founder of Costa Rican democracy and a leading democratic figure in Central America. Although he cooperated enthusiastically with the CIA, had called the United States "the standard-bearer of our cause" and was regarded by the US ambassador to Costa Rica as "the best advertising agency that the United Fruit Company could find in Latin America," Figueres had an independent streak and was therefore not considered as reliable as Somoza or other gangsters in our employ.

In the political rhetoric of the United States, this made him possibly a "Communist." So if Guatemala gave him money to help him win an election, that showed Guatemala supported Communists.

Worse yet, the same CIA memorandum continued, the "radical and nationalist policies" of the democratic capitalist government, including the "persecution of foreign economic interests,

37

especially the United Fruit Company," had gained "the support or acquiescence of almost all Guatemalans." The government was proceeding "to mobilize the hitherto politically inert peasantry" while undermining the power of large landholders.

Furthermore, the 1944 revolution had aroused "a strong national movement to free Guatemala from the military dictatorship, social backwardness, and 'economic colonialism' which had been the pattern of the past," and "inspired the loyalty and conformed to the self-interest of most politically conscious Guatemalans." Things became still worse after a successful land reform began to threaten "stability" in neighboring countries where suffering people did not fail to take notice.

In short, the situation was pretty awful. So the CIA carried out a successful coup. Guatemala was turned into the slaughterhouse it remains today, with regular US intervention whenever things threaten to get out of line.

By the late 1970s, atrocities were again mounting beyond the terrible norm, eliciting verbal protests. And yet, contrary to what many people believe, military aid to Guatemala continued at virtually the same level under the Carter "human rights" administration. Our allies have been enlisted in the cause as well—notably Israel, which is regarded as a "strategic asset" in part because of its success in guiding state terrorism.

Under Reagan, support for near-genocide in Guatemala became positively ecstatic. The most extreme of the Guatemalan Hitlers we've backed there, Rios Montt, was lauded by Reagan as a man totally dedicated to democracy. In the early 1980s, Washington's friends slaughtered tens of thousands of Guatemalans, mostly Indians in the highlands, with countless others tortured and raped. Large regions were decimated.

In 1988, a newly opened Guatemalan newspaper called *La Epoca* was blown up by government terrorists. At the time, the media here were very much exercised over the fact that the US-funded journal in Nicaragua, *La Prensa*, which was openly calling for the overthrow of the government and supporting the US-run terrorist army, had been forced to miss a couple of issues due to a shortage of newsprint. That led to a torrent of outrage and abuse, in the *Washington Post* and elsewhere, about Sandinista totalitarianism.

On the other hand, the destruction of *La Epoca* aroused no interest whatsoever and was not reported here, although it was well-known to US journalists. Naturally the US media couldn't be expected to notice that US-funded security forces had silenced the one, tiny independent voice that had tried, a few weeks earlier, to speak up in Guatemala.

A year later, a journalist from *La Epoca*, Julio Godoy, who had fled after the bombing, went back to Guatemala for a brief visit. When he returned to the US, he contrasted the situation in Central America with that in Eastern Europe. Eastern Europeans are "luckier than Central Americans," Godoy wrote, because

> while the Moscow-imposed government in Prague would degrade and humiliate reformers, the Washington-made government in Guatemala would kill them. It still does, in a virtual genocide that has taken more than 150,000 victims [in what Amnesty International calls] "a government program of political murder.

The press either conforms or, as in the case of *La Epoca*, disappears.

"One is tempted to believe," Godoy continued, "that some people in the White House worship Aztec gods—with the offering of Central American blood." And he quoted a Western European diplomat who said: "As long as the Americans don't change their attitude towards the region, there's no space here for the truth or for hope."

The invasion of Panama

Panama has been traditionally controlled by its tiny European elite, less than 10% of the population. That changed in 1968, when Omar Torrijos, a populist general, led a coup that allowed the black and mestizo [mixed-race] poor to obtain at least a share of the power under his military dictatorship.

In 1981, Torrijos was killed in a plane crash. By 1983, the effective ruler was Manuel Noriega, a criminal who had been a cohort of Torrijos and US intelligence.

The US government knew that Noriega was involved in drug trafficking since at least 1972, when the Nixon administration considered assassinating him. But he stayed on the CIA payroll. In 1983, a US Senate committee concluded that Panama was a major center for the laundering of drug funds and drug trafficking.

The US government continued to value Noriega's services. In May 1986, the director of the Drug Enforcement Agency praised Noriega for his "vigorous anti-drug-trafficking policy." A year later, the director "welcomed our close association" with Noriega, while Attorney-General Edwin Meese stopped a US Justice Department investigation of Noriega's criminal activities. In August 1987, a Senate resolution condemning Noriega was opposed by Elliott Abrams, the State Department official in charge of US policy in Central America and Panama.

And yet, when Noriega was finally indicted in Miami in 1988, all the charges except one were related to activities that took place *before* 1984—back when he was our boy, helping with the US war against Nicaragua, stealing elections with US approval and generally serving US interests satisfactorily. It had nothing to do with suddenly discovering that he was a gangster and a drug-peddler—that was known all along.

It's all quite predictable, as study after study shows. A brutal tyrant crosses the line from admirable friend to "villain" and "scum" when he commits the crime of independence. One common mistake is to go beyond robbing the poor—which is just fine—and to start interfering with the privileged, eliciting opposition from business leaders.

By the mid 1980s, Noriega was guilty of these crimes. Among other things, he seems to have been dragging his feet about helping the US in the Contra war. His independence also threatened our interests in the Panama Canal. On January 1, 1990, most of the administration of the Canal was due to go over to Panama—in the year 2000, it goes completely to them. We had to make sure that Panama was in the hands of people we could control before that date.

Since we could no longer trust Noriega to do our bidding, he had to go. Washington imposed economic sanctions that virtually destroyed the economy, the main burden falling on the poor nonwhite majority. They too came to hate Noriega, not least because he was responsible for the economic warfare (which was illegal, if anyone cares) that was causing their children to starve.

Next, a military coup was tried, but failed. Then, in December 1989, the US celebrated the fall of the Berlin wall and the end of the Cold War by invading Panama outright, killing hundreds or perhaps thousands of civilians (no one knows, and few north of the Rio Grande care enough to inquire). This restored power to the rich white elite that had been displaced by the Torrijos coup—just in time to ensure a compliant government for the administrative changeover of the Canal on January 1, 1990 (as noted by the right-wing European press).

Throughout this process the US press followed Washington's lead, selecting villains in terms of current needs. Actions we'd formerly condoned became crimes. For example, in 1984, the Panamanian presidential election had been won by Arnulfo Arias. The election was stolen by Noriega, with considerable violence and fraud. But Noriega hadn't yet become disobedient. He was our man in Panama, and the Arias party was considered to have dangerous elements of "ultranationalism." The Reagan administration therefore applauded the violence and fraud, and sent Secretary of State George Shultz down to legitimate the stolen election and praise Noriega's version of "democracy" as a model for the errant Sandinistas.

The Washington-media alliance and the major journals refrained from criticizing the fraudulent elections, but dismissed as utterly worthless the Sandinistas' far more free and honest election in the same year—because it could not be controlled.

In May 1989, Noriega again stole an election, this time from a representative of the business opposition, Guillermo Endara. Noriega used less violence than in 1984, but the Reagan administration had given the signal that it had turned against Noriega. Following the predictable script, the press expressed outrage over his failure to meet our lofty democratic standards.

The press also began passionately denouncing human rights violations that previously didn't reach the threshold of their attention. By the time we invaded Panama in December 1989, the press had demonized Noriega, turning him into the worst monster since Attila the Hun. (It was basically a replay of the demonization of Qaddafi of Libya.) Ted Koppel was orating that "Noriega belongs to that special fraternity of international villains, men like Qaddafi, Idi Amin and the Ayatollah Khomeini, whom Americans just love to hate." Dan Rather placed him "at the top of the list of the world's drug thieves and scums." In fact, Noriega remained a very minor thug—exactly what he was when he was on the CIA payroll.

In 1988, for example, Americas Watch [a US-based human-rights monitoring organization] published a report on human rights in Panama, giving an unpleasant picture. But as their reports— and other inquiries—make clear, Noriega's human rights record was nothing remotely like that of other US clients in the region, and no worse than in the days when Noriega was still a favorite, following orders. Take Honduras, for example. Although it's not a murderous terrorist state like El Salvador or Guatemala, human rights abuses were probably worse there than in Panama. In fact, there's one CIA-trained battalion in Honduras that all by itself had carried out more atrocities than Noriega did.

Or consider US-backed dictators like Trujillo in the Dominican Republic, Somoza in Nicaragua, Marcos in the Philippines, Duvalier in Haiti and a host of Central American gangsters through the 1980s. They were all *much* more brutal than Noriega, but the United States supported them enthusiastically right through decades of horrifying atrocities—as long as the profits were flowing out of their countries and into the US. George Bush's administration continued to honor Mobutu, Ceausescu and Saddam Hussein, among others, all far worse criminals than Noriega. Suharto of Indonesia, arguably the worst killer of them all, remains a Washington-media "moderate."

In fact, at exactly the moment it invaded Panama because of its outrage over Noriega's professed abuses of human rights, the Bush administration announced new high-technology sales to China, noting that $300 million in business for US firms was at stake and that contacts had secretly resumed a few weeks after the Tiananmen Square massacre.

On the same day—the day Panama was invaded—the White House also announced plans (and implemented them shortly afterwards) to lift a ban on loans to Iraq. The State Department explained with a straight face that this was to achieve the "goal of increasing US exports and put us in a better position to deal with Iraq regarding its human rights record...." The Department continued with the pose as Bush rebuffed the Iraqi democratic opposition (bankers, professionals, etc.) and blocked congressional efforts to condemn the atrocious crimes of his old friend

Saddam Hussein. Compared to Bush's buddies in Baghdad and Beijing, Noriega looked like Mother Teresa.

After the invasion, Bush announced a billion dollars in aid to Panama. Of this, $400 million consisted of incentives for US business to export products to Panama, $150 million was to pay off bank loans and $65 million went to private sector loans and guarantees to US investors. In other words, about half the aid was a gift from the American taxpayer to American businesses.

The US put the bankers back in power after the invasion. Noriega's involvement in drug trafficking had been trivial compared to theirs. Drug trafficking there has always been conducted primarily by the banks—the banking system is virtually unregulated, so it's a natural outlet for criminal money. This has been the basis for Panama's highly artificial economy and remains so—possibly at a higher level—after the invasion. The Panamanian Defense Forces have also been reconstructed with basically the same officers.

In general, everything's pretty much the same, only now more reliable servants are in charge. (The same is true of Grenada, which has become a major center of drug money laundering since the US invasion. Nicaragua, too, has become a significant conduit for drugs to the US market, after Washington's victory in the 1990 election. The pattern is standard—as is the failure to notice it.)

Inoculating Southeast Asia

The US wars in Indochina fall into the same general pattern. By 1948, the State Department recognized quite clearly that the Viet Minh, the anti-French resistance led by Ho Chi Minh, was *the* national movement of Vietnam. But the Viet Minh did not cede control to the local oligarchy. It favored independent development and ignored the interests of foreign investors.

There was fear that the Viet Minh might succeed, in which case "the rot would spread" and the "virus" would "infect" the region, to adopt the language the planners used year after year after year. (Except for a few madmen and nitwits, none feared conquest—they were afraid of a positive example of successful development.)

What do you do when you have a virus? First you destroy it, then you inoculate potential victims, so that the disease does not spread. That's basically the US strategy in the Third World.

If possible, it's advisable to have the local military destroy the virus for you. If they can't, you have to move your own forces in. That's more costly, and it's ugly, but sometimes you have to do it. Vietnam was one of those places where we had to do it.

Right into the late 1960s, the US blocked all attempts at political settlement of the conflict, even those advanced by the Saigon generals. If there were a political settlement, there might be progress toward successful development outside of our influence—an unacceptable outcome.

Instead, we installed a typical Latin American-style terror state in South Vietnam, subverted the only free elections in the history of Laos because the wrong side won, and blocked elections in Vietnam because it was obvious the wrong side was going to win there too.

The Kennedy administration escalated the attack against South Vietnam from massive state terror to outright aggression. Johnson sent a huge expeditionary force to attack South Vietnam and expanded the war to all of Indochina. That destroyed the virus, all right—Indochina will be lucky if it recovers in a hundred years.

While the United States was extirpating the disease of independent development at its source in Vietnam, it also prevented its spread by supporting the Suharto takeover in Indonesia in 1965, backing the overthrow of Philippine democracy by Ferdinand Marcos in 1972, supporting martial law in South Korea and Thailand and so on.

Suharto's 1965 coup in Indonesia was particularly welcome to the West, because it destroyed the only mass-based political party there. That involved the slaughter, in a few months, of about 700,000 people, mostly landless peasants—"a gleam of light in Asia," as the leading thinker of the *New York Times*, James Reston, exulted, assuring his readers that the US had a hand in this triumph.

The West was very pleased to do business with Indonesia's new "moderate" leader, as the *Christian Science Monitor* described General Suharto, after he had washed some of the blood off his hands—meanwhile adding hundreds of thousands of corpses in East Timor and elsewhere. This spectacular mass murderer is "at heart benign," the respected *Economist* [a British newsweekly based in London] assures us—doubtless referring to his attitude towards Western corporations.

After the Vietnam war was ended in 1975, the major policy goal of the US has been to maximize repression and suffering in the countries that were devastated by our violence. The degree of the cruelty is quite astonishing.

When the Mennonites tried to send pencils to Cambodia, the State Department tried to stop them. When Oxfam tried to send ten solar pumps, the reaction was the same. The same was true when religious groups tried to send shovels to Laos to dig up some of the unexploded shells left by American bombing.

When India tried to send 100 water buffalo to Vietnam to replace the huge herds that were destroyed by the American attacks—and remember, in this primitive country, water buffalo mean fertilizer, tractors, survival—the United States threatened to cancel Food for Peace aid. (That's one Orwell would have appreciated.) No degree of cruelty is too great for Washington sadists. The educated classes know enough to look the other way.

In order to bleed Vietnam, we've supported the Khmer Rouge indirectly through our allies, China and Thailand. The Cambodians have to pay with their blood so we can make sure there isn't any recovery in Vietnam. The Vietnamese have to be punished for having resisted US violence.

Contrary to what virtually everyone—left or right—says, the United States achieved its major objectives in Indochina. Vietnam was demolished. There's no danger that successful development there will provide a model for other nations in the region.

Of course, it wasn't a total victory for the US. Our larger goal was to reincorporate Indochina into the US-dominated global system, and that has not yet been achieved.

But our basic goal—the crucial one, the one that really counted—was to destroy the virus, and we did achieve that. Vietnam is a basket case, and the US is doing what it can to keep it that way. In October 1991, the US once again overrode the strenuous objections of its allies in Europe and Japan, and renewed the embargo and sanctions against Vietnam. The

Third World must learn that no one dare raise their head. The global enforcer will persecute them relentlessly if they commit this unspeakable crime.

The Gulf War

The Gulf War illustrated the same guiding principles, as we see clearly if we lift the veil of propaganda.

When Iraq invaded Kuwait in August, 1990, the UN Security Council immediately condemned Iraq and imposed severe sanctions on it. Why was the UN response so prompt and so unprecedently firm? The US government-media alliance had a standard answer.

First, it told us that Iraq's aggression was a unique crime, and thus merited a uniquely harsh reaction. "America stands where it always has—against aggression, against those who would use force to replace the rule of law"—so we were informed by [the first] President Bush, the invader of Panama and the only head of state condemned by the World Court for the "unlawful use of force" (in the Court's ruling on the US attack against Nicaragua). The media and the educated classes dutifully repeated the lines spelled out for them by their Leader, collapsing in awe at the magnificence of his high principles.

Second, these same authorities proclaimed—in a litany—that the UN was now at last functioning as it was designed to do. They claimed that this was impossible before the end of the Cold War, when the UN was rendered ineffective by Soviet disruption and the shrill anti-Western rhetoric of the Third World.

Neither of these claims can withstand even a moment's scrutiny. The US wasn't upholding any high principle in the Gulf, nor was any other state. The reason for the unprecedented response to Saddam Hussein wasn't his brutal aggression—it was because he stepped on the wrong toes.

Saddam Hussein is a murderous gangster—exactly as he was before the Gulf War, when he was our friend and favored trading partner. His invasion of Kuwait was certainly an atrocity, but well within the range of many similar crimes conducted by the US and its allies, and nowhere near as terrible as some.

For example, Indonesia's invasion and annexation of East Timor reached near-genocidal proportions, thanks to the decisive support of the US and its allies. Perhaps one-fourth of its population of 700,000 was killed, a slaughter exceeding that of Pol Pot, relative to the population, in the same years.

Our ambassador to the UN at the time (and now Senator from New York), Daniel Moynihan, explained his achievement at the UN concerning East Timor: "The United States wished things to turn out as they did, and worked to bring this about. The Department of State desired that the United Nations prove utterly ineffective in whatever measures it undertook. This task was given to me, and I carried it forward with no inconsiderable success."

The Australian Foreign Minister justified his country's acquiescence to the invasion and annexation of East Timor (and Australia's participation with Indonesia in robbing Timor's rich oil reserves) by saying simply that "the world is a pretty unfair place, littered with examples of acquisition by force." When Iraq invaded Kuwait, however, his government issued a ringing declaration that "big countries cannot invade small neighbors and get away with it." No heights of cynicism trouble the equanimity of Western moralists.

As for the UN finally functioning as it was designed to, the facts are clear—but absolutely barred by the guardians of political correctness who control the means of expression with an iron hand. For many years, the UN has been blocked by the great powers, primarily the United States—not the Soviet Union or the Third World. Since 1970, the United States has vetoed *far* more Security Council resolutions than any other country (Britain is second, France a distant third and the Soviet Union fourth). Our record in the General Assembly is similar. And the "shrill, anti-Western rhetoric" of the Third World commonly turns out to be a call to observe international law, a pitifully weak barrier against the depredations of the powerful.

The UN was able to respond to Iraq's aggression because—for once—the United States *allowed* it to. The unprecedented severity of the UN sanctions was the result of intense US pressure and threats. The sanctions had an unusually good chance of working, both because of their harshness and because the usual sanctions-busters—the United States, Britain and France—would have abided by them for a change.

But even after allowing sanctions, the US immediately moved to close off the diplomatic option by dispatching a huge military force to the Gulf, joined by Britain and backed by the family dictatorships that rule the Gulf's oil states, with only nominal participation by others.

A smaller, deterrent force could have been kept in place long enough for the sanctions to have had a significant effect; an army of half a million couldn't. The purpose of the quick military build-up was to ward off the danger that Iraq might be forced out of Kuwait by peaceful means.

Why was a diplomatic resolution so unattractive? Within a few weeks after the invasion of Kuwait on August 2, the basic outlines for a possible political settlement were becoming clear. Security Council Resolution 660, calling for Iraq's withdrawal from Kuwait, also called for simultaneous negotiations of border issues. By mid-August, the National Security Council considered an Iraqi proposal to withdraw from Kuwait in that context.

There appear to have been two issues: first, Iraqi access to the Gulf, which would have entailed a lease or other control over two uninhabited mudflats assigned to Kuwait by Britain in its imperial settlement (which had left Iraq virtually landlocked); second, resolution of a dispute over an oil field that extended two miles into Kuwait over an unsettled border.

The US flatly rejected the proposal, or any negotiations. On August 22, without revealing these facts about the Iraqi initiative (which it apparently knew), the *New York Times* reported that the Bush administration was determined to block the "diplomatic track" for fear that it might "defuse the crisis" in very much this manner. (The basic facts were published a week later by the Long Island daily *Newsday,* but the media largely kept their silence.)

The last known offer before the bombing, released by US officials on January 2, 1991, called for total Iraqi withdrawal from Kuwait. There were no qualifications about borders, but the offer was made in the context of unspecified agreements on other "linked" issues: weapons of mass destruction in the region and the Israel-Arab conflict. The latter issues include Israel's illegal occupation of southern Lebanon, in violation of

Security Council Resolution 425 of March 1978, which called for its immediate and unconditional withdrawal from the territory it had invaded. The US response was that there would be no diplomacy. The media suppressed the facts, Newsday aside, while lauding Bush's high principles.

The US refused to consider the "linked" issues because it was opposed to diplomacy on all the "linked" issues. This had been made clear months before Iraq's invasion of Kuwait, when the US had rejected Iraq's offer of negotiations over weapons of mass destruction. In the offer, Iraq proposed to destroy all such chemical and biological weapons, if other countries in the region also destroyed their weapons of mass destruction.

Saddam Hussein was then Bush's friend and ally, so he received a response, which was instructive. Washington said it welcomed Iraq's proposal to destroy its own weapons, but didn't want this linked to "other issues or weapons systems."

There was no mention of the "other weapons systems," and there's a reason for that. Israel not only may have chemical and biological weapons—it's also the only country in the Mideast with nuclear weapons (probably about 200 of them). But "Israeli nuclear weapons" is a phrase that can't be written or uttered by any official US government source. That phrase would raise the question of why all aid to Israel is not illegal, since foreign aid legislation from 1977 bars funds to any country that secretly develops nuclear weapons.

Independent of Iraq's invasion, the US had also always blocked any "peace process" in the Middle East that included an international conference and recognition of a Palestinian right of self-determination. For 20 years, the US has been virtually alone in this stance. UN votes indicate the regular annual pattern; once again in December 1990, right in the midst of the Gulf crisis, the call for an international conference was approved by a vote of 144 to 2 (the US and Israel). This had nothing to do with Iraq vs. Kuwait.

The US also adamantly refused to allow a reversal of Iraq's aggression by the peaceful means prescribed by international law. Instead it preferred to avoid diplomacy and to restrict the conflict to the arena of violence, in which a superpower facing no deterrent is bound to prevail over a Third World adversary.

As already discussed, the US regularly carries out or supports aggression, even in cases far more criminal than Iraq's invasion of Kuwait. Only the most dedicated commissar can fail to understand these facts, or the fact that in the rare case when the US happens to oppose some illegal act by a client or ally, it's quite happy with "linkage."

Take the South African occupation of Namibia, declared illegal by the World Court and the UN in the 1960s. The US pursued "quiet diplomacy" and "constructive engagement" for years, brokering a settlement that gave South Africa ample reward (including Namibia's major port) for its aggression and atrocities, with "linkage" extending to the Caribbean and welcome benefits for international business interests.

The Cuban forces that had defended Namibia's neighbor Angola from South African attack were withdrawn. Much as in Nicaragua after the 1987 "peace accords," the US is continuing to support the terrorist army backed by the US and its allies (South Africa and Zaire) and is preparing the ground for a 1992 Nicaragua-style "democratic election," where people will go to the polls under the threat of economic strangulation and terrorist attack if they vote the wrong way.

Meanwhile, South Africa was looting and destroying Namibia, and using it as a base for violence against its neighbors. In the Reagan-Bush years (1980–88) alone, South African violence led to about $60 billion in damage and over a million and a half people killed in the neighboring countries (excluding Namibia and South Africa). But the commissar class was unable to see these facts, and hailed George Bush's amazing display of principle as he opposed "linkage"—when someone steps on our toes.

More generally, opposing "linkage" amounts to little more than rejecting diplomacy, which always involves broader issues. In the case of Kuwait, the US position was particularly flimsy. After Saddam Hussein stepped out of line, the Bush administration insisted that Iraq's capacity for aggression be eliminated (a correct position, in contrast to its earlier support for Saddam's aggression and atrocities) and called for a regional settlement guaranteeing security. Well, that's linkage. The simple fact is that the US feared that diplomacy might "defuse the crisis," and therefore blocked diplomacy "linkage" at every turn during the build-up to the war.

50

By refusing diplomacy, the US achieved its major goals in the Gulf. We were concerned that the incomparable energy resources of the Middle East remain under our control, and that the enormous profits they produce help support the economies of the US and its British client.

The US also reinforced its dominant position, and taught the lesson that the world is to be ruled by force. Those goals having been acheived, Washington proceeded to maintain "stability," barring any threat of democratic change in the Gulf tyrannies and lending tacit support to Saddam Hussein as he crushed the popular uprising of the Shi'ites in the South, a few miles from US lines, and then the Kurds in the North.

But the Bush administration had not yet succeeded in achieving what its spokesman at the *New York Times,* chief diplomatic correspondent Thomas Friedman, calls "the best of all worlds: an iron-fisted Iraqi junta without Saddam Hussein." This, Friedman writes, would be a return to the happy days when Saddam's "iron fist...held Iraq together, much to the satisfaction of the American allies Turkey and Saudi Arabia," not to speak of the boss in Washington. But the current situation in the Gulf reflects the priorities of the superpower that held all the cards, another truism that must remain invisible to the guardians of the faith.

The Iran/Contra cover-up

The major elements of the Iran/Contra story were well known long before the 1986 exposures, apart from one fact: that the sale of arms to Iran via Israel and the illegal Contra war run out of Ollie North's White House office were connected.

The shipment of arms to Iran through Israel didn't begin in 1985, when the congressional inquiry and the special prosecuter pick up the story. It began almost immediately after the fall of the Shah in 1979. By 1982, it was public knowledge that Israel was providing a large part of the arms for Iran—you could read it on the front page of the *New York Times.*

In February 1982, the main Israeli figures whose names later appeared in the Iran/Contra hearings appeared on BBC television [the British Broadcasting Company, Britain's national broadcasting service] and described how they had helped organize an arms flow to the Khomeini regime. In October 1982, the Israeli ambassador to the US stated publicly that Israel was sending arms to the Khomeini regime, "with the cooperation of the United States...at almost the highest level." The high Israeli officials involved also gave the reasons: to establish links with elements of the military in Iran who might overthrow the regime, restoring the arrangements that prevailed under the Shah— standard operating procedure.

As for the Contra war, the basic facts of the illegal North-CIA operations were known by 1985 (over a year before the story broke, when a US supply plane was shot down and a US agent, Eugene Hasenfus, was captured). The media simply chose to look the other way.

So what finally generated the Iran/Contra scandal? A moment came when it was just impossible to suppress it any longer. When Hasenfus was shot down in Nicaragua while flying arms to the Contras for the CIA, and the Lebanese press reported that the US National Security Adviser was handing out Bibles and chocolate cakes in Teheran, the story just couldn't be kept under wraps. After that, the connection between the two well-known stories emerged.

We then move to the next phase: damage control. That's what the follow-up was about. For more on all of this, see my *Fateful Triangle* (1983), *Turning the Tide* (1985), and *Culture of Terrorism* (1987).

The prospects for Eastern Europe

What was remarkable about the events in Eastern Europe in the 1980s was that the imperial power simply backed off. Not only did the USSR permit popular movements to function, it actually encouraged them. There are few historical precedents for that.

It didn't happen because the Soviets are nice guys—they were driven by internal necessities. But it *did* happen and, as a result, the popular movements in Eastern Europe didn't have to face anything remotely like what they would have faced on our turf. The journal of the Salvadoran Jesuits quite accurately pointed out that in their country Vaclav Havel (the former political prisoner who became president of Czechoslovakia) wouldn't have been put in jail—he might well have been hacked to pieces and left by the side of the road somewhere.

The USSR even apologized for its past use of violence, and this too was unprecedented. US newspapers concluded that, because the Russians admitted that the invasion of Afghanistan was a crime that violated international law, they were finally joining the civilized world. That's an interesting reaction. Imagine someone in the US media suggesting that maybe the United States ought to try to rise to the moral level of the Kremlin and admit that the attacks against Vietnam, Laos and Cambodia violated international law.

The one country in Eastern Europe where there was extensive violence as the tyrannies collapsed was the very one where the Soviets had the least amount of influence and where we had the most: Romania. Nicolae Ceausescu, the dictator of Romania, was given the royal treatment when he visited England. The US gave him favored nation treatment, trade advantages and the like.

Ceausescu was just as brutal and crazed then as he was later, but because he'd largely withdrawn from the Warsaw Pact and was following a somewhat independent course, we felt he was partially on our side in the international struggle. (We're in favor of independence as long as it's in *other* people's empires, not in our own.)

Elsewhere in Eastern Europe, the uprisings were remarkably peaceful. There was some repression, but historically, 1989 was unique. I can't think of another case that comes close to it.

I think the prospects are pretty dim for Eastern Europe. The West has a plan for it—they want to turn large parts of it into a new, easily exploitable part of the Third World.

There used to be a sort of colonial relationship between Western and Eastern Europe; in fact, the Russians' blocking of that relationship was one of the reasons for the Cold War. Now it's being reestablished and there's a serious conflict over who's going to win the race for robbery and exploitation. Is it going to be German-led Western Europe (currently in the lead) or Japan (waiting in the wings to see how good the profits look) or the United States (trying to get into the act)?

There are a lot of resources to be taken, and lots of cheap labor for assembly plants. But first we have to impose the capitalist model on them. We don't accept it for *ourselves*—but for the Third World, we insist on it. That's the IMF system. If we can get them to accept that, they'll be very easily exploitable, and will move toward their new role as a kind of Brazil or Mexico.

In many ways, Eastern Europe is more attractive to investors than Latin America. One reason is that the population is white and blue-eyed, and therefore easier to deal with for investors who come from deeply racist societies like Western Europe and the United States.

More significantly, Eastern Europe has much higher general health and educational standards than Latin America—which, except for isolated sectors of wealth and privilege, is a total disaster area. One of the few exceptions in this regard is Cuba, which does approach Western standards of health and literacy, but its prospects are very grim.

One reason for this disparity between Eastern Europe and Latin America is the vastly greater level of state terror in the latter after the Stalin years. A second reason is economic policy.

According to US intelligence, the Soviet Union poured about 80 billion dollars into Eastern Europe in the 1970s. The situation has been quite different in Latin America. Between 1982 and 1987, about $150 billion were transferred *from* Latin America to the West. The *New York Times* cites estimates that "hidden transactions" (including drug money, illegal profits, etc.) might be in the $700 billion range. The effects in Central America have been particularly awful, but the same is true throughout Latin America—there's rampant poverty, malnutrition, infant mortality, environmental destruction, state terror, and a collapse of living standards to the levels of decades ago.

The situation in Africa is even worse. The catastrophe of capitalism was particularly severe in the 1980s, an "unrelenting nightmare" in the domains of the Western powers, in the accurate terms of the head of the Organization of African Unity. Illustrations provided by the World Health Organization estimate that eleven million children die every year in "the developing world," a "silent genocide" that could be brought to a quick end if resources were directed to human needs rather than enrichment of a few.

In a global economy designed for the interests and needs of international corporations and finance, and sectors that serve them, most people become superfluous. They will be cast aside if the institutional structures of power and privilege function without popular challenge or control.

The world's rent-a-thug

For most of this century, the United States was far-and-away the world's dominant economic power, and that made economic warfare an appealing weapon, including measures ranging from illegal embargoes to enforcement of IMF rules (for the weak). But in the last twenty years or so, the US has declined relative to Japan and German-led Europe (thanks in part to the economic mismanagement of the Reagan administration, which threw a party for the rich, with costs paid by the majority of the population and future generations). At the same time, however, US military power has become absolutely pre-eminent.

As long as the Soviet Union was in the game, there was a limit to how much force the US could apply, particularly in more remote areas where we didn't have a big conventional force advantage. Because the USSR used to support governments and political movements the US was trying to destroy, there was a danger that US intervention in the Third World might explode into a nuclear war. With the Soviet deterrent gone, the US is much more free to use violence around the world, a fact that has been recognized with much satisfaction by US policy analysts in the past several years.

In any confrontation, each participant tries to shift the battle to a domain in which it's most likely to succeed. You want to lead with your strength, play your strong card. The strong card of the United States is force—so if we can establish the principle that force rules the world, that's a victory for us. If, on the other hand, a conflict is settled through peaceful means, that benefits us less, because our rivals are just as good or better in that domain.

Diplomacy is a particularly unwelcome option, unless it's pursued under the gun. The US has very little popular support for its goals in the Third World. This isn't surprising, since it's trying to impose structures of domination and exploitation. A diplomatic settlement is bound to respond, at least to some degree, to the interests of the other participants in the negotiation, and that's a problem when your positions aren't very popular. As a result, negotiations are something the US commonly tries to avoid. Contrary to much propaganda, that has been true in southeast Asia, the Middle East and Central America for many years.

Against this background, it's natural that the [first] Bush administration should regard military force as a major policy instrument, preferring it to sanctions and diplomacy (as in the Gulf crisis). But since the US now lacks the economic base to impose "order and stability" in the Third World, it must rely on others to pay for the exercise—a necessary one, it's widely assumed, since someone must ensure a proper respect for the masters. The flow of profits from Gulf oil production helps, but Japan and German-led continental Europe must also pay their share as the US adopts the "mercenary role," following the advice of the international business press.

The financial editor of the conservative *Chicago Tribune* has been stressing these themes with particular clarity. We must be "willing mercenaries," paid for our ample services by our rivals, using our "monopoly power" in the "security market" to maintain "our control over the world economic system." We should run a global protection racket, he advises, selling "protection" to other wealthy powers who will pay us a "war premium." This is Chicago, where the words are understood: if someone bothers you, you call on the Mafia to break their bones. And if you fall behind in your premium, your health may suffer too.

To be sure, the use of force to control the Third World is only a last resort. The IMF is a more cost-effective instrument than the Marines and the CIA if it can do the job. But the "iron fist" must be poised in the background, available when needed.

Our rent-a-thug role also causes suffering at home. All of the successful industrial powers have relied on the state to protect and enhance powerful domestic economic interests, to direct public resources to the needs of investors, and so on—one reason why they are successful. Since 1950, the US has pursued these ends largely through the Pentagon system (including NASA and the Department of Energy, which produces nuclear weapons). By now we are locked into these devices for maintaining electronics, computers and high-tech industry generally.

Reaganite military Keynesian excesses added further problems. ["Keynesian" refers to the theories of the British economist John Maynard Keynes, 1883–1946, who recommended government spending to pull societies out of depressions.] The transfer of resources to wealthy minorities and other government policies led to a vast wave of financial manipulations and a consumption binge. But there was little in the way of productive investment, and the country was saddled with huge debts: government, corporate, household and the incalculable debt of unmet social needs as the society drifts towards a Third World pattern, with islands of great wealth and privilege in a sea of misery and suffering.

When a state is committed to such policies, it must somehow find a way to divert the population, to keep them from seeing what's happening around them. There are not many ways to do this. The standard ones are to inspire fear of terrible enemies about to overwhelm us, and awe for our grand leaders who rescue us from disaster in the nick of time.

That has been the pattern right through the 1980s, requiring no little ingenuity as the standard device, the Soviet threat, became harder to take seriously. So the threat to our existence has been Qaddafi and his hordes of international terrorists, Grenada and its ominous air base, Sandinistas marching on Texas, Hispanic narcotraffickers led by the arch-maniac Noriega, and crazed Arabs generally. Most recently it's Saddam Hussein, after he committed his sole crime—the crime of disobedience—in August 1990. It has become more necessary to recognize what has always been true: that the prime enemy is the Third World, which threatens to get "out of control."

These are not laws of nature. The processes, and the institutions that engender them, could be changed. But that will require cultural, social and institutional changes of no little moment, including democratic structures that go far beyond periodic selection of representatives of the business world to manage domestic and international affairs.

BRAINWASHING AT HOME

How the Cold War worked

Despite much pretense, national security has not been a major concern of US planners and elected officials. The historical record reveals this clearly. Few serious analysts took issue with George Kennan's position that "it is not Russian military power which is threatening us, it is Russian political power" (October 1947); or with President Eisenhower's consistent view that the Russians intended no military conquest of Western Europe and that the major role of NATO was to "convey a feeling of confidence to exposed populations, a confidence which will make them sturdier, politically, in their opposition to Communist inroads."

Similarly, the US dismissed possibilities for peaceful resolution of the Cold War conflict, which would have left the "political threat" intact. In his history of nuclear weapons, McGeorge Bundy writes that he is "aware of no serious contemporary proposal...that ballistic missiles should somehow be banned by agreement before they were ever deployed," even though these were the only potential military threat to the US. It was always the "political" threat of so-called "Communism" that was the primary concern.

(Recall that "Communism" is a broad term, and includes all those with the "ability to get control of mass movements.... something we have no capacity to duplicate," as Secretary of State John Foster Dulles privately complained to his brother Allen (who was director of the CIA). "The poor people are the ones they appeal to," he added, "and they have always wanted to plunder the rich." So they must be overcome, to protect our doctrine that the *rich* should plunder the *poor.)*

Of course, both the US and USSR would have preferred that the other simply disappear. But since this would obviously have involved mutual annihilation, a system of global management called the Cold War was established.

According to the conventional view, the Cold War was a conflict between two superpowers, caused by Soviet aggression, in which we tried to contain the Soviet Union and protect the world from it. If this view is a doctrine of theology, there's no need to discuss it. If it is intended to shed some light on history, we can easily put it to the test, bearing in mind a very simple point: if you want to understand the Cold War, you should look at the *events* of the Cold War. If you do so, a very different picture emerges.

On the Soviet side, the events of the Cold War were repeated interventions in Eastern Europe: tanks in East Berlin and Budapest and Prague. These interventions took place along the route that was used to attack and virtually destroy Russia three times in this century alone. The invasion of Afghanistan is the one example of an intervention outside that route, though also on the Soviet border. On the US side, intervention was world-wide, reflecting the status attained by the US as the first truly global power in history.

On the domestic front, the Cold War helped the Soviet Union entrench its military-bureaucratic ruling class in power, and it gave the US a way to compel its population to subsidize high-tech industry. It isn't easy to sell all that to the domestic populations. The technique used was the old stand-by—fear of a great enemy.

The Cold War provided that too. No matter how outlandish the idea that the Soviet Union and its tentacles were strangling the West, the "Evil Empire" *was* in fact evil, *was* an empire and *was* brutal. Each superpower controlled its primary enemy—its own population—by terrifying it with the (quite real) crimes of the other.

In crucial respects, then, the Cold War was a kind of tacit arrangement between the Soviet Union and the United States under which the US conducted its wars against the Third World and controlled its allies in Europe, while the Soviet rulers kept an iron grip on their own internal empire and their satellites in Eastern Europe—each side using the other to justify repression and violence in its own domains.

So why did the Cold War end, and how does its end change things? By the 1970s, Soviet military expenditures were leveling off and internal problems were mounting, with economic stagnation and increasing pressures for an end to tyrannical rule. Soviet power internationally had, in fact, been declining for some 30 years, as a study by the Center for Defense Information showed in 1980. A few years later, the Soviet system had collapsed. The Cold War ended with the victory of what had always been the far richer and more powerful contestant. The Soviet collapse was part of the more general economic catastrophe of the 1980s, more severe in most of the Third World domains of the West than in the Soviet empire.

As we've already seen, the Cold War had significant elements of North-South conflict (to use the contemporary euphemism for the European conquest of the world). Much of the Soviet empire had formerly been quasi-colonial dependencies of the West. The Soviet Union took an independent course, providing assistance to targets of Western attack and deterring the worst of Western violence. With the collapse of Soviet tyranny, much of the region can be expected to return to its traditional status, with the former higher echelons of the bureaucracy playing the role of the Third World elites that enrich themselves while serving the interests of foreign investors.

But while this particular phase has ended, North-South conflicts continue. One side may have called off the game, but the US is proceeding as before—more freely, in fact, with Soviet deterrence a thing of the past. It should have surprised no one that [the first] President Bush celebrated the symbolic end of the Cold War, the fall of the Berlin Wall, by immediately invading Panama and announcing loud and clear that the US would subvert Nicaragua's election by maintaining its economic stranglehold and military attack unless "our side" won.

Nor did it take great insight for Elliott Abrams to observe that the US invasion of Panama was unusual because it could be conducted without fear of a Soviet reaction anywhere, or for numerous commentators during the Gulf crisis to add that the US and Britain were now free to use unlimited force against its Third World enemy, since they were no longer inhibited by the Soviet deterrent.

Of course, the end of the Cold War brings its problems too. Notably, the technique for controlling the domestic population has had to shift, a problem recognized through the 1980s, as we've already seen. New enemies have to be invented. It becomes harder to disguise the fact that the real enemy has always been "the poor who seek to plunder the rich"—in particular, Third World miscreants who seek to break out of the service role.

The war on (certain) drugs

One substitute for the disappearing Evil Empire has been the threat of drug traffickers from Latin America. In early September 1989, a major government-media blitz was launched by the president. That month the AP wires carried more stories about drugs than about Latin America, Asia, the Middle East and Africa combined. If you looked at television, every news program had a big section on how drugs were destroying our society, becoming the greatest threat to our existence, etc.

The effect on public opinion was immediate. When Bush won the 1988 election, people said the budget deficit was the biggest problem facing the country. Only about 3% named drugs. After the media blitz, concern over the budget was way down and drugs had soared to about 40% or 45%, which is highly unusual for an open question (where no specific answers are suggested).

Now, when some client state complains that the US government isn't sending it enough money, they no longer say, "we need it to stop the Russians"—rather, "we need it to stop drug trafficking." Like the Soviet threat, this enemy provides a good excuse for a US military presence where there's rebel activity or other unrest.

So internationally, "the war on drugs" provides a cover for intervention. Domestically, it has little to do with drugs but a lot to do with distracting the population, increasing repression in the inner cities, and building support for the attack on civil liberties.

That's not to say that "substance abuse" isn't a serious problem. At the time the drug war was launched, deaths from tobacco were estimated at about 300,000 a year, with perhaps another 100,000 from alcohol. But these aren't the drugs the Bush administration targeted. It went after illegal drugs, which had caused many fewer deaths—3500+ a year—according to official figures. One reason for going after these drugs was that their use had been declining for some years, so the Bush administration could safely predict that its drug war would "succeed" in lowering drug use.

The administration also targeted marijuana, which hadn't caused any known deaths among some 60 million users. In fact, that crackdown has exacerbated the drug problem—many marijuana users have turned from this relatively harmless drug to more dangerous drugs like cocaine, which are easier to conceal.

Just as the drug war was launched with great fanfare in September 1989, the US Trade Representative (USTR) panel held a hearing in Washington to consider a tobacco industry request that the US impose sanctions on Thailand in retaliation for its efforts to restrict US tobacco imports and advertising. Such US government actions had already rammed this lethal addictive narcotic down the throats of consumers in Japan, South Korea and Taiwan, with human costs of the kind already indicated. The US Surgeon General, Everett Koop, testified at the USTR panel that "when we are pleading with foreign governments to stop the flow of cocaine, it is the height of hypocrisy for the United States to export tobacco." He added, "years from now, our nation will look back on this application of free-trade policy and find it scandalous."

Thai witnesses also protested, predicting that the consequence of US sanctions would be to reverse a decline in smoking achieved by their government's campaign against tobacco use. Responding to the US tobacco companies' claim that their product is the best in the world, a Thai witness said: "Certainly in the Golden Triangle we have some of the best products, but we never ask the principle of free trade to govern such products. In fact we suppressed [them]." Critics recalled the Opium War 150 years earlier, when the British government compelled China to open its doors to opium from British India, sanctimoniously pleading the virtues of free trade as they forcefully imposed large-scale drug addiction on China.

Here we have the biggest drug story of the day. Imagine the screaming headlines: *US Government World's Leading Drug Peddler.* It would surely sell papers. But the story passed virtually unreported, and with not a hint of the obvious conclusions.

Another aspect of the drug problem, which also received little attention, is the leading role of the US government in stimulating drug trafficking since World War II. This happened in part when the US began its postwar task of undermining the antifascist resistance and the labor movement became an important target. In France, the threat of the political power and influence of the labor movement was enhanced by its steps to impede the flow of arms to French forces seeking to reconquer their former colony of Vietnam with US aid. So the CIA undertook to weaken and split the French labor movement—with the aid of top American labor leaders, who were quite proud of their role.

The task required strikebreakers and goons. There was an obvious supplier: the Mafia. Of course, they didn't take on this work just for the fun of it—they wanted a return for their efforts. And it was given to them: They were authorized to reestablish the heroin racket that had been suppressed by the fascist governments, the famous "French connection" that dominated the drug trade until the 1960s.

By then, the center of the drug trade had shifted to Indochina, particularly Laos and Thailand. The shift was again a by-product of a CIA operation—the "secret war" fought in those countries during the Vietnam War by a CIA mercenary army. They also wanted a payoff for their contributions. Later, as the CIA shifted its activities to Pakistan and Afghanistan, the drug racket boomed there.

The clandestine war against Nicaragua also provided a shot in the arm to drug traffickers in the region, as illegal CIA arms flights to the US mercenary forces offered an easy way to ship drugs back to the US, sometimes through US Air Force bases, traffickers report.

The close correlation between the drug racket and international terrorism (sometimes called "counterinsurgency," "low intensity conflict" or some other euphemism) is not surprising. Clandestine operations need plenty of money, which should be undetectable. And they need criminal operatives as well. The rest follows.

War is peace.
Freedom is slavery.
Ignorance is strength.

The terms of political discourse typically have two meanings. One is the dictionary meaning, and the other is a meaning that is useful for serving power—the doctrinal meaning.

Take *democracy*. According to the common-sense meaning, a society is democratic to the extent that people can participate in a meaningful way in managing their affairs. But the doctrinal meaning of *democracy* is different—it refers to a system in which decisions are made by sectors of the business community and related elites. The public are to be only "spectators of action," not "participants," as leading democratic theorists (in this case, Walter Lippmann) have explained. They are permitted to ratify the decisions of their betters and to lend their support to one or another of them, but not to interfere with matters—like public policy—that are none of their business.

If segments of the public depart from their apathy and begin to organize and enter the public arena, that's not democracy. Rather, it's a *crisis of democracy* in proper technical usage, a threat that has to be overcome in one or another way: in El Salvador, by death squads; at home, by more subtle and indirect means.

Or take *free enterprise,* a term that refers, in practice, to a system of public subsidy and private profit, with massive government intervention in the economy to maintain a welfare state for the rich. In fact, in acceptable usage, just about any phrase containing the word "free" is likely to mean the opposite of its actual, literal meaning.

Or take defense *against aggression,* a phrase that's used—predictably—to refer to aggression. When the US attacked South Vietnam in the early 1960s, the liberal hero Adlai Stevenson (among others) explained that we were defending South Vietnam against "internal aggression"—that is, the aggression of South Vietnamese peasants against the US air force and a US-run mercenary army, which were driving them out of their homes and into concentration camps where they could be "protected" from the southern guerrillas. In fact, these peasants willingly supported the guerillas, while the US client regime was an empty shell, as was agreed on all sides.

So magnificently has the doctrinal system risen to its task that to this day, 30 years later, the idea that the US attacked South Vietnam is unmentionable, even unthinkable, in the mainstream. The essential issues of the war are, correspondingly, beyond any possibility of discussion now. The guardians of political correctness (the real PC) can be quite proud of an achievement that would be hard to duplicate in a well-run totalitarian state.

Or take the term *peace process*. The naive might think that it refers to efforts to seek peace. Under this meaning, we would say that the peace process in the Middle East includes, for example, the offer of a full peace treaty to Israel by President Sadat of Egypt in 1971, along lines advocated by virtually the entire world, including US official policy; the Security Council resolution of January 1976, introduced by the major Arab states with the backing of the PLO [the Palestine Liberation Organization], which called for a two-state settlement of the Arab-Israel conflict in the terms of a near-universal international consensus; PLO offers through the 1980s to negotiate with Israel for mutual recognition; and annual votes at the UN General Assembly, most recently in December 1990 (approved by a vote of 144 to 2), calling for an international conference on the Israel-Arab problem; etc.

But the sophisticated understand that these efforts do not form part of the peace process. The reason is that in the PC meaning, the term peace process refers to what the US government is doing—in the cases mentioned, this is to block international efforts to seek peace. The cases cited do not fall within the peace process, because the US backed Israel's rejection of Sadat's offer, vetoed the Security Council resolution, opposed negotiations and mutual recognition of the PLO and Israel, and regularly joins with Israel in opposing—thereby, in effect, vetoing—any attempt to move towards a peaceful diplomatic settlement at the UN or elsewhere.

The peace process is restricted to US initiatives, which call for a unilateral US-determined settlement with no recognition of Palestinian national rights. That's the way it works. Those who cannot master these skills must seek another profession.

There are many other examples. Take the term *special interest*. The well-oiled Republican PR systems of the 1980s regularly accused the Democrats of being the party of the special interests: women, labor, the elderly, the young, farmers—in short, the general population. There was only one sector of the population never listed as a special interest: corporations (and business generally). That makes sense. In PC discourse their (special) interests are the "national interest," to which all must bow.

The Democrats plaintively retorted that they were *not* the party of the special interests: they served the national interest too. That was correct, but their problem has been that they lack the single-minded class consciousness of their Republican opponents. The latter are not confused about their role as representatives of the owners and managers of the society, who are fighting a bitter class war against the general population—often adopting vulgar Marxist rhetoric and concepts, resorting to jingoist hysteria, fear and terror, awe of great leaders and the other standard devices of population control. The Democrats are less clear about their allegiances, hence less effective in the propaganda wars.

Finally, take the term *conservative,* which has come to refer to advocates of a powerful state that interferes massively in the economy and in social life. They advocate huge state expenditures and a postwar peak of protectionist measures and insurance against market risk, narrowing individual liberties through legislation and court-packing, protecting the Holy State from unwarranted inspection by the irrelevant citizenry—in short, those programs that are the precise opposite of traditional conservatism. Their allegiance is to "the people who own the country" and therefore "ought to govern it," in the words of Founding Father John Jay.

It's really not that hard, once one understands the rules.

To make sense of political discourse, it's necessary to give a running translation into English, decoding the doublespeak of the media, academic social scientists and the secular priesthood generally. Its function is not obscure: the effect is to make it impossible to find words to talk about matters of human significance in a coherent way. We can then be sure that little will be understood about how our society works and what is happening in the world—a major contribution to "democracy," in the PC sense of the word.

Socialism, real and fake

One can debate the meaning of the term *socialism,* but if it means anything, it means control of production by the workers themselves, not owners and managers who rule them and control all decisions, whether in capitalist enterprises or an absolutist state.

To refer to the Soviet Union as "socialist" is an interesting case of doctrinal doublespeak. The Bolshevik coup of October 1917 placed state power in the hands of Lenin and Trotsky, who moved quickly to dismantle the incipient socialist institutions that had grown up during the popular revolution of the preceding months—the factory councils, the soviets [popularly elected legislative assembies], in fact any organ of popular control—and to convert the workforce into what they called a "labor army" under the command of the leader. In any meaningful sense of the term "socialism," the Bolsheviks moved at once to destroy its existing elements. No socialist deviation has been permitted since.

These developments came as no surprise to leading Marxist intellectuals, who had criticized Lenin's doctrines for years (as had Trotsky) because they would centralize authority in the hands of the Vanguard Party and its leaders. In fact, decades earlier, the anarchist thinker Bakunin had predicted that the emerging intellectual class would follow one of two paths: either they would try to exploit popular struggles to take state power themselves, becoming a brutal and oppressive Red bureaucracy; or they would become the managers and ideologists of the state capitalist societies, if popular revolution failed. It was a perceptive insight, on both counts.

The world's two major propaganda systems did not agree on much, but they did agree on using the term *socialism* to refer to the immediate destruction of every element of socialism by the Bolsheviks. That's not too surprising. The Bolsheviks called their system socialist so as to exploit the moral prestige of socialism. The West adopted the same usage for the opposite reason: to defame the feared libertarian ideals by associating them with the Bolshevik dungeon, to undermine the popular belief that there really might be progress towards a more just society, with democratic control over its basic institutions and concern for human needs and rights.

If socialism is the tyranny of Lenin and Stalin, then sane people will say: *not for me.* And if that's the only alternative to corporate state capitalism, then many will submit to its authoritarian structures as the only reasonable choice.

With the collapse of the Soviet system, there's an opportunity to revive the lively and vigorous libertarian socialist thought that was not able to withstand the doctrinal and repressive assaults of the major systems of power. How large a hope that is, we cannot know. But at least one roadblock has been removed. In that sense, the disappearance of the Soviet Union is a small victory for socialism, much as the defeat of the fascist powers was.

The media

Whether they're called "liberal" or "conservative," the major media are large corporations, owned by and interlinked with even larger conglomerates. Like other corporations, they sell a product to a market. The market is advertisers—that is, other businesses. The product is audiences. For the elite media that set the basic agenda to which others adapt, the product is, furthermore, relatively privileged audiences.

So we have major corporations selling fairly wealthy and privileged audiences to other businesses. Not surprisingly, the picture of the world presented reflects the narrow and biased interests and values of the sellers, the buyers and the product.

Other factors reinforce the same distortion. The cultural managers (editors, leading columnists, etc.) share class interests and associations with state and business managers and other privileged sectors. There is, in fact, a regular flow of high-level people among corporations, government and media. Access to state authorities is important to maintain a competitive position; "leaks," for example, are often fabrications and deceit produced by the authorities with the cooperation of the media, who pretend they don't know.

In return, state authorities demand cooperation and submissiveness. Other power centers also have devices to punish departures from orthodoxy, ranging from the stock market to an effective vilification and defamation apparatus.

The outcome is not, of course, entirely uniform. To serve the interests of the powerful, the media must present a tolerably realistic picture of the world. And professional integrity and honesty sometimes interfere with the overriding mission. The best journalists are, typically, quite aware of the factors that shape the media product, and seek to use such openings as are provided. The result is that one can learn a lot by a critical and skeptical reading of what the media produce.

The media are only one part of a larger doctrinal system; other parts are journals of opinion, the schools and universities, academic scholarship and so on. We're much more aware of the media, particularly the prestige media, because those who critically analyze ideology have focused on them. The larger system hasn't been studied as much because it's harder to investigate systematically. But there's good reason to believe that it represents the same interests as the media, just as one would anticipate.

The doctrinal system, which produces what we call "propaganda" when discussing enemies, has two distinct targets. One target is what's sometimes called the "political class," the roughly 20% of the population that's relatively educated, more or less articulate, playing some role in decision-making. Their acceptance of doctrine is crucial, because they're in a position to design and implement policy.

Then there's the other 80% or so of the population. These are Lippmann's "spectators of action," whom he referred to as the "bewildered herd." They are supposed to follow orders and keep out of the way of the important people. They're the target of the real *mass* media: the tabloids, the sitcoms, the Super Bowl and so on.

These sectors of the doctrinal system serve to divert the unwashed masses and reinforce the basic social values: passivity, submissiveness to authority, the overriding virtue of greed and personal gain, lack of concern for others, fear of real or imagined enemies, etc. The goal is to keep the bewildered herd bewildered. It's unnecessary for them to trouble themselves with what's happening in the world. In fact, it's undesirable—if they see too much of reality they may set themselves to change it.

That's not to say that the media can't be influenced by the general population. The dominant institutions—whether political, economic or doctrinal—are not immune to public pressures. Independent (alternative) media can also play an important role. Though they lack resources, almost by definition, they gain significance in the same way that popular organizations do: by bringing together people with limited resources who can multiply their effectiveness, and their own understanding, through their interactions—precisely the democratic threat that's so feared by dominant elites.

THE FUTURE

Things have changed

It's important to recognize how much the scene has changed in the past 30 years as a result of the popular movements that organized in a loose and chaotic way around such issues as civil rights, peace, feminism, the environment and other issues of human concern.

Take the Kennedy and Reagan administrations, which were similar in a number of ways in their basic policies and commitments. When Kennedy launched a huge international terrorist campaign against Cuba after his invasion failed, and then escalated the murderous state terror in South Vietnam to outright aggression, there was no detectable protest.

It wasn't until hundreds of thousands of American troops were deployed and all of Indochina was under devastating attack, with hundreds of thousands slaughtered, that protest became more than marginally significant. In contrast, as soon as the Reagan administration hinted that they intended to intervene directly in Central America, spontaneous protest erupted at a scale sufficient to compel the state terrorists to turn to other means.

Leaders may crow about the end of the "Vietnam syndrome," but they know better. A National Security Policy Review of the Bush administration, leaked at the moment of the ground attack in the Gulf, noted that, "In cases where the US confronts much weaker enemies"—the only ones that the true statesman will agree to fight—"our challenge will be not simply to defeat them, but to defeat them decisively and rapidly." Any other outcome would be "embarrassing" and might "undercut political support," understood to be very thin.

By now, classical intervention is not even considered an option. The means are limited to clandestine terror, kept secret from the domestic population, or "decisive and rapid" demolition of "much weaker enemies"—after huge propaganda campaigns depicting them as monsters of indescribable power.

Much the same is true across the board. Take 1992. If the Columbus quincentenary had been in 1962, it would have been a celebration of the liberation of the continent. In 1992, that response no longer has a monopoly, a fact that has aroused much hysteria among the cultural managers who are used to near-totalitarian control. They now rant about the "fascist excesses" of those who urge respect for other people and other cultures.

In other areas too, there's more openness and understanding, more skepticism and questioning of authority.

Of course, these tendencies are double-edged. They may lead to independent thought, popular organizing and pressures for much-needed institutional change. Or they may provide a mass base of frightened people for new authoritarian leaders. These possible outcomes are not a matter for speculation, but for action, with stakes that are very large.

What you can do

In any country, there's some group that has the real power. It's not a big secret where power is in the United States. It basically lies in the hands of the people who determine investment decisions—what's produced, what's distributed. They staff the government, by and large, choose the planners, and set the general conditions for the doctrinal system.

One of the things they want is a passive, quiescent population. So one of the things that you can do to make life uncomfortable for them is *not* be passive and quiescent. There are lots of ways of doing that. Even just asking questions can have an important effect.

Demonstrations, writing letters and voting can all be meaningful—it depends on the situation. But the main point is—it's got to be sustained and organized.

If you go to one demonstration and then go home, that's something, but the people in power can live with that. What they can't live with is sustained pressure that keeps building, organizations that keep doing things, people that keep learning lessons from the last time and doing it better the next time.

71

Any system of power, even a fascist dictatorship, is responsive to public dissidence. It's certainly true in a country like this, where—fortunately—the state doesn't have a lot of force to coerce people. During the Vietnam War, direct resistance to the war was quite significant, and it was a cost that the government had to pay.

If elections are just something in which some portion of the population goes and pushes a button every couple of years, they don't matter. But if the citizens organize to press a position, and pressure their representatives about it, elections can matter.

Members of the House of Representatives can be influenced much more easily than senators, and senators somewhat more easily than the president, who is usually immune. When you get to that level, policy is decided almost totally by the wealthy and powerful people who own and manage the country.

But you can organize on a scale that will influence representatives. You can get them to come to your homes to be yelled at by a group of neighbors, or you can sit in at their offices—whatever works in the circumstances. It can make a difference—often an important one.

You can also do your own research. Don't just rely on the conventional history books and political science texts—go back to specialist monographs and to original sources: National Security Memoranda and similar documents. Most good libraries have reference departments where you can find them.

It does require a bit of effort. Most of the material is junk, and you have to read a ton of stuff before you find anything good. There are guides that give you hints about where to look, and sometimes you'll find references in secondary sources that look intriguing. Often they're misinterpreted, but they suggest places to search.

It's no big mystery, and it's not intellectually difficult. It involves some work, but anybody can do it as a spare-time job. And the results of that research can change people's minds. Real research is always a collective activity, and its results can make a large contribution to changing consciousness, increasing insight and understanding, and leading to constructive action.

The struggle continues

The struggle for freedom is never over. The people of the Third World need our sympathetic understanding and, much more than that, they need our help. We can provide them with a margin of survival by internal disruption in the United States. Whether they can succeed against the kind of brutality we impose on them depends in large part on what happens here.

The courage they show is quite amazing. I've personally had the privilege—and it is a privilege—of catching a glimpse of that courage at first hand in Southeast Asia, in Central America and on the occupied West Bank. It's a very moving and inspiring experience, and invariably brings to my mind some contemptuous remarks of Rousseau's on Europeans who have abandoned freedom and justice for the peace and repose "they enjoy in their chains." He goes on to say:

> When I see multitudes of entirely naked savages scorn European voluptuousness and endure hunger, fire, the sword and death to preserve only their independence, I feel that it does not behoove slaves to reason about freedom.

People who think that these are mere words understand very little about the world.

And that's just a part of the task that lies before us. There's a growing Third World at home. There are systems of illegitimate authority in every corner of the social, political, economic and cultural worlds. For the first time in human history, we have to face the problem of protecting an environment that can sustain a decent human existence. We don't know that honest and dedicated effort will be enough to solve or even mitigate such problems as these. We can be quite confident, however, that the lack of such efforts will spell disaster.

THE PROSPEROUS FEW AND THE RESTLESS MANY

FIRST PUBLISHED 1993

12 US PRINTINGS

5 FOREIGN EDITIONS

146,000 COPIES IN PRINT

TABLE OF CONTENTS

The new global economy

I was on Brattle Street [in Cambridge, Massachusetts] just last night. There were panhandlers, people asking for money, people sleeping in the doorways of buildings. This morning, in the subway station at Harvard Square, there was more of the same.

The spectre of poverty and despair has become increasingly obvious to the middle and upper class. You just can't avoid it as you could years ago, when it was limited to a certain section of town. This has a lot to do with the pauperization (the internal Third Worldization, I think you call it) of the United States.

There are several factors involved. About twenty years ago there was a big change in the world order, partly symbolized by Richard Nixon's dismantling of the postwar economic system. He recognized that US dominance of the global system had declined, and that in the new "tripolar" world order (with Japan and German-based Europe playing a larger role), the US could no longer serve—in effect—as the world's banker.

That led to a lot more pressure on corporate profits in the US and, consequently, to a big attack on social welfare gains. The crumbs that were permitted to ordinary people had to be taken away. Everything had to go to the rich.

There was also a tremendous expansion of unregulated capital in the world. In 1971, Nixon dismantled the Bretton Woods system, thereby deregulating currencies. That, and a number of other changes, tremendously expanded the amount of unregulated capital in the world, and accelerated what's called the globalization (or the internationalization) of the economy.

That's a fancy way of saying that you export jobs to high-repression, low-wage areas—which undercuts the opportunities for productive labor at home. It's a way of increasing corporate profits, of course. And it's much easier to do with a free flow of capital, advances in telecommunications, etc.

There are two important consequences of globalization. First, it extends the Third World model to industrial countries. In the Third World, there's a two-tiered society—a sector of extreme wealth and privilege, and a sector of huge misery and despair among useless, superfluous people.

That division is deepened by the policies dictated by the West. It imposes a neoliberal "free market" system that directs resources to the wealthy and to foreign investors, with the idea that something will trickle down by magic, some time after the Messiah comes.

You can see this happening everywhere in the industrial world, but most strikingly in the three English-speaking countries. In the 1980s, England under Thatcher, the United States under the Reaganites and Australia under a Labor government adopted some of the doctrines they preached for the Third World.

Of course, they would never really play this game completely. It would be too harmful to the rich. But they flirted with it. And they suffered. That is, the general population suffered.

Take, for example, South Central Los Angeles. It had factories once. They moved to Eastern Europe, Mexico, Indonesia— where you can get peasant women flocking off the land. But the rich did fine, just as they do in the Third World.

The second consequence, which is also important, has to do with governing structures. Throughout history, the structures of government have tended to coalesce around other forms of power—in modern times, primarily around economic power. So, when you have national economies, you get national states. We now have an international economy and we're moving towards an international state—which means, finally, an international executive.

To quote the business press, we're creating "a new imperial age" with a "de facto world government." It has its own institutions—like the International Monetary Fund (IMF) and the World Bank, trading structures like NAFTA and GATT [the North American Free Trade Agreement and the General Agreement on Tariffs and Trade, both discussed in more detail below], executive meetings like the G-7 [the seven richest industrial countries—the US, Canada, Japan, Germany, Britain, France and Italy—who meet regularly to discuss economic policy] and the European Community bureaucracy.

As you'd expect, this whole structure of decision making answers basically to the transnational corporations, international banks, etc. It's also an effective blow against democracy. All these structures raise decision making to the executive level, leaving what's called a "democratic deficit"—parliaments and populations with less influence.

Not only that, but the general population doesn't know what's happening, and it doesn't even know that it doesn't know. One result is a kind of alienation from institutions. People feel that nothing works for them.

Sure it doesn't. They don't even know what's going on at that remote and secret level of decision making. That's a real success in the long-term task of depriving formal democratic structures of any substance.

At Clinton's Little Rock economic conference and elsewhere, there was much talk of economic recovery and restoring competitiveness. Political economist Gar Alperovitz wrote in the *New York Times* that what's being proposed is "not likely to make a dent in our deeper economic problems. We may simply be in for a long, painful era of unresolved economic decay." Would you agree?

I haven't seen that piece yet, but the *Financial Times* [of London, the world's leading business newspaper] has been talking with some pleasure of the fiscal conservatism shown by Clinton and his advisors.

There are serious issues here. First of all, we have to be careful in the use of terms. When someone says America is in for a long period of decline, we have to decide what we mean by "America." If we mean the geographical area of the United States, I'm sure that's right. The policies now being discussed will have only a cosmetic effect. There has been decline and there will be further decline. The country is acquiring many of the characteristics of a Third World society.

But if we're talking about US-based corporations, then it's probably not right. In fact, the indications are to the contrary— their share in manufacturing production, for example, has been stable or is probably even increasing, while the share of the US itself has declined. That's an automatic consequence of sending productive labor elsewhere.

General Motors, as the press constantly reports, is closing some 24 factories in North America. But in the small print you read that it's opening new factories—including, for example, a $700 million high-tech factory in East Germany. That's an area of huge unemployment where GM can pay 40% of the wages of Western Europe and none of the benefits.

There was a nice story on the front page of the *Financial Times,* in which they described what a great idea this was. As they put it, GM doesn't have to worry about the "pampered" Western European workers any longer—they can just get highly exploited workers now that East Germany is being pushed back to its traditional Third World status. It's the same in Mexico, Thailand, etc.

The prescription for our economic problems is more of the same—"leave it to the market." There's such endless trumpeting of the free market that it assumes almost a myth-like quality. "It'll correct the problems." Are there any alternatives?

We have to first separate ideology from practice, because to talk about a free market at this point is something of a joke. Outside of ideologues, the academy and the press, no one thinks that capitalism is a viable system, and nobody has thought that for sixty or seventy years—if ever.

Herman Daly and Robert Goodland, two World Bank economists, circulated an interesting study recently. In it they point out that received economic theory—the standard theory on which decisions are supposed to be based—pictures a free market sea with tiny little islands of individual firms. These islands, of course, aren't internally free—they're centrally managed.

But that's okay, because these are just tiny little islands on the sea. We're supposed to believe that these firms aren't much different than a mom-and-pop store down the street.

Daly and Goodland point out that by now the islands are approaching the scale of the sea. A large percentage of cross-border transactions are within a single firm, hardly "trade" in any meaningful sense. What you have are centrally managed transactions, with a very visible hand—major corporate structures—directing it. And we have to add a further point—that the sea itself bears only a partial resemblance to free trade.

So you could say that one alternative to the free market system is the one we already have, because we often don't rely on the market where powerful interests would be damaged. Our actual economic policy is a mixture of protectionist, interventionist, free-market and liberal measures. And it's directed primarily to the needs of those who implement social policy, who are mostly the wealthy and the powerful.

For example, the US has always had an active state industrial policy, just like every other industrial country. It's been understood that a system of private enterprise can survive only if there is extensive government intervention. It's needed to regulate disorderly markets and protect private capital from the destructive effects of the market system, and to organize a public subsidy for targeting advanced sectors of industry, etc.

But nobody *called* it industrial policy, because for half a century it has been masked within the Pentagon system. Internationally, the Pentagon was an intervention force, but domestically it was a method by which the government could coordinate the private economy, provide welfare to major corporations, subsidize them, arrange the flow of taxpayer money to research and development, provide a state-guaranteed market for excess production, target advanced industries for development, etc. Just about every successful and flourishing aspect of the US economy has relied on this kind of government involvement.

At the Little Rock conference I heard Clinton talking about structural problems and rebuilding the infrastructure. One attendee, Ann Markusen, a Rutgers economist and author of the book *Dismantling the Cold War Economy,* talked about the excesses of the Pentagon system and the distortions and damages that it has caused to the US economy. So it seems that there's at least some discussion of these issues, which is something I don't recall ever before.

The reason is that they can't maintain the Pentagon-based system as readily as before. They've got to start talking about it, because the mask is dropping. It's very difficult now to get people to lower their consumption or their aspirations in order to divert investment funds to high-technology industry on the pretext that the Russians are coming.

So the system is in trouble. Economists and bankers have been pointing out openly for some time that one of the main reasons why the current recovery is so sluggish is that the government hasn't been able to resort to increased military spending with all of its multiplier effects—the traditional pump-priming mechanism of economic stimulation. Although there are various efforts to continue this (in my opinion, the current operation in Somalia is one such effort to do some public relations work for the Pentagon), it's just not possible the way it used to be.

There's another fact to consider. The cutting edge of technology and industry has for some time been shifting in another direction, away from the electronics-based industry of the postwar period and towards biology-based industry and commerce.

Biotechnology, genetic engineering, seed and drug design (even designing animal species), etc. is expected to be a huge growth industry with enormous profits. It's potentially vastly more important than electronics—in fact, compared to the potential of biotechnology (which may extend to the essentials of life), electronics is sort of a frill.

But it's hard to disguise government involvement in these areas behind the Pentagon cover. Even if the Russians were still there, you couldn't do that.

There are differences between the two political parties about what should be done. The Reagan-Bush types, who are more fanatically ideological, have their heads in the sand about it to some extent. They are a bit more dogmatic. The Clinton people are more up front about these needs. That's one of the main reasons why Clinton had substantial business support.

Take the question of "infrastructure" or "human capital"—a kind of vulgar way of saying keep people alive and allow them to have an education. By now the business community is well aware that they've got problems with that.

The *Wall Street Journal*, for example, was the most extreme advocate of Reaganite lunacies for ten years. They're now publishing articles in which they're bemoaning the consequences—without, of course, conceding that they're consequences.

They had a big news article on the collapse of California's educational system, which they're very upset about. Businessmen in the San Diego area have relied on the state system—on a public subsidy—to provide them with skilled workers, junior managers, applied research, etc. Now the system is in collapse.

The reason is obvious—the large cutbacks in social spending in the federal budget, and the fiscal and other measures that greatly increased the federal debt (which the *Wall Street Journal* supported), simply transferred the burden of keeping people alive and functioning to the states. The states are unable to support that burden. They're in serious trouble and have tried to hand down the problem to the municipalities, which are also in serious trouble.

The same is true if you're a rich businessman living in a rich suburb here in the Boston area. You would like to be able to get into your limousine and drive downtown and have a road. But the road has potholes. That's no good. You also want to be able to walk around the city and go to the theater without getting knifed.

So now businessmen are complaining. They want the government to get back into the business of providing them with what they need. That's going to mean a reversal of the fanaticism that the *Wall Street Journal* and others like it have been applauding all these years.

Talking about it is one thing, but do they really have a clue about what to do?

I think they do have a clue. If you listen to smart economists like Bob Solow, who started the Little Rock conference off, they have some pretty reasonable ideas.

What they want to do is done openly by Japan and Germany and every functioning economy—namely, rely on government initiatives to provide the basis for private profit. In the periphery of Japan—for example in South Korea and Taiwan—we've been seeing a move out of the Third World pattern to an industrial society through massive state intervention.

Not only is the state there powerful enough to control labor, but it's powerful enough to control capital. In the 1980s, Latin America had a huge problem of capital flight because they're open to international capital markets. South Korea has no such problem—they have the death penalty for capital flight. Like any sane planners, they use market systems for allocating resources, but very much under planned central direction.

The US has been doing it indirectly through the Pentagon system, which is kind of inefficient. It won't work as well anymore anyway, so they'd like to do it openly. The question is whether that can be done. One problem is that the enormous debt created during the Reagan years—at the federal, state, corporate, local and even household levels—makes it extremely difficult to launch constructive programs.

There's no capital available.

That's right. In fact, that was probably part of the purpose of the Reaganite borrow-and-spend program.

To eliminate capital?

Recall that about ten years ago, when David Stockman [director of the Office of Management and Budget in the early Reagan years] was kicked out, he had some interviews with economic journalist William Greider. There Stockman pretty much said that the idea was to try to put a cap on social spending, simply by debt. There would always be plenty to subsidize the rich. But they wouldn't be able to pay aid to mothers with dependent children—only aid to dependent corporate executives.

Incidentally, the debt itself, just the numbers, may not be such a huge problem. We've had bigger debts than that—not in numbers, but relative to the GNP [the Gross National Product]—in the past. The exact amount of the debt is a bit of a statistical artifact. You can make it different things depending on how you count. Whatever it is, it's not something that couldn't be dealt with.

The question is—what was done with the borrowing? If the borrowing in the last ten years had been used for constructive purposes—say for investment or infrastructure—we'd be quite well off. But the borrowing was used for enrichment of the rich—for consumption (which meant lots of imports, building up the trade deficit), financial manipulation and speculation. All of these are very harmful to the economy.

There's another problem, a cultural and ideological problem. The government has for years relied on a propaganda system that denies these truths. It's other countries that have government involvement and social services—we're rugged individualists. So IBM doesn't get anything from the government. In fact, they get plenty, but it's through the Pentagon.

The propaganda system has also whipped up hysteria about taxation (though we're undertaxed by comparative standards) and about bureaucracies that interfere with profits—say, by protecting worker and consumer interests. Pointy-headed bureaucrats who funnel a public subsidy to industry and banks are just fine, of course.

Propaganda aside, the population *is*, by comparative standards, pretty individualistic and kind of dissident and doesn't take orders very well, so it's not going to be easy to sell state industrial policy to people. These cultural factors are significant.

In Europe there's been a kind of social contract. It's now declining, but it has been largely imposed by the strength of the unions, the organized work force and the relative weakness of the business community (which, for historical reasons, isn't as dominant in Europe as it has been here). European governments do see primarily to the needs of private wealth, but they also have created a not-insubstantial safety net for the rest of the population. They have general healthcare, reasonable services, etc.

We haven't had that, in part because we don't have the same organized work force, and we have a much more class-conscious and dominant business community.

Japan achieved pretty much the same results as Europe, but primarily because of the highly authoritarian culture. People just do what they're told. So you tell them to cut back consumption—they have a very low standard of living, considering their wealth—work hard, etc. and people just do it. That's not so easy to do here.

Given the economic situation, it would seem to be a propitious moment for the left, the progressive movement, to come forward with some concrete proposals. Yet the left seems to be either bogged down in internecine warfare or in a reactive mode. It's not proactive.

What people call the "left" (the peace and justice movements, whatever they are) has expanded a lot over the years. They tend to be very localized. On particular issues they focus and achieve things.

But there's not much of a broader vision, or of institutional structure. The left can't coalesce around unions because the unions are essentially gone. To the extent that there's any formal structure, it's usually something like the church.

There's virtually no functioning left intelligentsia [intellectuals viewed as a distinct group or class]. Nobody's talking much about what should be done, or is even available to give talks. The class warfare of the last decades has been fairly successful in weakening popular organizations. People are isolated.

I should also say that the policy issues that have to be faced are quite deep. It's always nice to have reforms. It would be nice to have more money for starving children. But there are some objective problems which you and I would have to face if we ran the country.

One problem was kindly pointed out to the Clinton administration by a front page article in the *Wall Street Journal* the other day. It mentioned what might happen if the administration gets any funny ideas about taking some of their own rhetoric seriously—like spending money for social programs. (Granted, that's not very likely, but just in case anybody has some funny ideas.)

The United States is so deeply in hock to the international financial community (because of the debt) that they have a lock on US policy. If something happens here—say, increasing workers' salaries—that the bondholders don't like and will cut down their short-term profit, they'll just start withdrawing from the US bond market.

That will drive interest rates up, which will drive the economy down, which will increase the deficit. The *Journal* points out that Clinton's $20-billion spending program could be turned into a $20-billion cost to the government, to the debt, just by slight changes in the purchase and sale of bonds.

So social policy, even in a country as rich and powerful as the United States (which is the richest and most powerful of them all), is mortgaged to the international wealthy sectors here and abroad. Those are issues that have to be dealt with—and that means facing problems of revolutionary change.

There are doubtless many debates over this issue. All those debates assume that investors have the right to decide what happens. So we have to make things as attractive as possible to them. But as long as the investors have the right to decide what happens, nothing much is going to change.

It's like trying to decide whether to change from proportional representation to some other kind of representation in the state-run parliament of a totalitarian state. That might change things a little, but it's not going to matter much.

Until you get to the source of power, which ultimately is investment decisions, other changes are cosmetic and can only take place in a limited way. If they go too far, the investors will just make other choices, and there's nothing much you can do about it.

To challenge the right of investors to determine who lives, who dies, and how they live and die—that would be a significant move toward Enlightenment ideals (actually the classical liberal ideal). That would be revolutionary.

I'd like you to address another factor at work here. Psychologically, it's a lot easier to criticize something than to promote something constructive. There's a completely different dynamic at work.

You can see a lot of things that are wrong. Small changes you can propose. But to be realistic, substantial change (which will really alter the large-scale direction of things and overcome major problems) will require profound democratization of the society and the economic system.

A business or a big corporation is a fascist structure internally. Power is at the top. Orders go from top to bottom. You either follow the orders or get out.

The concentration of power in such structures means that everything in the ideological or political domains is sharply constrained. It's not totally controlled, by any means. But it's sharply constrained. Those are just facts.

The international economy imposes other kinds of constraints. You can't overlook those things—they're just true. If anybody bothered to read [the Scottish moral philosopher] Adam Smith instead of prating about him, they'd see he pointed out that social policy is class-based. He took the class analysis for granted.

If you studied the canon properly at the University of Chicago [home of Milton Friedman and other right-wing economists], you learned that Adam Smith denounced the mercantilist system and colonialism because he was in favor of free trade. That's only half the truth. The other half is that he pointed out that the mercantilist system and colonialism were very beneficial to the "merchants and manufacturers...the principal architects of policy" but were harmful to the people of England.

In short, it was a class-based policy which worked for the rich and powerful in England. The people of England paid the costs. He was opposed to that because he was an enlightened intellectual, but he recognized it. Unless you recognize it, you're just not in the real world.

NAFTA and GATT—who benefits?

The last US-based typewriter company, Smith Corona, is moving to Mexico. There's a whole corridor of *maquiladoras* [factories where parts made elsewhere are assembled at low wages] along the border. People work for five dollars a day, and there are incredible levels of pollution, toxic waste, lead in the water, etc.

One of the major issues before the country right now is the North American Free Trade Agreement. There's no doubt that NAFTA's going to have very large effects on both Americans and Mexicans. You can debate what the effect will be, but nobody doubts that it'll be significant.

Quite likely the effect will be to accelerate just what you've been describing—a flow of productive labor to Mexico. There's a brutal and repressive dictatorship there, so it's guaranteed wages will be low.

During what's been called the "Mexican economic miracle" of the last decade, their wages have dropped 60%. Union organizers get killed. If the Ford Motor Company wants to toss out its work force and hire super cheap labor, they just do it. Nobody stops them. Pollution goes on unregulated. It's a great place for investors.

One might think that NAFTA, which includes sending pro - ductive labor down to Mexico, might improve their real wages, maybe level the two countries. But that's most unlikely. One reason is that the repression there prevents organizing for higher wages. Another reason is that NAFTA will flood Mexico with industrial agricultural products from the United States.

These products are all produced with big public subsidies, and they'll undercut Mexican agriculture. Mexico will be flooded with American crops, which will contribute to driving an estimated thirteen million people off the land to urban areas or into the *maquiladora* areas—which will again drive down wages.

NAFTA will very likely be quite harmful for American workers too. We may lose hundreds of thousands of jobs, or lower the level of jobs. Latino and black workers are the ones who are going to be hurt most.

But it'll almost certainly be a big bonanza for investors in the United States and for their counterparts in the wealthy sectors in Mexico. They're the ones—along with the professional classes who work for them—who are applauding the agreement.

Will NAFTA and GATT essentially formalize and institutionalize relations between the North [prosperous, industrialized, mostly northern nations] and the South [poorer, less industrialized, mostly southern nations]?

That's the idea. NAFTA will also almost certainly degrade environmental standards. For example, corporations will be able to argue that EPA [the Environmental Protection Agency] standards are violations of free-trade agreements. This is already happening in the Canada-US part of the agreement. Its general effect will be to drive life down to the lowest level while keeping profits high.

It's interesting to see how the issue has been handled. The public hasn't the foggiest idea what's going on. In fact, they can't know. One reason is that NAFTA is effectively a secret—it's an executive agreement that isn't publicly available.

In 1974, the Trade Act was passed by Congress. One of its provisions was that the Labor Advisory Committee—which is based in the unions—had to have input and analysis on any trade-related issue. Obviously that committee had to report on NAFTA, which was an executive agreement signed by the president.

The Labor Advisory Committee was notified in mid-August 1992 that their report was due on September 9, 1992. However, they weren't given a text of the agreement until about 24 hours before the report was due. That meant they couldn't even convene, and they obviously couldn't write a serious report in time.

Now these are conservative labor leaders, not the kind of guys who criticize the government much. But they wrote a very acid report. They said that, to the extent that we can look at this in the few hours given to us, it looks like it's going to be a disaster for working people, for the environment, for Mexicans—and a great boon for investors.

The committee pointed out that although treaty advocates said it won't hurt many American workers, maybe just unskilled workers, their definition of "unskilled worker" would include 70% of the workforce. The committee also pointed out that property rights were being protected all over the place, but workers' rights were scarcely mentioned. The committee then bitterly condemned the utter contempt for democracy that was demonstrated by not giving the committee the complete text ahead of time.

GATT is the same—nobody knows what's going on there unless they're some kind of specialist. And GATT is even more far-reaching. One of the things being pressed very hard in those negotiations is what's called "intellectual property rights." That means protection for patents—also things like software, records, etc. The idea is to guarantee that the technology of the future remains in the hands of multinational corporations, for whom the world government works.

You want to make sure, for example, that India can't produce drugs for its population at 10% the cost of drugs produced by Merck Pharmaceutical, which is government supported and subsidized. Merck relies extensively on research that comes out of university biology laboratories (which are supported by public funds) and on all sorts of other forms of government intervention.

Have you seen details of these treaties?

By now it's theoretically possible to get a text. But what I've seen is the secondary comment on the text, like the Labor Advisory Committee report, and the report of the Congressional Office of Technology Assessment, which is fairly similar.

The crucial point is that even if you and I could get a text, what does that mean for American democracy? How many people even know that this is going on? The Labor Advisory

Committee report, and the fact that the treaty was withheld from the Committee, was never even reported by the press (to my knowledge).

I just came back from a couple of weeks in Europe, where GATT is a pretty big issue for the people in the countries of the European Community. They're concerned about the gap that's developing between executive decisions (which are secret) and democratic (or at least partially democratic) institutions like parliaments, which are less and less able to influence decisions made at the EU level.

It seems that the Clinton-Gore administration is going to be in a major conflict. It supports NAFTA and GATT, while at the same time talking—at least rhetorically—about its commitment to environmental protection and creating jobs for Americans.

I would be very surprised if there's a big conflict over that. I think your word "rhetorically" is accurate. Their commitment is to US-based corporations, which means transnational corporations. They approve of the form NAFTA is taking—special protection for property rights, but no protection for workers' rights—and the methods being developed to undercut environmental protection. That's in their interests. I doubt that there'll be a conflict in the administration unless there's a lot of public pressure.

Food and Third World "economic miracles"

Talk about the political economy of food, its production and distribution, particularly within the framework of IMF and World Bank policies. These institutions extend loans under very strict conditions to the nations of the South: they have to promote the market economy, pay back the loans in hard currency and increase exports—like coffee, so that we can drink cappuccino, or beef, so that we can eat hamburgers—at the expense of indigenous agriculture.

You've described the basic picture. It's also interesting to have a close look at the individual cases. Take Bolivia. It was in trouble. There'd been brutal, highly repressive dictators, huge debt—the whole business.

The West went in—Jeffrey Sachs, a leading Harvard expert, was the advisor—with the IMF rules: stabilize the currency, increase agro-export, cut down production for domestic needs, etc. It worked. The figures, the macroeconomic statistics, looked quite good. The currency has been stabilized. The debt has been reduced. The GNP has been increasing.

But there are a few flies in the ointment. Poverty has rapidly increased. Malnutrition has increased. The educational system has collapsed. But the most interesting thing is what's stabilized the economy—exporting coca [the plant from which cocaine is made]. It now accounts for about two-thirds of Bolivian exports, by some estimates.

The reason is obvious. Take a peasant farmer somewhere and flood his area with US-subsidized agriculture—maybe through a Food for Peace program—so he can't produce or compete. Set up a situation in which he can only function as an agricultural exporter. He's not an idiot. He's going to turn to the most profitable crop, which happens to be coca.

The peasants, of course, don't get much money out of this, and they also get guns and DEA [the US Drug Enforcement Agency] helicopters. But at least they can survive. And the world gets a flood of coca exports.

The profits mostly go to big syndicates or, for that matter, to New York banks. Nobody knows how many billions of dollars of cocaine profits pass through New York banks or their offshore affiliates, but it's undoubtedly plenty.

Plenty of it also goes to US-based chemical companies which, as is well known, are exporting the chemicals used in cocaine production to Latin America. So there's plenty of profit. It's probably giving a shot in the arm to the US economy as well. And it's contributing nicely to the international drug epidemic, including here in the US.

That's the economic miracle in Bolivia. And that's not the only case. Take a look at Chile. There's another big economic miracle. The poverty level has increased from about 20% during the Allende years [Salvador Allende, a democratically elected Socialist president of Chile, was assassinated in a US-backed military coup in 1973] up to about 40% now, after the great miracle. And that's true in country after country.

These are the kinds of consequences that will follow from what has properly been called "IMF fundamentalism." It's having a disastrous effect everywhere it's applied.

But from the point of view of the perpetrators, it's quite successful. As you sell off public assets, there's lots of money to be made, so much of the capital that fled Latin America is now back. The stock markets are doing nicely. The professionals and businessmen are very happy with it. And they're the ones who make the plans, write the articles, etc.

And now the same methods are being applied in Eastern Europe. In fact, the same people are going. After Sachs carried through the economic miracle in Bolivia, he went off to Poland and Russia to teach them the same rules.

You hear lots of praise for this economic miracle in the US too, because it's just a far more exaggerated version of what's happening here. The wealthy sector is doing fine, but the general public is in deep trouble. It's mild compared with the Third World, but the structure is the same.

Between 1985 and 1992, Americans suffering from hunger rose from twenty to thirty million. Yet novelist Tom Wolfe described the 1980s as one of the "great golden moments that humanity has ever experienced."

A couple of years ago, Boston City Hospital—that's the hospital for the poor and the general public in Boston, not the fancy Harvard teaching hospital—had to institute a malnutrition clinic, because they were seeing it at Third World levels.

Most of the deep starvation and malnutrition in the US had pretty well been eliminated by the Great Society programs in the 1960s. But by the early 1980s it was beginning to creep up again, and now the latest estimates are thirty million or so in deep hunger.

It gets much worse over the winter because parents have to make an agonizing decision between heat and food, and children die because they're not getting water with some rice in it.

The group World Watch says that one of the solutions to the shortage of food is control of population. Do you support efforts to limit population?

First of all, there's no shortage of food. There are serious problems of distribution. That aside, I think there should be efforts to control population. There's a well-known way to do it—increase the economic level.

Population is declining very sharply in industrial societies. Many of them are barely reproducing their own population. Take Italy, which is a late industrializing country. The birth rate now doesn't reproduce the population. That's a standard phenomenon.

Coupled with education?

Coupled with education and, of course, the means for birth control. The United States has had a terrible role. It won't even help fund international efforts to provide education about birth control.

Photo ops in Somalia

Does Operation Restore Hope in Somalia represent a new pattern of US intervention in the world?

I don't think it really should be classified as an intervention. It's more of a public relations operation for the Pentagon.

In fact, it's intriguing that it was almost openly stated this time. Colin Powell, the [former] Chairman of the Joint Chiefs, made a statement about how this was a great public relations job for the military. A *Washington Post* editorial described it as a bonanza for the Pentagon.

The reporters could scarcely fail to see what was happening. After all, when the Pentagon calls up all the news bureaus and major television networks and says: "Look, be at such-and-such a beach at such-and-such an hour with your cameras aiming in this direction because you're going to watch Navy Seals climbing out of the water and it will be real exciting," nobody can fail to see that this is a PR job. That would be a level of stupidity that's too much for anyone.

The best explanation for the "intervention," in my opinion, was given in an article in the *Financial Times* on the day of the landing. It didn't mention Somalia—it was about the US recession and why the recovery is so sluggish.

It quoted various economists from investment firms and banks—guys that really care about the economy. The consensus was that the recovery is slow because the standard method of government stimulation—pump-priming through the Pentagon system—simply isn't available to the extent that it's been in the past.

Bush put it pretty honestly in his farewell address when he explained why we intervened in Somalia and not Bosnia. What it comes down to is that in Bosnia somebody might shoot at us. In Somalia it's just a bunch of teenaged kids. We figure thirty thousand Marines can handle that.

The famine was pretty much over and fighting had declined. So it's photo opportunities, basically. One hopes it will help the Somalis more than harm them, but they're more or less incidental. They're just props for Pentagon public relations.

This has to be finessed by the press at the moment, because Somalia is not a pretty story. The US was the main support for Siad Barre, a kind of Saddam Hussein clone, from 1978 through 1990 (so it's not ancient history). He was tearing the country apart.

He destroyed the civil and social structures—in fact, laid the basis for what's happening now—and, according to Africa Watch [a human rights monitoring group based in Washington DC], probably killed fifty or sixty thousand people. The US was, and may well be still, supporting him. The forces, mostly loyal to him, are being supported through Kenya, which is very much under US influence.

The US was in Somalia for a reason—the military bases there are part of the system aimed at the Gulf region. However, I doubt that that's much of a concern at this point. There are much more secure bases and more stable areas. What's needed now, desperately needed, is some way to prevent the Pentagon budget from declining.

When the press and commentators say the US has no interests there, that's taking a very narrow and misleading view. Maintaining the Pentagon system is a major interest for the US economy.

A Navy and Marine White Paper in September 1992 discussed the military's shift in focus from global threats to "regional challenges and opportunities," including "humanitarian assistance and nation-building efforts in the Third World."

That's always been the cover, but the military budget is mainly for intervention. In fact, even strategic nuclear forces were basically for intervention.

The US is a global power. It isn't like the Soviet Union, which used to carry out intervention right around its borders, where they had an overwhelming conventional force advantage. The US carried out intervention everywhere—in Southeast Asia, in the Middle East and in places where it had no such dominance. So the US had to have an extremely intimidating posture to make sure that nobody got in the way.

That required what was called a "nuclear umbrella"—powerful strategic weapons to intimidate everybody, so that conventional forces could be an instrument of political power. In fact, almost the entire military system—its military aspect, not its economic aspect—was geared for intervention. But that was often covered as "nation-building." In Vietnam, in Central America—we're always humanitarian.

So when the Marine Corps documents say we now have a new mission—"humanitarian nation-building"—that's just the old cover story. We now have to emphasize it more because the traditional pretext—the conflict with the Russians—is gone, but it's the same as it's always been.

What kind of impact will the injection of US armed forces into Somalia have on the civil society? Somalia has been described by one US military official as "Dodge City" and the Marines as "Wyatt Earp." What happens when the marshal leaves town?

First of all, that description has little to do with Somalia. One striking aspect of this intervention is that there's no concern for Somalia. No one who knew anything about Somalia was involved in planning it, and there's no interaction with Somalis as far as we know (so far, at least).

Since the Marines have gone in, the only people they've dealt with are the so-called "warlords," and they're the biggest gangsters in the country. But Somalia is a *country.* There are people who know and care about it, but they don't have much of a voice here.

One of the most knowledgeable is a Somali woman named Rakiya Omaar, who was the executive director of Africa Watch. She did most of the human rights work, writing, etc. up until the intervention. She strongly opposed the intervention and was fired from Africa Watch.

Another knowledgeable voice is her co-director, Alex de Waal, who resigned from Africa Watch in protest after she was fired. In addition to his human rights work, he's an academic specialist on the region. He's written many articles and has published a major book on the Sudan famine with Oxford University Press. He knows not only Somalia but the region very well. And there are others. Their picture is typically quite different from the one we get here.

Siad Barre's main atrocities were in the northern part of Somalia, which had been a British colony. They were recovering from his US-backed attack and were pretty well organized (although they could, no doubt, have used aid). Their own civil society was emerging—a rather traditional one, with traditional elders, but with lots of new groups. Women's groups, for example, emerged in this crisis.

The area of real crisis was one region in the south. In part, that's because of General Mohammed Hersi's forces, which are supported from Kenya. (Hersi, who's known as *Morgan,* is Siad Barre's son-in-law.) His forces, as well as those of General Mohammed Farah Aidid and Ali Mahdi, were carrying out some of the worst atrocities. This led to a serious breakdown in which people just grabbed guns in order to survive. There was lots of looting, and teenaged gangsters.

By September–October [1992], that region was already re-covering. Even though groups like US Care and the UN operations were extremely incompetent, other aid groups—like the International Red Cross, Save The Children, and smaller groups like the American Friends Service Committee or Australian Care—were getting most of the aid through.

By early November, 80–90% of their aid was reportedly getting through; by late November the figures were up to 95%. The reason was that they were working with the reconstituting Somalian society. In this southern corner of real violence and starvation, things were already recovering, just as they had in the north.

A lot of this had been under the initiative of a UN negotiator, Mohammed Sahnoun of Algeria, who was extremely successful and highly respected on all sides. He was working with traditional elders and the newly emerging civic groups, especially the women's groups, and they were coming back together under his guidance, or at least his initiative.

But Sahnoun was kicked out by [UN Secretary General] Boutros-Ghali in October because he publicly criticized the incompetence and corruption of the UN effort. The UN put in an Iraqi replacement, who apparently achieved very little.

A US intervention was apparently planned for shortly after the election. The official story is that it was decided upon at the end of November, when George Bush saw heart-rending pictures on television. But, in fact, US reporters in Baidoa in early November saw Marine officers in civilian clothes walking around and scouting out the area, planning for where they were going to set up their base.

This was rational timing. The worst crisis was over, the society was reconstituting and you could be pretty well guaranteed a fair success at getting food in, since it was getting in anyway. Thirty thousand troops would only expedite it in the short term. There wouldn't be too much fighting, because that was subsiding. So it wasn't Dodge City.

Bush got the photo opportunities and left somebody else to face the problems that were bound to arise later on. Nobody cared what happened to the Somalis. If it works, great, we'll applaud and cheer ourselves and bask in self-acclaim. If it turns into a disaster, we'll treat it the same as other interventions that turn into disasters.

After all, there's a long series of them. Take Grenada. That was a humanitarian intervention. We were going to save the people from tragedy and turn it into what Reagan called a "showplace for democracy" or a "showplace for capitalism."

The US poured aid in. Grenada had the highest per capita aid in the world the following year—next to Israel, which is in another category. And it turned into a complete disaster.

The society is in total collapse. About the only thing that's functioning there is money-laundering for drugs. But nobody hears about it. The television cameras were told to look somewhere else.

So if the Marine intervention turns out to be a success, which is conceivable, then there'll be plenty of focus on it and how marvelous we are. If it turns into a disaster, it's off the map—forget about it. Either way we can't lose.

Slav vs. Slav

Would you comment on the events in the former Yugoslavia, which constitute the greatest outburst of violence in Europe in fifty years— tens of thousands killed, hundreds of thousands of refugees. This isn't some remote place like East Timor we're talking about—this is Europe—and it's on the news every night.

In a certain sense, what's happening is that the British and American right wings are getting what they asked for. Since the 1940s they've been quite bitter about the fact that Western support turned to Tito and the partisans, and against Mikailhovich and his Chetniks, and the Croatian anti-Communists, including the Ustasha, who were outright Nazis. The Chetniks were also playing with the Nazis and were trying to overcome the partisans.

The partisan victory imposed a communist dictatorship, but it also federated the country. It suppressed the ethnic violence that had accompanied the hatreds and created the basis of some sort of functioning society in which the parts had their role. We're now essentially back to the 1940s, but without the partisans.

Serbia is the inheritor of the Chetniks and their ideology. Croatia is the inheritor of the Ustasha and its ideology (less ferocious than the Nazi original, but similar). It's possible that they're now carrying out pretty much what they would've done if the partisans hadn't won.

Of course, the leadership of these elements comes from the Communist party, but that's because every thug in the region went into the ruling apparatus. (Yeltsin, for example, was a Communist party boss.)

It's interesting that the right wing in the West—at least its more honest elements—defend much of what's happening. For example, Nora Beloff, a right-wing British commentator on Yugoslavia, wrote a letter to the *Economist* condemning those who denounce the Serbs in Bosnia. She's saying it's the fault of the Muslims. They're refusing to accommodate the Serbs, who are just defending themselves.

She's been a supporter of the Chetniks from way back, so there's no reason why she shouldn't continue to support Chetnik violence (which is what this amounts to). Of course there may be another factor. She's an extremist Zionist, and the fact that the Muslims are involved already makes them guilty.

Some say that, just as the Allies should have bombed the rail lines to Auschwitz to prevent the deaths of many people in concentration camps, so we should now bomb the Serbian gun positions surrounding Sarajevo that have kept that city under siege. Would you advocate the use of force?

First of all, there's a good deal of debate about how much effect bombing the rail lines to Auschwitz would have had. Putting that aside, it seems to me that a judicious threat and use of force, not by the Western powers but by some international or multinational group, might, at an earlier stage, have suppressed a good deal of the violence and maybe blocked it. I don't know if it would help now.

If it were possible to stop the bombardment of Sarajevo by threatening to bomb some emplacements (and perhaps even carrying the threat out), I think you could give an argument for it. But that's a very big *if*. It's not only a moral issue—you have to ask about the consequences, and they could be quite complex.

What if a Balkan war were set off? One consequence is that conservative military forces within Russia could move in. They're already there, in fact, to support their Slavic brothers in Serbia. They might move in *en masse*. (That's traditional, incidentally. Go back to Tolstoy's novels and read about how Russians were going to the south to save their Slavic brothers from attacks. It's now being reenacted.)

At that point you're getting fingers on nuclear weapons involved. It's also entirely possible that an attack on the Serbs, who feel that they're the aggrieved party, could inspire them to move more aggressively in Kosovo, the Albanian area. That could set off a large-scale war, with Greece and Turkey involved. So it's not so simple.

Or what if the Bosnian Serbs, with the backing of both the Serbian and maybe even other Slavic regions, started a guerrilla war? Western military "experts" have suggested it could take a hundred thousand troops just to more or less hold the area. Maybe so.

So one has to ask a lot of questions about consequences. Bombing Serbian gun emplacements sounds simple, but you have to ask how many people are going to end up being killed. That's not so simple.

Zeljko Raznjatovic, known as Arkan, a fugitive wanted for bank robbery in Sweden, was elected to the Serb Parliament in December 1992. His Tigers' Militia is accused of killing civilians in Bosnia. He's among ten people listed by the US State Department as a possible war criminal. Arkan dismissed the charges and said, "There are a lot of people in the United States I could list as war criminals."

That's quite correct. By the standards of Nuremberg, there are plenty of people who could be listed as war criminals in the West. It doesn't absolve him in any respect, of course.

The chosen country

The conditions of the US-Israel alliance have changed, but have there been any structural changes?

There haven't been any significant structural changes. It's just that the capacity of Israel to serve US interests, at least in the short term, has probably increased.

The Clinton administration has made it very clear that it intends to persist in the extreme pro-Israeli bias of the Bush administration. They've appointed Martin Indyk, whose background is in AIPAC [the American Israel Public Affairs Committee, the main pro-Israel lobbying group in the US], to the Middle East desk of the National Security Council.

He's headed a fraudulent research institute, the Washington Institute for Near East Studies. It's mainly there so that journalists who want to publish Israeli propaganda, but want to do it "objectively," can quote somebody who'll express what they want said.

The United States has always had one major hope from the so-called peace negotiations—that the traditional tacit alliance between Israel and the family dictatorships ruling the Gulf states will somehow become a little more overt or solidified. And it's conceivable.

There's a big problem, however. Israel's plans to take over and integrate what they want of the occupied territories—plans which have never changed—are running into some objective problems. Israel has always hoped that in the long run they would be able to expel much of the Palestinian population.

Many moves were made to accelerate that. One of the reasons they instituted an educational system on the West Bank was in hopes that more educated people would want to get out because there weren't any job opportunities.

For a long time it worked—they were able to get a lot of people to leave—but they now may well be stuck with the population. This is going to cause some real problems, because Israel intends to take the water and the usable land. That may not be so pretty or so easy.

What's Israel's record of compliance with the more than twenty Security Council resolutions condemning its policies?

It's in a class by itself.

No sanctions, no enforcement?

None. Just to pick one at random—Security Council Resolution 425 of March 1978. It called on Israel to withdraw immediately and unconditionally from southern Lebanon. Israel is still there, even though the request was renewed by the government of Lebanon in February 1991, when everyone was going at Iraq.

The United States will block any attempt to change things. Many of the large number of Security Council resolutions vetoed by the US have to do with Israeli aggression or atrocities.

For example, take the invasion of Lebanon in 1982. At first the United States went along with the Security Council condemnations. But within a few days the US had vetoed the major Security Council resolution that called on everyone to withdraw and stop fighting, and later vetoed another, similar one.

The US has gone along with the last few UN resolutions or deportations.

The US has gone along, but has refused to allow them to have any teeth. The crucial question is: Does the US do anything about it? For example, the United States went along with the Security Council resolution condemning the annexation of the Golan Heights. But when the time came to do something about it, they refused.

International law transcends state law, but Israel says these resolutions are not applicable. How are they not applicable?

Just as international law isn't applicable to the United States, which has even been condemned by the World Court. States do what they feel like—though of course small states have to obey.

Israel's not a small state. It's an appendage to the world superpower, so it does what the United States allows. The United States tells it: You don't have to obey any of these resolutions, therefore they're null and void—just as they are when the US gets condemned.

The US never gets condemned by a Security Council resolution, because it vetoes them. Take the invasion of Panama. There were two resolutions in the Security Council condemning the United States for that invasion. We vetoed them both.

You can find repeated Security Council resolutions that never passed that condemn the US, ones which would have passed if they were about a defenseless country. And the General Assembly passes resolutions all the time, but they have no standing—they're just recommendations.

I remember talking to Mona Rishmawi, a lawyer for the human rights organization Al Haq in Ramallah on the West Bank. She told me that when she would go to court, she wouldn't know whether the Israeli prosecutor would prosecute her clients under British mandate emergency law, Jordanian law, Israeli law or Ottoman law.

Or their own laws. There are administrative regulations, some of which are never published. As any Palestinian lawyer will tell you, the legal system in the territories is a joke. There's no law—just pure authority.

Most of the convictions are based on confessions, and everybody knows what it means when people confess. Finally, after about sixteen years, a Druze Israeli army veteran who'd confessed and was sentenced was later proven to be innocent. Then it became a scandal.

There was an investigation, and the Supreme Court stated that for sixteen years the secret services had been lying to them. The secret services had been torturing people—as everybody knew—but telling the Court they weren't.

There was a big fuss about the fact that they'd been lying to the Supreme Court. How could you have a democracy when they lie to the Supreme Court? But the torture wasn't a big issue—everyone knew about it all along.

Amnesty International interviewed Supreme Court Justice Moshe Etzioni in London in 1977. They asked him to explain why such an extremely high percentage of Arabs confessed. He said, "It's part of their nature."

That's the Israeli legal system in the territories.

Explain these Orwellisms of "security zone" and "buffer zone."

In southern Lebanon? That's what Israel calls it, and that's how it's referred to in the media.

Israel invaded southern Lebanon in 1978. It was all in the context of the Camp David agreements. It was pretty obvious that those agreements would have the consequence they did—namely, freeing up Israel to attack Lebanon and inte - grate the occupied territories, now that Egypt was eliminated as a deterrent.

Israel invaded southern Lebanon and held onto it through clients—at the time it was Major Sa'ad Haddad's militia, basically an Israeli mercenary force. That's when Security Council Resolution 425 was passed.

When Israel invaded in 1982, there'd been a lot of recent violence across the border, all from Israel north. There had been an American-brokered ceasefire which the PLO had held to scrupulously, initiating no cross-border actions. But Israel carried out thousands of provocative actions, including bombing of civilian targets—all to try to get the PLO to do something, thus giving Israel an excuse to invade.

It's interesting the way that period has been reconstructed in American journalism. All that remains is tales of the PLO's bombardment of Israeli settlements, a fraction of the true story (and in the year leading up to the 1982 Israeli invasion, not even that).

The truth was that Israel was bombing and invading north of the border, and the PLO wasn't responding. In fact, they were trying to move towards a negotiated settlement. (The truth about earlier years also has only a limited resemblance to the standard picture, as I've documented several times—uselessly, of course.)

We know what happened after Israel invaded Lebanon. They were driven out by what they call "terrorism"—meaning resistance by people who weren't going to be cowed. Israel succeeded in awakening a fundamentalist resistance, which it couldn't control. They were forced out.

They held on to the southern sector, which they call a "security zone"—although there's no reason to believe that it has the slightest thing to do with security. It's Israel's foothold in Lebanon. It's now run by a mercenary army, the South Lebanon Army, which is backed up by Israeli troops. They're very brutal. There are horrible torture chambers.

We don't know the full details, because they refuse to allow inspections by the Red Cross or anyone else. But there have been investigations by human rights groups, journalists and others. Independent sources—people who got out, plus some Israeli sources—overwhelmingly attest to the brutality. There was even an Israeli soldier who committed suicide because he couldn't stand what was going on. Some others have written about it in the Hebrew press.

Ansar is the main camp. They very nicely put it in the town of Khiyam. There was a massacre there by the Haddad militia under Israeli eyes in 1978, after years of Israeli bombing, that drove out most of the population. That's mainly for Lebanese who refuse to cooperate with the South Lebanon Army.

So that's the "security zone."

Israel dumped scores of deportees in Lebanon in the 1970s and 1980s. Why has that changed now? Why has Lebanon refused?

It's not so much that it has refused. If Israel dropped some deportees by helicopter into the outskirts of Sidon, Lebanon couldn't refuse. But this time I think Israel made a tactical error. The deportation of 415 Palestinians [in December 1992] is going to be very hard for them to deal with.

According to the Israeli press, this mass deportation was fairly random, a brutal form of collective punishment. I read in *Ha'aretz* [the leading Israeli newspaper] that the Shabak [the Israeli secret police] leaked the information that they had only given six names of security risks, adding a seventh when the Rabin Labor government wanted a larger number. The other four hundred or so were added by Rabin's government, without intelligence information.

So there's no reason to believe that those who were deported were Hamas [Islamic fundamentalist] activists. In fact, Israel deported virtually the whole faculty of one Islamic university. They essentially deported the intellectuals, people involved in welfare programs and so on.

But to take this big class of people and put them in the mountains of southern Lebanon, where it's freezing now and boiling hot in the summer—that's not going to look pretty in front of the TV cameras. And that's the only thing that matters. So there may be some problems, because Israel's not going to let them back in without plenty of pressure.

I heard Steven Solarz [former Democratic congressman from Brooklyn] on the BBC. He said the world has a double standard: 700,000 Yemenis were expelled from Saudi Arabia and no one said a word (which is true); 415 Palestinians get expelled from Gaza and the West Bank and everybody's screaming.

Every Stalinist said the same thing: "We sent Sakharov into exile and everyone was screaming. What about this or that other atrocity—which is worse?" There is always somebody who has committed a worse atrocity. For a Stalinist mimic like Solarz, why not use the same line?

Incidentally, there is a difference—the Yemenis were deported *to* their country, the Palestinians *from* their country. Would Solarz claim that we all should be silent if he and his family were dumped into a desert in Mexico?

Israel's record and its attitude toward Hamas have evolved over the years. Didn't Israel once favor it?

They not only favored it, they tried to organize and stimulate it. Israel was sponsoring Islamic fundamentalists in the early days of the [first] *intifada* [an uprising of Palestinians within Israel against the Israeli government]. If there was a strike of students at some West Bank university, the Israeli army would sometimes bus in Islamic fundamentalists to break up the strike.

Sheikh Yaseen, an antisemitic maniac in Gaza and the leader of the Islamic fundamentalists, was protected for a long time. They liked him. He was saying, "Let's kill all the Jews." It's a standard thing, way back in history. Seventy years ago Chaim Weizmann was saying: Our danger is Arab moderates, not the Arab extremists.

The invasion of Lebanon was the same thing. Israel wanted to destroy the PLO because it was secular and nationalist, and was calling for negotiations and a diplomatic settlement. That was the threat, not the terrorists. Israeli commentators have been quite frank about that from the start.

Israel keeps making the same mistake, with the same predictable results. In Lebanon, they went in to destroy the threat of moderation and ended up with Hezbollah [Iranian-backed fundamentalists] on their hands. In the West Bank, they also wanted to destroy the threat of moderation—people who wanted to make a political settlement. There Israel's ending up with Hamas, which organizes effective guerrilla attacks on Israeli security forces.

It's important to recognize how utterly incompetent secret services are when it comes to dealing with people and politics. Intelligence agencies make the most astonishing mistakes—just as academics do.

In a situation of occupation or domination, the occupier, the dominant power, has to justify what it's doing. There is only one way to do it—become a racist. You have to blame the victim. Once you become a raving racist in self-defense, you've lost your capacity to understand what's happening.

The US in Indochina was the same. They never could understand—there are some amazing examples in the internal record. The FBI is the same; they make the most astonishing mistakes, for similar reasons.

In a letter to the *New York Times,* the director of the [B'nai Brith's] Anti-Defamation League, Abraham Foxman, wrote that the Rabin government has "unambiguously demonstrated its commitment to the peace process" since assuming leadership. "Israel is the last party that has to prove its desire to make peace." What's been the record of Rabin's Labor government?

It's perfectly true that Israel wants peace. So did Hitler. Everybody wants peace. The question is, On what terms?

The Rabin government, exactly as was predicted, harshened the repression in the territories. Just this afternoon I was speaking to a woman who's spent the last couple of years in Gaza doing human rights work. She reported what everyone reports, and what everybody with a brain knows—as soon as Rabin came, it got tougher. He's the iron-fist man—that's his record.

Likud [the major irght-wing party in Israel] actually had a better record in the territories than Labor did. Torture and collective punishment stopped under Begin. There was one bad period when Sharon was in charge, but under Begin it was generally better. When the Labor party came back into the government in 1984, torture and collective punishment started again, and later the *intifada* came.

In February 1989, Rabin told a group of Peace Now leaders that the negotiations with the PLO didn't mean anything—they were going to give him time to crush the Palestinians by force. And they will be crushed, he said, they will be broken.

It hasn't happened.

It happened. The *intifada* was pretty much dead, and Rabin awakened it again with his own violence. He has also continued settlement in the occupied territories, exactly as everyone with their eyes open predicted. Although there was a very highly publicized settlement cutoff, it was clear right away that it was a fraud. Foxman knows that. He reads the Israeli press, I'm sure.

What Rabin stopped was some of the more extreme and crazy Sharon plans. Sharon was building houses all over the place, in places where nobody was ever going to go, and the country couldn't finance it. So Rabin eased back to a more rational settlement program. I think the current number is 11,000 new homes going up.

Labor tends to have a more rational policy than Likud—that's one of the reasons the US has always preferred Labor. They do pretty much the same things as Likud, but more quietly, less brazenly. They tend to be more modern in their orientation, better attuned to the norms of Western hypocrisy. Also, they're more realistic. Instead of trying to make seven big areas of settlement, they're down to four.

But the goal is pretty much the same—to arrange the settlements so that they separate the Palestinian areas. Big highway networks will connect Jewish settlements and surround some little Arab village way up in the hills. That's to make certain that any local autonomy will never turn into a form of meaningful self-government. All of this is continuing and the US is, of course, funding it.

Critics of the Palestinian movement point to what they call the *"intrafada,"* the fact that Palestinians are killing other Palestinians—as if this justifies Israeli rule and delegitimizes Palestinian aspirations.

You might look back at the Zionist movement—there were plenty of Jews killed by other Jews. They killed collaborators, traitors and people they thought were traitors. And they weren't under anything like the harsh conditions of the Palestinian occupation. As plenty of Israelis have pointed out, the British weren't nice, but they were gentlemen compared with us.

The Labor-based defense force Haganah had torture chambers and assassins. I once looked up their first recorded assassination in the official Haganah history. It's described there straight.

It was in 1921. A Dutch Jew named Jacob de Haan had to be killed, because he was trying to approach local Palestinians to see if things could be worked out between them and the new Jewish settlers. His murderer was assumed to be the woman who later became the wife of the first president of Israel. They said that another reason for assassinating him was that he was a homosexual.

Yitzhak Shamir became head of the Stern gang by killing the guy who was designated to be the head. He didn't like him for some reason. Shamir was supposed to take a walk with him on a beach. He never came back. Everyone knows Shamir killed him.

As the *intifada* began to self-destruct under tremendous repression, the killing got completely out of hand. It began to be a matter of settling old scores and gangsters killing anybody they saw. Originally the *intifada* was pretty disciplined, but it ended up with a lot of random killing, which Israel loves. Then they can point out how rotten the Arabs are.

It's a dangerous neighborhood.

Yes, it is. They help make it dangerous.

Gandhi, nonviolence and India

I've never heard you talk about Gandhi. Orwell wrote of him that, "Compared to other leading political figures of our times, how clean a smell he has managed to leave behind." What are your views on the Mahatma?

I'd hesitate to say without analyzing more closely what he did and what he achieved. There were some positive things— for example, his emphasis on village development, self-help and communal projects. That would have been very healthy for India. Implicitly, he was suggesting a model of development that could have been more successful and humane than the Stalinist model that was adopted (which emphasized the development of heavy industry, etc.).

But you really have to think through the talk about nonviolence. Sure, everybody's in favor of nonviolence rather than violence, but under what conditions and when? Is it an absolute principle?

You know what he said to Lewis Fisher in 1938 about the Jews in Germany—that German Jews ought to commit collective suicide, which would "have aroused the world and the people of Germany to Hitler's violence."

He was making a tactical proposal, not a principled one. He wasn't saying that they should have walked cheerfully into the gas chambers because that's what nonviolence dictates. He was saying, *If you do it, you may be better off.*

If you divorce his proposal from any principled concern other than how many people's lives can be saved, it's conceivable that it would have aroused world concern in a way that the Nazi slaughter didn't. I don't believe it, but it's not literally impossible. On the other hand, there's nothing much that the European Jews could have done anyway under the prevailing circumstances, which were shameful everywhere.

Orwell adds that after the war Gandhi justified his position, saying, *The Jews had been killed anyway and might as well have died significantly.*

Again, he was making a tactical, not a principled, statement. One has to ask what the consequences of the actions he recommended would have been. That's speculation based on little evidence. But for him to have made that recommendation at the time would have been grotesque.

What he should have been emphasizing is: "Look, powerless people who are being led to slaughter can't do anything. Therefore it's up to others to prevent them from being massacred." To give them advice on how they should be slaughtered isn't very uplifting—to put it mildly.

You can say the same about lots of other things. Take people being tortured and murdered in Haiti. You want to tell them: "The way you ought to do it is to walk up to the killers and put your head in front of their knife—and maybe people on the outside will notice." Could be. But it'd be a little more significant to tell the people who are giving the murderers the knives that they should do something better.

Preaching nonviolence is easy. One can take it seriously when it's someone like [long-time pacifist and activist] Dave Dellinger, who's right up front with the victims.

India today is torn asunder by various separatist movements. Kashmir [a far-northern province disputed by Inda and Pakistan] is an incredible mess, occupied by the Indian army, and there are killings, detentions and massive human-rights violations in the Punjab [a province that straddles Pakistan and India] and elsewhere.

I'd like you to comment on a tendency in the Third World to blame the colonial masters for all the problems that are besetting their countries today. They seem to say, "Yes, India has problems, but it's the fault of the British—before that, India was just one happy place."

It's difficult to assess blame for historical disasters. It's somewhat like trying to assess blame for the health of a starving and diseased person. There are lots of different factors. Let's say the person was tortured—that certainly had an effect. But maybe when the torture was over, that person ate the wrong diet, lived a dissolute life and died from the combined effects. That's the kind of thing we're talking about.

There's no doubt that imperial rule was a disaster. Take India. When the British first moved into Bengal, it was one of the richest places in the world. The first British merchant-warriors described it as a paradise. That area is now Bangladesh and Calcutta—the very symbols of despair and hopelessness.

There were rich agricultural areas producing unusually fine cotton. They also had advanced manufacturing, by the standards of the day. For example, an Indian firm built one of the flagships for an English admiral during the Napoleonic Wars. It wasn't built in British factories—it was the Indians' own manufacture.

You can read about what happened in Adam Smith, who was writing over two hundred years ago. He deplored the deprivations that the British were carrying out in Bengal. As he puts it, they first destroyed the agricultural economy and then turned "dearth into a famine." One way they did this was by taking the agricultural lands and turning them into poppy production (since opium was the only thing Britain could sell to China). Then there was mass starvation in Bengal.

The British also tried to destroy the existing manufacturing system in the parts of India they controlled. Starting from about 1700, Britain imposed harsh tariff regulations to prevent Indian manufacturers from competing with British textiles. They had to undercut and destroy Indian textiles because India had a comparative advantage. They were using better cotton and their manufacturing system was in many respects comparable to, if not better than, the British system.

The British succeeded. India deindustrialized, it ruralized. As the industrial revolution spread in England, India was turning into a poor, ruralized and agrarian country.

It wasn't until 1846, when their competitors had been destroyed and they were way ahead, that Britain suddenly discovered the merits of free trade. Read the British liberal historians, the big advocates of free trade—they were very well aware of it. Right through that period they say: "Look, what we're doing to India isn't pretty, but there's no other way for the mills of Manchester to survive. We have to destroy the competition."

And it continues. We can pursue this case by case through India. In 1944, Nehru wrote an interesting book *[The Discovery of India]* from a British prison. He pointed out that if you trace British influence and control in each region of India, and then compare that with the level of poverty in the region, they correlate. The longer the British have been in a region, the poorer it is. The worst, of course, was Bengal—now Bangladesh. That's where the British were first.

You can't trace these same things in Canada and North America, because there they just decimated the population. It's not only the current "politically correct" commentators that describe this—you can go right back to the founding fathers.

The first secretary of defense, General Henry Knox, said that what we're doing to the native population is worse than what the conquistadors did in Peru and Mexico. He said future historians will look at the "destruction" of these people—what would nowadays be called genocide—and paint the acts with "sable colors" [in other words, darkly].

This was known all the way through. Long after John Quincy Adams, the intellectual father of Manifest Destiny, left power, he became an opponent of both slavery and the policy toward the Indians. He said he'd been involved—along with the rest of them—in a crime of "extermination" of such enormity that surely God would punish them for these "heinous sins."

Latin America was more complex, but the initial population was virtually destroyed within a hundred and fifty years. Meanwhile, Africans were brought over as slaves. That helped devastate Africa even before the colonial period, then the conquest of Africa drove it back even further.

After the West had robbed the colonies—as they did, no question about that, and there's also no question that it contributed to their own development—they changed over to so-called "neo-colonial" relationships, which means domination without direct administration. After that it was generally a further disaster.

Divide and conquer

To continue with India: talk about the divide-and-rule policy of the British Raj, playing Hindus off against Muslims. You see the results of that today.

Naturally, any conqueror is going to play one group against another. For example, I think about 90% of the forces that the British used to control India were Indians.

There's that astonishing statistic that at the height of British power in India, they never had more than 150,000 people there.

That was true everywhere. It was true when the American forces conquered the Philippines, killing a couple hundred thousand people. They were being helped by Philippine tribes, exploiting conflicts among local groups. There were plenty who were going to side with the conquerors.

But forget the Third World—just take a look at the Nazi conquest of nice, civilized Western Europe, places like Belgium and Holland and France. Who was rounding up the Jews? Local people, often. In France they were rounding them up faster than the Nazis could handle them. The Nazis also used Jews to control Jews.

If the United States was conquered by the Russians, Ronald Reagan, George Bush, Elliott Abrams and the rest of them would probably be working for the invaders, sending people off to concentration camps. They're the right personality types.

That's the traditional pattern. Invaders quite typically use collaborators to run things for them. They very naturally play upon any existing rivalries and hostilities to get one group to work for them against others.

It's happening right now with the Kurds. The West is trying to mobilize Iraqi Kurds to destroy Turkish Kurds, who are by far the largest group and historically the most oppressed. Apart from what we might think of those guerrillas, there's no doubt that they had substantial popular support in southeastern Turkey.

(Turkey's atrocities against the Kurds haven't been covered much in the West, because Turkey is our ally. But right into the Gulf War they were bombing in Kurdish areas, and tens of thousands of people were driven out.)

Now the Western goal is to use the Iraqi Kurds as a weapon to try and restore what's called "stability"—meaning their own kind of system—in Iraq. The West is using the Iraqi Kurds to destroy the Turkish Kurds, since that will extend Turkey's power in the region, and the Iraqi Kurds are cooperating.

In October 1992, there was a very ugly incident in which there was a kind of pincers movement between the Turkish army and the Iraqi Kurdish forces to expel and destroy Kurdish guerrillas from Turkey.

Iraqi Kurdish leaders, and some sectors of the population, cooperated because they thought they could gain something by it. You could understand their position—not necessarily approve of it, that's another question—but you could certainly understand it.

These are people who are being crushed and destroyed from every direction. If they grasp at some straw for survival, it's not surprising—even if grasping at that straw means helping to kill people like their cousins across the border.

That's the way conquerors work. They've always worked that way. They worked that way in India.

It's not that India was a peaceful place before—it wasn't. Nor was the western hemisphere a pacifist utopia. But there's no doubt that almost everywhere the Europeans went they raised the level of violence to a significant degree. Serious military historians don't have any doubts about that—it was already evident by the eighteenth century. Again, you can read it in Adam Smith.

One reason for that is that Europe had been fighting vicious, murderous wars internally. So it had developed an unsurpassed culture of violence. That culture was even more important than the technology, which was not all that much greater than other cultures.

The description of what the Europeans did is just monstrous. The British and Dutch merchants—actually merchant-warriors—moved into Asia and broke into trading areas that had been functioning for long, long periods, with pretty well-established rules. They were more or less free, fairly pacific—sort of like free-trade areas.

The Europeans destroyed what was in their way. That was true over almost the entire world, with very few exceptions. European wars were wars of extermination. If we were to be honest about that history, we would describe it simply as a barbarian invasion.

The natives had never seen anything like it. The only ones who were able to fend it off for a while were Japan and China. China sort of made the rules and had the technology and was powerful, so they were able to fend off Western intervention for a long time. But when their defenses finally broke down in the nineteenth century, China collapsed.

Japan fended it off almost entirely. That's why Japan is the one area of the Third World that developed. That's striking. The one part of the Third World that wasn't colonized is the one part that's part of the industrialized world. That's not by accident.

To strengthen the point, you need only look at the parts of Europe that were colonized. Those parts—like Ireland—are much like the Third World. The patterns are striking. So when people in the Third World blame the history of imperialism for their plight, they have a very strong case to make.

It's interesting to see how this is treated in the West these days. There was an amazing article in the *Wall Street Journal* [of January 7, 1993] criticizing the intervention in Somalia. It was by Angelo Codevilla, a so-called scholar at the Hoover Institute at Stanford, who says: Look, the problem in the world is that Western intellectuals hate their culture and therefore they terminated colonialism. Only civilizations of great generosity can undertake tasks as noble as colonialism, which tries to rescue barbarians all over the world from their miserable fate. The Europeans did it—and of course gave them enormous gifts and benefits. But then these Western intellectuals who hate their own cultures forced them to withdraw. The result is what you now see.

You really have to go to the Nazi archives to find anything comparable to that. Apart from the stupendous ignorance—ignorance so colossal that it can only appear among respected intellectuals—the moral level is so low you'd have to go to the Nazi archives. And yet this is an op-ed in the *Wall Street Journal*. It probably won't get much criticism.

It was interesting to read the right-wing papers in England—the *Sunday Telegraph* and the *Daily Telegraph*—after Rigoberta Menchu [a Guatemalan Indian activist and author] won the Nobel Prize. They, especially their Central America correspondent, were infuriated. Their view is: True, there were atrocities in Guatemala. But either they were carried out by the left-wing guerrillas or they were an understandable response by the respectable sectors of the society to the violence and atrocities of these Marxist priests. So to give a Nobel Prize to the person who's been torturing the Indians all these years, Rigoberta Menchu....

It's hard for me to reproduce this. You have to read the original. Again, it's straight out of the Stalinist and Nazi archives—at their worst. But it's very typical of elements of British and American culture.

The roots of racism

All over the world—from LA to the Balkans to the Caucasus to India—there's a surge of tribalism, nationalism, religious fanaticism, racism. Why now?

First of all, let's remember that it's always been going on.

I grant you that, but it seems more pronounced.

In parts of the world it's more pronounced. Take Eastern Europe. Europe is altogether a very racist place, even worse than the US, but Eastern Europe is particularly ugly. That society traditionally had very bitter ethnic hatreds. One of the reasons why many of us are here is that our grandparents fled from that.

Up until a couple of years ago, Eastern Europe was under the control of a very harsh tyranny—the Soviet system. It immobilized the civil society, which meant that it eliminated what was good, but it also suppressed what was bad. Now that the tyranny is gone, the civil society is coming back—including its warts, of which there are plenty.

Elsewhere in the world, say in Africa, there are all kinds of atrocities. They were always there. One of the worst atrocities was in the 1980s. From 1980 to 1988, US-backed South Africa was responsible for about a million and a half killings, plus about sixty billion dollars worth of damage—and that's only in the region surrounding South Africa.

Nobody here cared about that, because the US was backing it. If you go back to the 1970s in Burundi, there was a huge massacre, tens of thousands of people killed. Nobody cared.

In Western Europe, there's an increase in regionalism. This in part reflects the decline of their democratic institutions. As the European Community slowly consolidates towards executive power, reflecting big economic concentrations, people are trying to find other ways to preserve their identity. That leads to a lot of regionalism, with both positive and negative aspects. That's not the whole story, but a lot of it.

Germany had the most liberal asylum policies in the world—now they want to limit civil liberties, and ban political parties.

There's a lot of talk about German racism, and it's bad enough. For example, kicking out the Gypsies and sending them off to Romania is a scandal you can't even describe. The Gypsies were treated just like the Jews in the Holocaust, but nobody's batting an eyelash about that because nobody gives a damn about the Gypsies.

But we should remember that there are other things going on too, which are getting less publicity. Take Spain. It was admitted into the European Community with some conditions. One was that it's to be a barrier to the hordes of North Africans whom the Europeans are afraid will flock up to Europe.

There are plenty of boat people trying to get across the narrow distance between North Africa to Spain—kind of like Haiti and the Dominican Republic. If they make it, the boat people are immediately expelled by the Spanish police and navy. It's very ugly.

There are, of course, reasons why people are going from Africa to Europe and not the other direction. There are five hundred years of reasons for that. But it's happening, and Europe doesn't want it. They want to preserve their wealth and keep the poor people out.

The same problem is occurring in Italy. The Lombard League, which includes a kind of neofascist element, won a recent electoral victory. It reflects northern Italian interests. They don't want to be saddled with the poor people in the south of Italy. And they're concerned about the North Africans coming up from the south, drifting up through Sicily into Italy. The north Italians don't want them—they want rich white people.

That brings in the whole question of race and racism and how that factored into the relationship between the North and the South.

There has always been racism. But it developed as a leading principle of thought and perception in the context of colonialism. That's understandable. When you have your boot on someone's neck, you have to justify it. The justification has to be their depravity.

It's very striking to see this in the case of people who aren't very different from one another. Take a look at the British conquest of Ireland, the earliest of the Western colonial conquests. It was described in the same terms as the conquest of Africa. The Irish were a different race. They weren't human. They weren't like us. We had to crush and destroy them.

Some Marxists say racism is a product of the economic system, of capitalism. Would you accept that?

No. It has to do with conquest, with oppression. If you're robbing somebody, oppressing them, dictating their lives, it's a very rare person who can say: *Look, I'm a monster. I'm doing this for my own good.* Even Himmler didn't say that.

A standard technique of belief formation goes along with oppression, whether it's throwing them in gas chambers or charging them too much at a corner store, or anything in between. The standard reaction is to say: *It's their depravity. That's why I'm doing it. Maybe I'm even doing them good.*

118

If it's their depravity, there's got to be something about them that makes them different from me. What's different about them will be whatever you can find.

And that's the justification.

Then it becomes racism. You can always find something—they have a different color hair or eyes, they're too fat, or they're gay. You find something that's different enough. Of course you can lie about it, so it's easier to find.

Take the Serbs and the Croats. They're indistinguishable. They use a different alphabet, but they speak the same language. They belong to different branches of the Catholic Church. That's about it. But many of them are perfectly ready to murder and destroy each other. They can imagine no higher task in life.

The unmentionable five-letter word

It's a given that ideology and propaganda are phenomena of other cultures. They don't exist in the United States. *Class* is in the same category. You've called it the "unmentionable five-letter word."

It's kind of interesting the way it works. Statistics about things like quality of life, infant mortality, life expectancy, etc. are usually broken down by race. It always turns out that blacks have horrible statistics as compared with whites.

But an interesting study was done by Vicente Navarro, a professor at Johns Hopkins who works on public health issues. He decided to reanalyze the statistics, separating out the factors of race and class. For example, he looked at white workers and black workers versus white executives and black executives. He discovered that much of the difference between blacks and whites was actually a class difference. If you look at poor white workers and white executives, the gap between them is enormous.

The study was obviously relevant to epidemiology and public health, so he submitted it to the major American medical journals. They all rejected it. He then sent it to the world's leading medical journal, the *Lancet,* in Britain. They accepted it right away.

The reason is very clear. In the United States you're not allowed to talk about class differences. In fact, only two groups are allowed to be class-conscious in the United States. One of them is the business community, which is rabidly class-conscious. When you read their literature, it's all full of the danger of the masses and their rising power and how we have to defeat them. It's kind of vulgar, inverted Marxism.

The other group is the high planning sectors of the government. They talk the same way—how we have to worry about the rising aspirations of the common man and the impoverished masses who are seeking to improve standards and harming the business climate.

So they can be class-conscious. They have a job to do. But it's extremely important to make other people, the rest of the population, believe that there is no such thing as class. We're all just equal, we're all Americans, we live in harmony, we all work together, everything is great.

Take, for example, the book *Mandate for Change*, put out by the Progressive Policy Institute, the Clinton think tank. It was a book you could buy at airport newsstands, part of the campaign literature describing the Clinton administration's program. It has a section on "entrepreneurial economics," which is economics that's going to avoid the pitfalls of the right and the left.

It gives up these old-fashioned liberal ideas about entitlement and welfare mothers having a right to feed their children—that's all passé. We're not going to have any more of that stuff. We now have "enterprise economics," in which we improve investment and growth. The only people we want to help are workers and the firms in which they work.

According to this picture, we're all workers. There are firms in which we work. We would like to improve the firms in which we work, like we'd like to improve our kitchens, get a new refrigerator.

There's somebody missing from this story—there are no managers, no bosses, no investors. They don't exist. It's just workers and the firms in which we work. All the administration's interested in is helping us folks out there.

The word *entrepreneurs* shows up once, I think. They're the people who assist the workers and the firms in which they work.

The word *profits* also appears once, if I recall. I don't know how that sneaked in—that's another dirty word, like *class*.

Or take the word *jobs*. It's now used to mean *profits*. So when, say, George [H. W.] Bush took off to Japan with Lee Iacocca and the rest of the auto executives, his slogan was *Jobs, jobs, jobs*. That's what he was going for.

We know exactly how much George Bush cares about jobs. All you have to do is look at what happened during his presidency, when the number of unemployed and underemployed officially reached about seventeen million or so—a rise of eight million during his term of office.

He was trying to create conditions for exporting jobs overseas. He continued to help out with the undermining of unions and the lowering of real wages. So what does he mean when he and the media shout, "Jobs, jobs, jobs"? It's obvious: "Profits, profits, profits." Figure out a way to increase profits.

The idea is to create a picture among the population that we're all one happy family. We're America, we have a national interest, we're working together. There are us nice workers, the firms in which we work and the government who works for us. We pick them—they're our servants.

And that's all there is in the world—no other conflicts, no other categories of people, no further structure to the system beyond that. Certainly nothing like class—unless you happen to be in the ruling class, in which case you're very well aware of it.

So then equally exotic issues like class oppression and class warfare occur only in obscure books and on Mars?

Or in the business press and the business literature, where it's written about all the time. It exists there because they have to worry about it.

You use the term *elite*. The political economist and economic historian Samir Amin says it confers too much dignity upon them. He prefers *ruling class*. Incidentally, a more recent invention is *the ruling crass*.

The only reason I don't use the word *class* is that the terminology of political discourse is so debased it's hard to find any words at all. That's part of the point—to make it impossible to talk. For one thing, *class* has various associations. As soon as you say the word *class*, everybody falls down dead. They think, *There's some Marxist raving again.*

But the other thing is that to do a really serious class analysis, you can't just talk about the ruling class. Are the professors at Harvard part of the ruling class? Are the editors of the *New York Times* part of the ruling class? Are the bureaucrats in the State Department? There are lots of different categories of people. So you can talk vaguely about *the establishment* or *the elites* or the people in *the dominant sectors.*

But I agree, you can't get away from the fact that there are sharp differences in power which in fact are ultimately rooted in the economic system. You can talk about *the masters,* if you like. It's Adam Smith's word, and he's now in fashion. The elite are the masters, and they follow what Smith called their "vile maxim"—namely, *all for ourselves and nothing for anyone else.*

You say that class transcends race, essentially.

It certainly does. For example, the United States *could* become a color-free society. It's possible. I don't think it's going to happen, but it's perfectly possible that it would happen, and it would hardly change the political economy at all—just as women could pass through the "glass ceiling" and that wouldn't change the political economy at all.

That's one of the reasons why you commonly find the business sector reasonably willing to support efforts to overcome racism and sexism. It doesn't matter that much for them. You lose a little white-male privilege in the executive suite, but that's not all that important as long as the basic institutions of power and domination survive intact.

And you can pay the women less.

Or you can pay them the same amount. Take England. They just went through ten pleasant years with the Iron Lady running things. Even worse than Reaganism.

Lingering in the shadows of the liberal democracies—where there's this pyramid of control and domination, where there's class and race and gender bias—is coercion, force.

That comes from the fact that objective power is concentrated. It lies in various places, like in patriarchy, in race. Crucially it also lies in ownership.

If you think about the way the society generally works, it's pretty much the way the founding fathers said. As John Jay put it, "The country should be governed by those who own it" and the owners intend to follow Adam Smith's vile maxim. That's at the core of things. That can remain even if lots of other things change.

On the other hand, it's certainly worth overcoming the other forms of oppression. For people's lives, racism and sexism may be much worse than class oppression. When a kid was lynched in the South, that was worse than being paid low wages. So when we talk about the roots of the system of oppression, that can't be spelled out simply in terms of suffering. Suffering is an independent dimension, and you want to overcome suffering.

Human nature and self-image

Is racism something that's learned, or is it innately endowed?

I don't think either of those is the right answer. There's no doubt that there's a rich, complex human nature. We're not rocks. Anybody sane knows that an awful lot about us is genetically determined, including aspects of our behavior, our attitudes. That's not even a question among sane people.

When you go beyond that and ask what it is, you're entering into general ignorance. We know there's something about human nature that forces us to grow arms, not wings, and undergo puberty at roughly a certain age. And by now we know that acquisition of language, growth of the visual system and so on, are part of human nature in fundamental respects.

When you get to cultural patterns, belief systems and the like, the guess of the next guy you meet at the bus stop is about as good as that of the best scientist. Nobody knows anything. People can rant about it if they like, but they basically know almost nothing.

In this particular area we can at best make some reasonable speculations. I think the one I've outlined may be a reasonable guess. It's not so much that racism is in our genes; what *is* in our genes is the need for protecting our self-image. It's probably in our nature to find a way to recast anything that we do in some way that makes it possible for us to live with it.

It's the same in the broader social sphere, where there are institutions functioning, and systems of oppression and domination. The people who are in control, who are harming others—those people will construct justifications for themselves. They may do it in sophisticated ways or nonsophisticated ways, but they're going to do it. That much is in human nature. One of the consequences of that can turn out to be racism. It can turn out to be other things too.

Take the sophisticated ones. One of the intellectual gurus of the modern period in the United States was Reinhold Niebuhr. He was called the "theologian of the establishment." He was revered by the Kennedy liberal types, by people like George Kennan. He was considered a moral teacher of the contemporary generation.

It's interesting to look at why he was so revered. I went through his stuff once. (There was supposed to be a chapter about him in one of my books—but the publisher thought it would be too arcane for the audience, so I didn't include it.) The intellectual level is depressingly low—you can hardly keep a straight face.

But something made him appealing—his concept of the "paradox of grace." What it comes down to is this: No matter how much you try to do good, you're always going to do harm. Of course, he's an intellectual, so he had to dress it up with big words, but that's what it comes down to.

That's very appealing advice for people who are planning to enter a life of crime—to say, *No matter how much I try to do good, I'm always going to harm people. I can't get out of it.* It's a wonderful idea for a Mafia don. He can go ahead and do whatever he feels like. If he harms people, *Oh my God, the paradox of grace.*

That may well explain why Niebuhr was so appealing to American intellectuals in the post-World War II period. They were preparing to enter a life of major crime. They were going to be either the managers or the apologists for a period of global conquest.

Running the world is obviously going to entail enormous crimes. So they think, "Isn't it nice to have this doctrine behind us? Of course we're superbenevolent and humane, but the paradox of grace...."

Again, if you're an intellectual, you dress it up and write articles about it. The mechanisms, however, are quite simple.

I suppose all of that is, if you like, part of our nature, but in such a transparent way that we can't seriously call this a theory. Everybody knows from their own experience just about everything that's understood about human beings—how they act and why—if they stop to think about it. It's not quantum physics.

What about the so-called "competitive ethic?" Is there any evidence that we are naturally competitive? Many proponents of free market theory and market capitalism say you've got to give people the ability to compete—it's a natural thing.

There are certainly conditions under which people will compete, and there are also conditions under which people will cooperate. For example, take a family. Suppose that whoever is providing the money for the family loses his or her job, so they don't have enough food to eat.

The father is probably the strongest one in the family. Does he steal all the food and eat it, so all the kids starve? (I guess there are people who do that, but then you lock them up. There's a pathological defect there somewhere.) No, what you do is share.

Does that mean they're not competitive? No. It means that in *that* circumstance, they share. Those circumstances can extend quite broadly—for example, they can extend to the whole working class. That's what happens in periods of working class solidarity, when people struggle together to create unions and decent working conditions.

That's true of the United States, after all. Take a look at the Homestead strike a century ago [when Andrew Carnegie locked striking workers out of a steel mill in Pennsylvania]. That was a period of enormous ethnic rivalry and racism, directed mostly against Eastern European immigrants. But during that conflict they worked together. It's one of the few periods of real ethnic harmony. They worked together with Anglo-Saxon Americans and the Germans and the rest of them.

Let me tell you a personal story. I'm not particularly violent, but when I was in college, we had to take boxing. So the way we did it was to spar with a friend, wait until the thing was over and go home. But we were all amazed to find that after doing this pushing around for a while, we really wanted to hurt that other guy, our best friend. We could feel it coming out—we wanted to kill each other.

Does that mean that the desire to kill people is innate? In certain circumstances that desire is going to come out, even if it's your best friend. There are circumstances under which this aspect of our personality will dominate. But there are other circumstances in which other aspects will dominate. If you want to create a humane world, you change the circumstances.

How crucial is social conditioning in all of this? Let's say you're a child growing up in Somalia today.

How about a child growing up two blocks from here in Cambridge? Just last summer a student at MIT was killed—knifed—by a couple of teenagers from the local high school. They were engaged in a sport that works like this: They walk around and find somebody walking the street. Then one of the teenagers is picked to knock the person down with a single blow. If he fails to do it, the other kids beat the kid who failed.

So they were walking along and saw this MIT student. The chosen kid knocked the student down with one blow. For unexplained reasons, they also knifed and killed him. The teenagers didn't see anything especially wrong with it. They walked off and went to a bar somewhere. They were later picked up by the police because somebody had seen them. They hadn't even tried to get away.

These kids are growing up in Cambridge—not in the wealthy sections, but probably in the slums. Those aren't Somali slums by any means, or even Dorchester slums, but surely kids in the more affluent suburbs wouldn't act like that.

Does that mean they're different genetically? No. There's something about the social conditions in which they're growing up that makes this acceptable behavior, even natural behavior. Anyone who has grown up in an urban area must be aware of this.

I can remember from childhood, that there were neighborhoods where if you went, you'd be beaten up. You weren't supposed to be there. The people who were doing it—kids—felt justified and righteous about it. They were defending their turf. What else did they have to defend?

It can't happen here...can it?

Huey Long [a populist Louisiana governor and senator in the early 1930s] once said that when fascism comes to this country, it's going to be wrapped in an American flag. You've commented on tendencies toward fascism in this country. You've even been quoting Hitler on the family and the role of women.

The Republican convention—fortunately I saved myself the pain of watching it on television, but I read about it—struck such chords that I began looking up some literature on fascism from the 1930s. I looked up Hitler's speeches to women's groups and big rallies. The rhetoric was very similar to that of the "God-and-country" rally the first night of the Republican convention.

But I don't really take that similarity too seriously, because the levers of power are firmly in the hands of the corporate sector. It'll permit rabid fundamentalists to scream about God and country and family, but they're very far from having any influence over major power decisions.

That was obvious in the way the campaign developed. They were given the first night to scream and yell. They were even given the party platform—it was pre-Enlightenment. But then when the campaign started, we were back to business as usual.

But that can change. When people grow more alienated and isolated, they begin to develop highly irrational and very self-destructive attitudes. They want something in their lives. They want to identify themselves somehow. They don't want to be just glued to the television set. If most of the constructive ways are cut off, they turn to other ways.

You can see that in the polls too. I was just looking at a study by an American sociologist (published in England) of comparative religious attitudes in various countries. The figures are shocking. Three quarters of the American population literally believe in religious miracles. The numbers who believe in the devil, in resurrection, in God doing this and that—it's astonishing.

These numbers aren't duplicated anywhere else in the industrial world. You'd have to maybe go to mosques in Iran or do a poll among old ladies in Sicily to get numbers like this. Yet this is the American population.

Just a couple of years ago, there was a study of what people thought of evolution. The percentage of the population that believed in Darwinian evolution at that point was 9%—not all that much above statistical error. About half the population believed in divinely-guided evolution, Catholic church doctrine. About 40% thought the world was created a few thousand years ago.

Again, you've got to go back to pre-technological societies, or devastated peasant societies, before you get numbers like that. Those are the kinds of belief systems that show up in things like the God-and-country rally.

Religious fundamentalism can be a very scary phenomenon. It could be the mass base for an extremely dangerous popular movement. These fundamentalist leaders aren't stupid. They have huge amounts of money, they're organizing, they're moving the way they should, beginning to take over local offices where nobody notices them.

There was a striking phenomenon in the last election—it even made the front pages of the national newspapers. It turned out that in many parts of the country ultraright fundamentalist extremists had been running candidates without identifying them. It doesn't take a lot of work to get somebody elected to the school board. Not too many people pay attention. You don't have to say who you are. You just appear with a friendly face and a smile and say, *I'm going to help your kids* and people will vote for you.

A lot of people got elected because of these organized campaigns to take over local structures. If that ties in with some charismatic power figure who says, *I'm your leader, follow me,* it could be very ugly. We could move back to real pre-Enlightenment times.

There's also a huge increase in fundamentalist media, particularly electronic media. You can't drive across the country without noticing it.

That was true years ago. I remember driving across the country, being bored out of my head and turning on the radio. Every station I found was some ranting minister. Now it's much worse, and of course now there's television.

Hume's paradox

You've said the real drama since 1776 has been the "relentless attack of the prosperous few upon the rights of the restless many." I want to ask you about the "restless many." Do they hold any cards?

Sure. They've won a lot of victories. The country is a lot more free than it was two hundred years ago. For one thing, we don't have slaves. That's a big change. Thomas Jefferson's goal, at the very left-liberal end of the spectrum, was to create a country "free of blot or mixture"—meaning no red Indians, no black people, just good white Anglo-Saxons. That's what the liberals wanted.

They didn't succeed. They did pretty much get rid of the native population—they almost succeeded in "exterminating" them (as they put it in those days)—but they couldn't get rid of the black population, and over time they've had to incorporate them in some fashion into society.

Freedom of speech has been vastly extended. Women finally received the franchise 150 years after the revolution. After a very bloody struggle, workers finally won some rights in the 1930s—about fifty years after they did in Europe. (They've been losing them ever since, but they won them to some extent.)

In many ways large parts of the general population have been integrated into the system of relative prosperity and relative freedom—almost always as a result of popular struggle. So the general population has lots of cards.

That's something that [English philosopher] David Hume pointed out a couple of centuries ago. In his work on political theory, he describes the paradox that, in any society, the population submits to the rulers, even though force is always in the hands of the governed.

Ultimately the governors, the rulers, can only rule if they control opinion—no matter how many guns they have. This is true of the most despotic societies and the most free, he wrote. If the general population won't accept things, the rulers are finished.

That underestimates the resources of violence, but expresses important truths nonetheless. There's a constant battle between people who refuse to accept domination and injustice and those who are trying to force people to accept them.

How to break from the system of indoctrination and propaganda? You've said that it's nearly impossible for individuals to do anything, that it's much easier and better to act collectively. What prevents people from getting associated?

There's a big investment involved. Everybody lives within a cultural and social framework which has certain values and certain opportunities. It assigns cost to various kinds of action and benefits to others. You just live in that—you can't help it.

We live in a society that assigns benefits to efforts to achieve individual gain. Let's say I'm the father or mother of a family. What do I do with my time? I've got 24 hours a day. If I've got children to take care of, a future to worry about, what do I do?

One thing I can do is try to play up to the boss and see if I can get a dollar more an hour. Or maybe I can kick somebody in the face when I walk past them (if not directly then indirectly, by the mechanisms that are set up within a capitalist society). That's one way.

The other way is to spend my evenings trying to organize other people, who will then spend their evenings at meetings, go out on a picket line and carry out a long struggle in which they'll be beaten up by the police and lose their jobs. Maybe they'll finally get enough people together so they'll ultimately achieve a gain, which may or may not be greater than the gain that they tried to achieve by following the individualist course.

In game theory, this kind of situation is called "prisoner's dilemma." You can set up things called "games"—interactions—in which each participant will gain more if they work together, but you only gain if the other person works with you. If the other person is trying to maximize his or her own gain, you lose.

Let me take a simple case—driving to work. It would take me longer to take the subway than to drive to work. If we all took the subway and put the money into that instead of into roads, we'd all get there faster by the subway. But we all have to do it. If other people are going to be driving and I'm taking the subway, then private transportation is going to be better for the people who are doing it.

It's only if we all do something a different way that we'll all benefit a lot more. The costs to you—an individual—to work to create the possibilities to do things together can be severe. It's only if lots of people begin to do it, and do it seriously, that you get real benefits.

The same has been true of every popular movement that ever existed. Suppose you were a twenty-year-old black kid at Spelman College in Atlanta in 1960. You had two choices. One was: *I'll try to get a job in a business somewhere. Maybe somebody will be willing to pick a black manager. I'll be properly humble and bow and scrape. Maybe I'll live in a middle-class home.*

The other was to join SNCC [the Student Nonviolent Coordinating Committee, a black civil rights group of the 1960s], in which case you might get killed. You were certainly going to get beaten and defamed. It would be a very tough life for a long time. Maybe you'd finally be able to create enough popular support so that people like you and your family could live better.

It would be hard to make that second choice, given the alternatives available. Society is very much structured to try to drive you toward the individualist alternative. It's a remarkable fact that many young people took that second choice, suffered for it and helped create a much better world.

You've noted polls that indicate that 83% of the population regard the entire economic system as "inherently unfair." But it doesn't translate into anything.

It can only translate into anything if people do something about it. That's true whether you're talking about general things—like the inherent unfairness of the economic system, which requires revolutionary change—or about small things.

Take, say, health insurance. In public, almost nobody calls for a "Canadian-style" system. (That's the kind of system they have everywhere in the world—an efficient, nationally organized public health system that guarantees health services for everyone and—if it's more serious than Canada's system—also provides preventive care.)

And yet according to some polls, a majority of the population is in favor of it anyway, even though they've scarcely heard anybody advocate it. Does it matter? No. There'll be some kind of insurance company-based, "managed" healthcare system—designed to ensure that insurance companies and the health corporations they run will make plenty of money.

There are only two ways we could get the healthcare that most of the population wants. There either needs to be a large-scale popular movement—which would mean moving towards democracy, and nobody in power wants that—or the business community must decide that it would be good for them. They might do that.

This highly bureaucratized, extremely inefficient system designed for the benefit of one sector of the private enterprise system happens to harm other sectors. Auto companies pay more in health benefits here than they would across the border. They notice that. They may press for a more efficient system that breaks away from the extreme inefficiencies and irrationalities of the capitalist-based system.

"Outside the pale
of intellectual responsibility"

Canadian journalist David Frum has called you the "great American crackpot." I think that ranks up there with the *New Republic's* Martin Peretz placing you "outside the pale of intellectual responsibility." Frum also says, "There was a time when the *New York Times* op-ed page was your stomping ground." Have I missed something here?

I guess I have too. I did have an op-ed once—it was in 1971, I think. This was the period when the corporate sector, and later the *New York Times,* had decided we'd better get out of Vietnam because it was costing us too much.

I had testified before the Senate Foreign Relations Committee. Senator Fulbright had in effect turned the Committee into a seminar. He was very turned off by the war and American foreign policy at that time. He invited me to testify. That was respectable enough. So they ran a segment of....

Excerpts of your comments. It wasn't an original piece you had written for the *Times*.

Maybe it was slightly edited, but it was essentially a piece of my testimony at the committee. So it's true, the *Times* did publish a piece of my testimony to the Foreign Relations Committee.

And that was your "stomping grounds." What about letters? How many letters of yours have they printed?

Occasionally, when an outlandish slander and lie about me has appeared there, I've written back to them. Sometimes they don't publish the letters. Once, maybe more, I was angry enough that I contacted a friend inside, who was able to put enough pressure on so they ran the letter.

But sometimes they just refused. In the *Times* book review section, there were a bunch of vicious lies about me and the Khmer Rouge. I wrote back a short letter responding, and they refused to publish it. I got annoyed and wrote back again—and actually got a response. They said they'd published a different letter—one they thought was better.

SECRETS,
LIES
AND
DEMOCRACY

FIRST PUBLISHED 1994

11 US PRINTINGS

7 FOREIGN EDITIONS

142,000 COPIES IN PRINT

TABLE OF CONTENTS

Defective democracy

Clinton's national security advisor, Anthony Lake, is encouraging the enlargement of democracy overseas. Should he extend that to the US?

I can't tell you what Anthony Lake has in mind, but the concept of democracy that's been advanced is a very special one, and the more honest people on the right describe it rather accurately. For example, Thomas Carothers, who was involved in what was called the "democracy assistance project" during the Reagan administration, has written a book and several articles about it.

He says the US seeks to create a form of top-down democracy that leaves traditional structures of power—basically corporations and their allies—in effective control. Any form of democracy that leaves the traditional structures essentially unchallenged is admissible. Any form that undermines their power is as intolerable as ever.

So there's a dictionary definition of *democracy* and then a real-world definition.

The real-world definition is more or less the one Carothers describes. The dictionary definition has lots of different dimensions, but, roughly speaking, a society is democratic to the extent that people in it have meaningful opportunities to take part in the formation of public policy. There are a lot of different ways in which that can be true, but insofar as it's true, the society is democratic.

A society can have the formal trappings of democracy and not be democratic at all. The Soviet Union, for example, had elections.

The US obviously has a formal democracy with primaries, elections, referenda, recalls, and so on. But what's the content of this democracy in terms of popular participation?

Over long periods, the involvement of the public in planning or implementation of public policy has been quite marginal. This is a business-run society. The political parties have reflected business interests for a long time.

One version of this view which I think has a lot of power behind it is what political scientist Thomas Ferguson calls "the investment theory of politics." He believes that the state is controlled by coalitions of investors who join together around some common interest. To participate in the political arena, you must have enough resources and private power to become part of such a coalition.

Since the early nineteenth century, Ferguson argues, there's been a struggle for power among such groups of investors. The long periods when nothing very major seemed to be going on are simply times when the major groups of investors have seen more or less eye to eye on what public policy should look like. Moments of conflict come along when groups of investors have differing points of view.

During the New Deal, for example, various groupings of private capital were in conflict over a number of issues. Ferguson identifies a high-tech, capital-intensive, export-oriented sector that tended to be quite pro-New Deal and in favor of the reforms. They wanted an orderly work force and an opening to foreign trade.

A more labor-intensive, domestically oriented sector, grouped essentially around the National Association of Manufacturers, was strongly anti-New Deal. They didn't want any of these reform measures. (Those groups weren't the only ones involved, of course. There was the labor movement, a lot of public ferment and so on.)

You view corporations as being incompatible with democracy, and you say that if we apply the concepts that are used in political analysis, corporations are fascist. That's a highly charged term. What do you mean?

I mean fascism pretty much in the traditional sense. So when a rather mainstream person like Robert Skidelsky, the biographer of [British economist John Maynard] Keynes, describes the early postwar systems as modeled on fascism, he simply means a system in which the state integrates labor and capital under the control of the corporate structure.

That's what a fascist system traditionally was. It can vary in the way it works, but the ideal state that it aims at is absolutist—top-down control with the public essentially following orders.

Fascism is a term from the political domain, so it doesn't apply strictly to corporations, but if you look at them, power goes strictly top-down, from the board of directors to managers to lower managers and ultimately to the people on the shop floor, typists, etc. There's no flow of power or planning from the bottom up. Ultimate power resides in the hands of investors, owners, banks, etc.

People can disrupt, make suggestions, but the same is true of a slave society. People who aren't owners and investors have nothing much to say about it. They can choose to rent their labor to the corporation, or to purchase the commodities or services that it produces, or to find a place in the chain of command, but that's it. That's the totality of their control over the corporation.

That's something of an exaggeration, because corporations are subject to some legal requirements and there is some limited degree of public control. There are taxes and so on. But corporations are more totalitarian than most institutions we call totalitarian in the political arena.

Is there anything large corporate conglomerates do that has beneficial effects?

A lot of what's done by corporations will happen to have, by accident, beneficial effects for the population. The same is true of the government or anything else. But what are they trying to achieve? Not a better life for workers and the firms in which they work, but profits and market share.

That's not a big secret—it's the kind of thing people should learn in third grade. Businesses try to maximize profit, power, market share and control over the state. Sometimes what they do helps other people, but that's just by chance.

There's a common belief that, since the Kennedy assassination, business and elite power circles control our so-called democracy. Has that changed at all with the Clinton administration?

First of all, Kennedy was very pro-business. He was essentially a business candidate. His assassination had no significant effect on policy that anybody has been able to detect. (There *was* a change in policy in the early 1970s, under Nixon, but that had to do with changes in the international economy.)

Clinton is exactly what he says he is, a pro-business candidate. The *Wall Street Journal* had a very enthusiastic, big, front-page article about him right after the NAFTA vote. They pointed out that the Republicans tend to be the party of business as a whole, but that the Democrats tend to favor big business over small business. Clinton, they said, is typical of this. They quoted executives from the Ford Motor Company, the steel industry, etc. who said that this is one of the best administrations they've ever had.

The day after the House vote on NAFTA, the *New York Times* had a very revealing front-page, pro-Clinton story by their Washington correspondent, R.W. Apple. It went sort of like this: People had been criticizing Clinton because he just didn't have any principles. He backed down on Bosnia, on Somalia, on his economic stimulus program, on Haiti, on the health program. He seemed like a guy with no bottom line at all.

Then he proved that he really was a man of principle and that he really does have backbone—by fighting for the corporate version of NAFTA. So he does have principles—he listens to the call of big money. The same was true of Kennedy.

Radio listener: I've often wondered about people who have a lot of power because of their financial resources. Is it possible to reach them with logic?

They're acting very logically and rationally in their own interests. Take the CEO of Aetna Life Insurance, who makes $23 million a year in salary alone. He's one of the guys who is going to be running our healthcare program if Clinton's plan passes.

Suppose you could convince him that he ought to lobby against having the insurance industry run the healthcare program, because that will be very harmful to the general population (as indeed it will be). Suppose you could convince him that he ought to give up his salary and become a working person.

What would happen then? He'd get thrown out and someone else would be put in as CEO. These are *institutional* problems.

Why is it important to keep the general population in line?

Any form of concentrated power doesn't want to be subjected to popular democratic control—or, for that matter, to market discipline. That's why powerful sectors, including corporate wealth, are naturally opposed to functioning democracy, just as they're opposed to functioning markets...for themselves, at least.

It's just natural. They don't want external constraints on their capacity to make decisions and act freely.

And has that been the case?

Always. Of course, the descriptions of the facts are a little more nuanced, because modern "democratic theory" is more articulate and sophisticated than in the past, when the general population was called "the rabble." More recently, Walter Lippmann called them "ignorant and meddlesome outsiders." He felt that "responsible men" should make the decisions and keep the "bewildered herd" in line.

Modern "democratic theory" takes the view that the role of the public—the "bewildered herd," in Lippmann's words—is to be spectators, not participants. They're supposed to show up every couple of years to ratify decisions made elsewhere, or to select among representatives of the dominant sectors in what's called an "election." That's helpful, because it has a legitimizing effect.

It's very interesting to see the way this idea is promoted in the slick PR productions of the right-wing foundations. One of the most influential in the ideological arena is the Bradley Foundation. Its director, Michael Joyce, recently published an article on this. I don't know whether he wrote it or one of his PR guys did, but I found it fascinating.

It starts off with rhetoric drawn, probably consciously, from the left. When left liberals or radical activists start reading it, they get a feeling of recognition and sympathy (I suspect it's directed at them and at young people). It begins by talking about how remote the political system is from us, how we're asked just to show up every once in a while and cast our votes and then go home.

This is meaningless, the article says—this isn't real participation in the world. What we need is a functioning and active civil society in which people come together and do important things, not just this business of pushing a button now and then.

Then the article asks, How do we overcome these inadequacies? Strikingly, you don't overcome them with more active participation in the political arena. You do it by abandoning the political arena and joining the PTA and going to church and getting a job and going to the store and buying something. That's the way to become a real citizen of a democratic society.

Now, there's nothing wrong with joining the PTA. But there are a few gaps here. What happened to the political arena? It disappeared from the discussion after the first few comments about how meaningless it is.

If you abandon the political arena, somebody is going to be there. Corporations aren't going to go home and join the PTA. They're going to run things. But that we don't talk about.

As the article continues, it talks about how we're being oppressed by the liberal bureaucrats, the social planners who are trying to convince us to do something for the poor. They're the ones who are really running the country. They're that impersonal, remote, unaccountable power that we've got to get off our backs as we fulfill our obligations as citizens at the PTA and the office.

This argument isn't quite presented step-by-step like that in the article—I've collapsed it. It's very clever propaganda, well designed, well crafted, with plenty of thought behind it. Its goal is to make people as stupid, ignorant, passive and obedient as possible, while at the same time making them feel that they're somehow moving towards higher forms of participation.

In your discussions of democracy, you often refer to a couple of comments of Thomas Jefferson's.

Jefferson died on July 4, 1826—fifty years to the day after the Declaration of Independence was signed. Near the end of his life, he spoke with a mixture of concern and hope about what had been achieved, and urged the population to struggle to maintain the victories of democracy.

He made a distinction between two groups—aristocrats and democrats. Aristocrats "fear and distrust the people, and wish to draw all powers from them into the hands of the higher classes." This view is held by respectable intellectuals in many different societies today, and is quite similar to the Leninist doctrine that the vanguard party of radical intellectuals should take power and lead the stupid masses to a bright future. Most liberals are aristocrats in Jefferson's sense. [Former Secretary of State] Henry Kissinger is an extreme example of an aristocrat.

Democrats, Jefferson wrote, "identify with the people, have confidence in them, cherish and consider them as the most honest and safe, although not the most wise, depository of the public interest." In other words, democrats believe the people should be in control, whether or not they're going to make the right decisions. Democrats do exist today, but they're becoming increasingly marginal.

Jefferson specifically warned against "banking institutions and monied incorporations" (what we would now call "corporations") and said that if they grow, the aristocrats will have won and the American Revolution will have been lost. Jefferson's worst fears were realized (although not entirely in the ways he predicted).

Later on, [the Russian anarchist Mikhail] Bakunin predicted that the contemporary intellectual classes would separate into two groups (both of which are examples of what Jefferson meant by aristocrats). One group, the "red bureaucracy," would take power into their own hands and create one of the most malevolent and vicious tyrannies in human history.

The other group would conclude that power lies in the private sector, and would serve the state and private power in what we now call state capitalist societies. They'd "beat the people with the people's stick," by which he meant that they'd profess democracy while actually keeping the people in line.

You also cite [the American philosopher and educator] John Dewey. What did he have to say about this?

Dewey was one of the last spokespersons for the Jeffersonian view of democracy. In the early part of this century, he wrote that democracy isn't an end in itself, but a means by which people discover and extend and manifest their fundamental human nature and human rights. Democracy is rooted in freedom, solidarity, a choice of work and the ability to participate in the social order. Democracy produces real people, he said. That's the major product of a democratic society—real people.

He recognized that democracy in that sense was a very withered plant. Jefferson's "banking institutions and monied incorporations" had of course become vastly more powerful by this time, and Dewey felt that "the shadow cast on society by big business" made reform very difficult, if not impossible. He believed that reform may be of some use, but as long as there's no democratic control of the workplace, reform isn't going to bring democracy and freedom.

Like Jefferson and other classical liberals, Dewey recognized that institutions of private power were absolutist institutions, unaccountable and basically totalitarian in their internal structure. Today, they're far more powerful than anything Dewey dreamed of.

This literature is all accessible. It's hard to think of more leading figures in American history than Thomas Jefferson and John Dewey. They're as American as apple pie. But when you read them today, they sound like crazed Marxist lunatics. That just shows how much our intellectual life has deteriorated.

In many ways, these ideas received their earliest—and often most powerful—formulation in people like [the German intellectual] Wilhelm von Humboldt, who inspired [the English philosopher] John Stuart Mill and was one of the founders of the classical liberal tradition in the late eighteenth century. Like Adam Smith and others, von Humboldt felt that at the root of human nature is the need for free creative work under one's own control. That must be at the basis of any decent society.

Those ideas, which run straight through to Dewey, are deeply anticapitalist in character. Adam Smith didn't call himself an anticapitalist because, back in the eighteenth century, he was basically precapitalist, but he had a good deal of skepticism about capitalist ideology and practice—even about what he called "joint stock companies" (what we call corporations today, which existed in quite a different form in his day). He worried about the separation of managerial control from direct participation, and he also feared that these joint stock companies might turn into "immortal persons."

This indeed happened in the nineteenth century, after Smith's death [under current law, corporations have even more rights than individuals, and can live forever]. It didn't happen through parliamentary decisions—nobody voted on it in Congress. In the

US, as elsewhere in the world, it happened through judicial decisions. Judges and corporate lawyers simply crafted a new society in which corporations have immense power.

Today, the top two hundred corporations in the world control over a quarter of the world's total assets, and their control is increasing. *Fortune* magazine's annual listing of the top American corporations found increasing profits, increasing concentration, and reduction of jobs—tendencies that have been going on for some years.

Von Humboldt's and Smith's ideas feed directly into the socialist-anarchist tradition, into the left-libertarian critique of capitalism. This critique can take the Deweyian form of a sort of workers'-control version of democratic socialism, or the left-Marxist form of people like [the Dutch astronomer and political theorist] Anton Pannekoek and [the Polish-German revolutionary] Rosa Luxemburg [1871–1919], or [the leading anarchist] Rudolf Rocker's anarcho-syndicalism (among others).

All this has been grossly perverted or forgotten in modern intellectual life but, in my view, these ideas grow straight out of classical, eighteenth-century liberalism. I even think they can be traced back to seventeenth-century rationalism.

Keeping the rich on welfare

A book called *America: Who Pays the Taxes?,* written by a couple of *Philadelphia Inquirer* reporters, apparently shows that the amount of taxes paid by corporations has dramatically declined in the US.

That's for sure. It's been very striking over the last fifteen years.

Some years ago, a leading specialist, Joseph Pechman, pointed out that despite the apparently progressive structure that's built into the income tax system (that is, the higher your income, the higher your tax rate), all sorts of other regressive factors end up making everyone's tax rate very near a fixed percentage.

An interesting thing happened in Alabama involving Daimler-Benz, the big German auto manufacturer.

Under Reagan, the US managed to drive labor costs way below the level of our competitors (except for Britain). That's produced consequences not only in Mexico and the US but all across the industrial world.

For example, one of the effects of the so-called free trade agreement with Canada was to stimulate a big flow of jobs from Canada to the southeast US, because that's an essentially nonunion area. Wages are lower; you don't have to worry about benefits; workers can barely organize. So that's an attack against Canadian workers.

Daimler-Benz, which is Germany's biggest conglomerate, was seeking essentially Third World conditions. They managed to get our southeastern states to compete against one another to see who could force the public to pay the largest bribe to bring them there. Alabama won. It offered hundreds of millions of dollars in tax benefits, practically gave Daimler-Benz the land on which to construct their plant, and agreed to build all sorts of infrastructure for them.

Some people will benefit—the small number who are employed at the plant, with some spillover to hamburger stands and so on, but primarily bankers, corporate lawyers, people involved in investment and financial services. They'll do very well, but the cost to most of the citizens of Alabama will be substantial.

Even the *Wall Street Journal,* which is rarely critical of business, pointed out that this is very much like what happens when rich corporations go to Third World countries, and it questioned whether there were going to be overall benefits for the state of Alabama. Meanwhile Daimler-Benz can use this to drive down the lifestyle of German workers.

German corporations have also set up factories in the Czech Republic, where they can get workers for about 10% the cost of German workers. The Czech Republic is right across the border; it's a Westernized society with high educational levels and nice white people with blue eyes. Since they don't believe in the free market any more than any other rich people do, they'll leave the Czech Republic to pay the social costs, pollution, debts and so on, while they pick up the profits.

It's exactly the same with the plants GM is building in Poland, where it's insisting on 30% tariff protection. The free market is for the poor. We have a dual system—protection for the rich and market discipline for everyone else.

I was struck by an article in the *New York Times* whose headline was, "Nation Considers Means to Dispose of Its Plutonium." So the nation has to figure out how to dispose of what was essentially created by private capital.

That's the familiar idea that profits are privatized but costs are socialized. The costs are the nation's, the people's, but the profits weren't for the people, nor did they make the decision to produce plutonium in the first place, nor are they making the decisions about how to dispose of it, nor do they get to decide what ought to be a reasonable energy policy.

One of the things I've learned from working with you is the importance of reading *Business Week, Fortune* and the *Wall Street Journal.* In the business section of the *New York Times,* I read a fascinating discussion by a bureaucrat from MITI [Japan's Ministry of International Trade and Industry] who trained at the Harvard Business School.

One of his classes was studying a failed airline that went out of business. They were shown a taped interview with the company's president, who noted with pride that through the whole financial crisis and eventual bankruptcy of the airline, he'd never asked for government help. To the Japanese man's astonishment, the class erupted into applause.

He commented, "There's a strong resistance to government intervention in America. I understand that. But I was shocked. There are many shareholders in companies. What happened to his employees, for example?" Then he reflects on what he views as America's blind devotion to a free-market ideology. He says, "It is something quite close to a religion. You cannot argue about it with most people. You believe it or you don't." It's interesting.

It's interesting, in part, because of the Japanese man's failure to understand what actually happens in the US, which apparently was shared by the students in his business class. If it was Eastern Airlines they were talking about, Frank Lorenzo, the director, was trying to put it out of business. He made a personal profit out of that.

He wanted to break the unions in order to support his other enterprises (he ripped off profits from Eastern Airlines to support them). He wanted to leave the airline industry less unionized and more under corporate control, and to leave himself wealthier. All of that happened. So naturally he didn't call on government intervention to save him—things were working the way he wanted.

On the other hand, the idea that corporations don't ask for government help is a joke. They demand an extraordinary amount of government intervention. That's largely what the whole Pentagon system is about.

Take the airline industry, which was created by government intervention. A large part of the reason for the huge growth in the Pentagon in the late 1940s was to salvage the collapsing aeronautical industry, which obviously couldn't survive in a civilian market. That's worked—it's now the United States' leading export industry, and Boeing is the leading exporter.

An interesting and important book on this by Frank Kofsky just came out. It describes the war scares that were manipulated in 1947 and 1948 to try to ram spending bills through Congress to save the aeronautical industry. (That wasn't the only purpose of these war scares, but it was a big factor.)

Huge industries were spawned, and are maintained, by massive government intervention. Many corporations couldn't survive without it. (For some, it's not a huge part of their profits at the moment, but it's a cushion.) The public also provides the basic technology—metallurgy, avionics or whatever—via the public subsidy system.

The same is true just across the board. You can hardly find a functioning sector of the US manufacturing or service economy which hasn't gotten that way and isn't sustained by government intervention.

The Clinton administration has been pouring new funds into the National Institute of Standards and Technology. It used to try to work on how long a foot is but it will now be more actively involved in serving the needs of private capital. Hundreds of corporations are beating on their doors asking for grants.

The idea is to try to replace the somewhat declining Pentagon system. With the end of the Cold War, it's gotten harder to maintain the Pentagon system, but you've got to keep the subsidy going to big corporations. The public has to pay the research and development costs.

The idea that a Japanese investigator could fail to see this is fairly remarkable. It's pretty well known in Japan.

Healthcare

I don't suppose you can see the Boston skyline from your home in Lexington. But if you could, what would be the two tallest buildings?

The John Hancock and the Prudential.

And they happen to be two types of what?

They're going to be running our healthcare program if Clinton has his way.

There's a general consensus that the US healthcare system needs to be reformed. How did that consensus evolve?

It evolved very simply. We have a relatively privatized healthcare system. As a result, it's geared towards high-tech intervention rather than public health and prevention. It's also hopelessly inefficient and extremely bureaucratic, with huge administrative expenses.

This has gotten just too costly for American business. In fact, a bit to my surprise, *Business Week*, the main business journal, has come out recently with several articles advocating a Canadian-style, single-payer program. Under this system, healthcare is individual, but the government is the insurer. Similar plans exist in every industrial country in the world, except the US.

The Clinton plan is called "managed competition." What is that, and why are the big insurance companies supporting it?

"Managed competition" means that big insurance companies will put together huge conglomerates of healthcare institutions, hospitals, clinics, labs and so on. Various bargaining units will be set up to determine which of these conglomerates to work with. That's supposed to introduce some kind of market forces.

But a very small number of big insurance conglomerates, in limited competition with one another, will be pretty much in charge of organizing your healthcare. (This plan will drive the little insurance companies out of the market, which is why they're opposed to it.)

Since they're in business for profit, not for your comfort, the big insurance companies will doubtlessly micromanage health-care, in an attempt to reduce it to the lowest possible level. They'll also tend away from prevention and public health meas-ures, which aren't their concern. Enormous inefficiencies will be involved—huge profits, advertising costs, big corporate salaries and other corporate amenities, big bureaucracies that control in precise detail what doctors and nurses do and don't do—and we'll have to pay for all that.

There's another point that ought to be mentioned. In a Canadian-style, government-insurance system, the costs are distributed in the same way that taxes are. If the tax system is progressive—that is, if rich people pay a higher percentage of their income in taxes (which all other industrial societies as-sume, correctly, to be the only ethical approach)—then the wealthy will also pay more of the costs of healthcare.

But the Clinton program, and all the others like it, are rad-ically regressive. A janitor and a CEO pay the same amount. It's as if they were both taxed the same amount, which is un-heard of in any civilized society.

Actually, it's even worse than that—the janitor will probably pay more. He'll be living in a poor neighborhood and the exec-utive will be living in a rich suburb or a downtown high-rise, which means they'll belong to different health groupings. Be-cause the grouping the janitor belongs to will include many more poor and high-risk people, the insurance companies will demand higher rates from it than the one the executive belongs to, which will include mostly wealthier, lower-risk people.

According to a Harris poll, Americans prefer the Canadian-style health-care system by a huge majority. That's kind of remarkable, given the min-imal amount of media attention the single-payer system has received.

The best work I know on this is by Vicente Navarro. He's dis-covered that there's been quite consistent support for some-thing like a Canadian-style system ever since polls began on this issue, which is now over forty years.

Back in the 1940s, Truman tried to put through such a program. It would have brought the US into line with the rest of the industrial world, but it was beaten back by a huge corporate offensive, complete with tantrums about how we were going to turn into a Bolshevik society and so on.

Every time the issue has come up, there's been a major corporate offensive. One of Ronald Reagan's great achievements back in the late 1960s was to give somber speeches (written for him by the AMA) about how if the legislation establishing Medicare was passed, we'd all be telling our children and grandchildren decades hence what freedom used to be like.

Steffie Woolhandler and David Himmelstein [both of Harvard Medical School] also cite another poll result: When Canadians were asked if they'd want a US-style system, only 5% said yes.

By now, even large parts of the business community don't want it. It's just too inefficient, too bureaucratic and too costly for them. The auto companies estimated a couple of years ago that it was costing them about $500 extra per car just because of the inefficiencies of the US healthcare system—as compared with, say, their Canadian operations.

When business starts to get hurt, then the issue moves into the public agenda. The public has been in favor of a big change for a long time, but what the public thinks doesn't matter much.

There was a nice phrase about this sort of thing in the *Economist*. It was concerned about the fact that Poland has degenerated into a system where they have democratic elections, which is sort of a nuisance.

The population in all of the East European countries is being smashed by the economic changes that are being rammed down their throats. (These changes are called "reforms," which is supposed to make them sound good.) In the last election, the Poles voted in an anti-"reform" government. The *Economist* pointed out that this really wasn't too troublesome because "policy is insulated from politics." In their view, that's a good thing.

In this country too, policy is insulated from politics. People can have their opinions; they can even vote if they like. But policy goes on its merry way, determined by other forces.

What the public wants is called "politically unrealistic." Translated into English, that means the major centers of power and privilege are opposed to it. A change in our healthcare system has now become politically more realistic because the corporate community wants a change, since the current system is harming them.

Vicente Navarro says that a universal and comprehensive healthcare program is "directly related to the strength of the working class and its political and economic instruments."

That's certainly been true in Canada and Europe. Canada had a system rather like ours up until the mid-1960s. It was changed first in one province, Saskatchewan, where the NDP [the New Democratic Party, a mildly reformist, umbrella political party with labor backing] was in power.

The NDP was able to put through a provincial insurance program, driving the insurance companies out of the healthcare business. It turned out to be very successful. It was giving good medical care and reducing costs and was much more progressive in payment. It was mimicked by other provinces, also under labor pressure, often using the NDP as an instrument. Pretty soon it was adopted across Canada nationally.

The history in Europe is pretty much the same. Working-class organizations have been one of the main (although not the only) mechanisms by which people with very limited power and resources can get together to participate in the public arena. That's one of the reasons unions are so hated by business and elites generally. They're just too democratizing in their character.

So Navarro is surely right. The strength and organization of labor and its ability to enter into the public arena is certainly related—maybe even decisively related—to the establishment of social programs of this kind.

There may be a parallel movement going on in California, where there's a ballot initiative to have single-payer healthcare.

The situation in the US is a little different from what Navarro described, because business still plays an inordinate role here in determining what kind of system will evolve. Unless there are significant changes in the US—that is, unless public pressure and organizations, including labor, do a lot more than they've done so far—the outcome will once again be determined by business interests.

Much more media attention has been paid to AIDS than to breast cancer, but a half a million women in the US will die from breast cancer in the 1990s. Many men will die from prostate cancer. These aren't considered political questions, are they?

Well, there's no vote taken on them, but if you're asking if there are questions of policy involved, of course there are. You might add to those cancers the number of children who will suffer or die because of extremely poor conditions in infancy and childhood.

Take, say, malnutrition. That decreases life span quite considerably. If you count that up in deaths, it outweighs anything you're talking about. I don't think many people in the public health field would question the conclusion that the major contribution to improving health, reducing mortality figures and improving the quality of life, would come from simple public-health measures like ensuring people adequate nutrition and safe and healthy conditions of life, clean water, effective sewage treatment, and so on.

You'd think that in a rich country like this, these wouldn't be big issues, but they are for a lot of the population. *Lancet,* the British medical journal—the most prestigious medical journal in the world—recently pointed out that 40% of children in New York City live below the poverty line. They suffer from malnutrition and other poor conditions that cause very high mortality rates—and, if they survive, they have very severe health problems all through their lives.

The *New England Journal of Medicine* pointed out a couple of years ago that black males in Harlem have about the same mortality rate as people in Bangladesh. That's essentially because of the extreme deterioration of the most elementary public health conditions, and social conditions.

Some people have linked the increase in breast cancer and prostate cancer to environmental degradation, to diet, and to the increase of additives and preservatives. What do you think about that?

It's doubtless some kind of a factor. How big or serious a factor it is I'm not sure.

Are you at all interested in the so-called natural or organic food movement?

Sure. I think there ought to be concerns about the quality of food. I would say this falls into the question of general public health. It's like having good water and good sewage and making sure that people have enough food and so on.

All these things are in roughly the same category—they don't have to do with high-technology medical treatment but with essential conditions of life. These general public-health issues, of which eating food that doesn't contain poisons is naturally a part, are the overwhelming factors in quality of life and mortality.

Crime and punishment

There's been a tendency over the last few years for local TV news programs to concentrate on crimes, rapes, kidnappings, etc. Now this is spilling over into the national network news programs.

That's true, but it's just a surface phenomenon. Why is there an increase in attention to violent crime? Is it connected to the fact that there's been a considerable decline in income for the large majority of the population, and a decline as well in the opportunity for constructive work?

But until you ask why there's an increase in social disintegration, and why more and more resources are being directed towards the wealthy and privileged sectors and away from the general population, you can't have even a concept of why there's rising crime or how you should deal with it.

Over the past twenty or thirty years, there's been a considerable increase in inequality. This trend accelerated during the Reagan years. The society has been moving visibly towards a kind of Third World model.

The result is an increasing crime rate, as well as other signs of social disintegration. Most of the crime is poor people attacking each other, but it spills over to more privileged sectors.

People are very worried—and quite properly, because the society is becoming very dangerous.

A constructive approach to the problem would require dealing with its fundamental causes, but that's off the agenda, because we must continue with a social policy that's aimed at strengthening the welfare state for the rich.

The only kind of responses the government can resort to under those conditions is pandering to the fear of crime with increasing harshness, attacking civil liberties and attempting to control the poor, essentially by force.

Do you know what "smash and grab" is? When your car is in traffic or at a stop light, people come along, smash in the window and grab your purse or steal your wallet.

The same thing is going on right around Boston. There's also a new form, called "Good Samaritan robbery." You fake a flat tire on the highway and when somebody stops to help, you jump them, steal their car, beat them up if they're lucky, kill them if they're not.

The causes are the increasing polarization of the society that's been going on for the past twenty-five years, and the marginalization of large sectors of the population. Since they're superfluous for wealth production (meaning profit production), and since the basic ideology is that a person's human rights depend on what they can get for themselves in the market system, they have no human value.

Larger and larger sectors of the population have no form of organization and no viable, constructive way of reacting, so they pursue the available options, which are often violent. To a large extent, those are the options that are encouraged in the popular culture.

You can tell a great deal about a society when you look at its system of justice. I was wondering if you'd comment on the Clinton crime bill, which authorizes hiring 100,000 more cops, boot camps for juveniles, more money for prisons, extending the death penalty to about fifty new offenses and making gang membership a federal crime—which is interesting, considering there's something about freedom of association in the Bill of Rights.

It was hailed with great enthusiasm by the far right as the greatest anticrime bill ever. It's certainly the most *extraordinary* crime bill in history. It's greatly increased, by a factor of five or six, federal spending for repression. There's nothing much constructive in it. There are more prisons, more police, heavier sentences, more death sentences, new crimes, three strikes and you're out.

It's unclear how much pressure and social decline and deterioration people will accept. One tactic is just drive them into urban slums—concentration camps, in effect—and let them prey on one another. But they have a way of breaking out and affecting the interests of wealthy and privileged people. So you have to build up the jail system, which is incidentally also a shot in the arm for the economy.

It's natural that Clinton picked up this crime bill as a major social initia tve, not only for a kind of ugly political reason—namely, that it's easy to whip up hysteria about it—but also because it reflects the general point of view of the so-called New Democrats, the business-oriented segment of the Democratic Party to which Clinton belongs.

What are your views on capital punishment?

It's a crime. I agree with Amnesty Interna ion d on that one, and indeed with most of the world. The state should have no right to take people's lives.

Radio listener: Does this country have a vested interest in supporting the drug trade?

It's complicated; I don't want to be too brief about it. For one thing, you can't talk about marijuana and cocaine in the same breath. Marijuana simply doesn't have the lethal effects of cocaine. You can debate about whether marijuana is good or bad, but out of about sixty million users, I don't think there's a known case of overdose. The criminalization of marijuana has motives other than concern about drugs.

On the other hand, hard drugs, to which people have been driven to a certain extent by the prohibitions against soft drugs, are very harmful—although nowhere near the harm of, say, tobacco and alcohol in terms of overall societal effects, including deaths.

There are sectors of American society that profit from the hard drug trade, like the big international banks that do the money laundering or the corporations that provide the chemicals for the industrial production of hard drugs. On the other hand, people who live in the inner cities are being devastated by them. So there are different interests.

Gun control

Advocates of free access to arms cite the Second Amendment. Do you believe that it permits unrestricted, uncontrolled possession of guns?

It's pretty clear that, taken literally, the Second Amendment doesn't permit people to have guns. But laws are never taken literally, including amendments to the Constitution or constitutional rights. Laws permit what the tenor of the times interprets them as permitting.

But underlying the controversy over guns are some serious questions. There's a feeling in the country that people are under attack. I think they're misidentifying the source of the attack, but they do feel under attack.

The government is the only power structure that's even partially accountable to the population, so naturally the business sectors want to make that the enemy—not the corporate system, which is totally unaccountable. After decades of intensive business propaganda, people feel that the government is some kind of enemy and that they have to defend themselves from it.

It's not that that doesn't have its justifications. The government *is* authoritarian and commonly hostile to much of the population. But it's partially influenceable—and potentially very influenceable—by the general population.

Many people who advocate keeping guns have fear of the government in the back of their minds. But that's a crazy response to a real problem.

Do the media foster the feeling people have that they're under attack?

At the deepest level, the media contribute to the sense that the government is the enemy, and they suppress the sources of real power in the society, which lie in the totalitarian institutions—the corporations, now international in scale—that control the economy and much of our social life. In fact, the corporations set the conditions within which the government operates, and control it to a large extent.

The picture presented in the media is constant, day after day. People simply have no awareness of the system of power under which they're suffering. As a result—as intended—they turn their attention against the government.

People have all kinds of motivations for opposing gun control, but there's definitely a sector of the population that considers itself threatened by big forces, ranging from the Federal Reserve to the Council on Foreign Relations to big government to who knows what, and they're calling for guns to protect themselves.

Radio listener: On the issue of gun control, I believe that the US is becoming much more like a Third World country, and nothing is necessarily going to put a stop to it. I look around and see a lot of Third World countries where, if the citizens had weapons, they wouldn't have the government they've got. So I think that maybe people are being a little short-sighted in arguing for gun control and at the same time realizing that the government they've got is not exactly a benign one.

Your point illustrates exactly what I think is a major fallacy. The government is far from benign—that's true. On the other hand, it's at least partially accountable, and it can become as benign as we make it.

What's not benign (what's extremely harmful, in fact) is something you didn't mention—business power, which is highly concentrated and, by now, largely transnational. Business power is very far from benign and it's completely unaccountable. It's a totalitarian system that has an enormous effect on our lives. It's also the main reason why the government isn't benign.

As for guns being the way to respond to this, that's outlandish. First of all, this is not a weak Third World country. If people have pistols, the government has tanks. If people get tanks, the government has atomic weapons. There's no way to deal with these issues by violent force, even if you think that that's morally legitimate.

Guns in the hands of American citizens are not going to make the country more benign. They're going to make it more brutal, ruthless and destructive. So while one can recognize the motivation that lies behind some of the opposition to gun control, I think it's sadly misguided.

Becoming a Third World country

A recent Census Bureau report stated that there's been a 50% increase in the working poor—that is, people who have jobs but are still below the poverty level.

That's part of the Third-Worldization of the society. It's not just unemployment, but also wage reduction. Real wages have been declining since the late 1960s. Since 1987, they've even been declining for college-educated people, which was a striking shift.

There's supposed to be a recovery going on, and it's true that a kind of recovery is going on. It's at about half the rate of preceding postwar recoveries from recession (there've been half a dozen of them) and the rate of job creation is less than a third. Furthermore—out of line with earlier recoveries—the jobs themselves are low-paying, and a huge number of them are temporary.

This is what's called "increasing flexibility of the labor market." *Flexibility* is a word like reform—it's supposed to be a good thing. Actually, *flexibility* means insecurity. It means you go to bed at night and don't know if you'll have a job in the morning. Any economist can explain that that's a good thing for the economy—that is, for profit-making, not for the way people live.

Low wages also increase job insecurity. They keep inflation low, which is good for people who have money—bondholders, say. Corporate profits are zooming, but for most of the population, things are grim. And grim circumstances, without much prospect for a future or for constructive social action, express themselves in violence.

It's interesting that you should say that. Most of the examples of mass murders are in the workplace. I'm thinking of the various killings in post offices and fast-food restaurants, where workers are disgruntled for one reason or another, or have been fired or laid off.

Not only have real wages stagnated or declined, but working conditions have gotten much worse. You can see that just in counting hours of work. Julie Schor, an economist at Harvard, brought out an important book on this a couple of years ago, called *The Overworked American.* If I remember her figures correctly, by around 1990, the time she was writing, workers had to put in about six weeks extra work a year to maintain something like a 1970 real wage level.

Along with the increasing hours of work comes increasing harshness of work conditions, increasing insecurity and, because of the decline of unions, reduced ability to protect oneself. In the Reagan years, even the minimal government programs for protecting workers against workplace accidents and the like were reduced, in the interest of maximizing profits. The absence of constructive options, like union organizing, leads to violence.

Labor

[Harvard professor] Elaine Bernard and [union official] Tony Mazzocchi have been talking about creating a new labor-based party. What are your views on that?

I think that's an important initiative. The US is becoming very depoliticized and negative. About half the population thinks both political parties should be disbanded. There's a real need for something that would articulate the concerns of that substantial majority of the population that's being left out of social planning and the political process.

Labor unions have often been a significant force—in fact, the main social force—for democratization and progress. On the other hand, when they aren't linked to the political system through a labor-based party, there's a limit on what they can do. Take healthcare, for example.

Powerful unions in the US were able to get fairly reasonable healthcare provisions for themselves. But since they were acting independently of the political system, they typically didn't attempt to bring about decent health conditions for the general population. Compare Canada, where the unions, being linked to labor-based parties, were able to implement healthcare for everybody.

That's an illustration of the kind of difference a politically oriented, popular movement like labor can achieve. We're not in the day any longer where the industrial workers are the majority or even the core of the labor force. But the same questions arise. I think Bernard and Mazzocchi are on the right track in thinking along those lines.

Yesterday was May 1. What's its historical significance?

It's May Day, which throughout the world has been a working-class holiday for more than a hundred years. It was initiated in solidarity with American workers who, back in the 1880s, were suffering unusually harsh conditions in their effort to achieve an eight-hour workday. The US is one of the few countries where this day of solidarity with US labor is hardly even known.

This morning, way in the back of the *Boston Globe*, there was a little item whose headline read, "May Day Celebration in Boston." I was surprised, because I don't think I've ever seen that here in the US. It turned out that there indeed was a May Day celebration, of the usual kind, but it was being held by Latin American and Chinese workers who've recently immigrated here.

That's a dramatic example of the efficiency with which business controls US ideology, of how effective its propaganda and indoctrination have been in depriving people of any awareness of their own rights and history. You have to wait for poor Latino and Chinese workers to celebrate an international holiday of solidarity with American workers.

In his *New York Times* column, Anthony Lewis wrote: "Unions in this country, sad to say, are looking more and more like the British unions...backward, unenlightened....The crude, threatening tactics used by unions to make Democratic members of the House vote against NAFTA underline the point."

That brings out Lewis's real commitments very clearly. What he called "crude, threatening tactics" were labor's attempt to get their representatives to represent their interests. By the standards of the elite, that's an attack on democracy, because the political system is supposed to be run by the rich and powerful.

Corporate lobbying vastly exceeded labor lobbying, but you can't even talk about it in the same breath. It wasn't considered raw muscle or antidemocratic. Did Lewis have a column denouncing corporate lobbying for NAFTA?

I didn't see it.

I didn't see it either.

Things reached the peak of absolute hysteria the day before the vote. The *New York Times* lead editorial was exactly along the lines of that quote from Lewis, and it included a little box that listed the dozen or so representatives in the New York region who were voting against NAFTA. It showed their contributions from labor and said that this raises ominous questions about the political influence of labor, and whether these politicians are being honest, and so on.

As a number of these representatives later pointed out, the *Times* didn't have a box listing corporate contributions to them or to other politicians—nor, we may add, was there a box listing advertisers of the *New York Times* and their attitudes towards NAFTA.

It was quite striking to watch the hysteria that built up in privileged sectors, like the *Times'* commentators and editorials, as the NAFTA vote approached. They even allowed themselves the use of the phrase "class lines." I've never seen that in the *Times* before. You're usually not allowed to admit that the US has class lines. But this was considered a really serious issue, and all bars were let down.

The end result is very intriguing. In a recent poll, about 70% of the respondents said they were opposed to the actions of the labor movement against NAFTA, but it turned out that they took pretty much the same position that labor took. So why were they opposed to it?

I think it's easy to explain that. The media scarcely reported what labor was actually *saying.* But there was plenty of hysteria about labor's alleged tactics.

The CIA

What about the role of the CIA in a democratic society? Is that an oxymoron?

You could imagine a democratic society with an organization that carries out intelligence-gathering functions. But that's a very minor part of what the CIA does. Its main purpose is to carry out secret and usually illegal activities for the executive branch, which wants to keep these activities secret because it knows that the public won't accept them. So even inside the US, it's highly undemocratic.

The activities that it carries out are quite commonly efforts to undermine democracy, as in Chile through the 1960s into the early 1970s. That's far from the only example. By the way, although most people focus on Nixon's and Kissinger's involvement with the CIA, Kennedy and Johnson carried out similar policies.

Is the CIA an instrument of state policy, or does it formulate policy on its own?

You can't be certain, but my own view is that the CIA is very much under the control of executive power. I've studied those records fairly extensively in many cases, and it's very rare for the CIA to undertake initiatives on its own.

It often looks as though it does, but that's because the executive wants to preserve deniability. The executive branch doesn't want to have documents lying around that say, *I told you to murder Lumumba*, or to overthrow the government of Brazil, or to assassinate Castro.

So the executive branch tries to follow policies of plausible deniability, which means that messages are given to the CIA to do things but without a paper trail, without a record. When the story comes out later, it looks as if the CIA is doing things on their own. But if you really trace it through, I think this almost never happens.

The media

Let's talk about the media and democracy. In your view, what are the communications requirements of a democratic society?

I agree with Adam Smith on this—we'd like to see a tendency toward equality. Not just equality of opportunity, but actual equality—the ability, at every stage of one's existence, to access information and make decisions on the basis of it. So a democratic communications system would be one that involves large-scale public participation, and that reflects both public interests and real values like truth, integrity and discovery.

Bob McChesney, in his recent book *Telecommunications, Mass Media and Democracy,* details the debate between 1928 and 1935 for control of radio in the US. How did that battle play out?

That's a very interesting topic, and he's done an important service by bringing it out. It's very pertinent today, because we're involved in a very similar battle over this so-called "information superhighway."

In the 1920s, the first major means of mass communication since the printing press came along—radio. It's obvious that radio is a bounded resource, because there's only a fixed bandwidth. There was no question in anyone's mind that the government was going to have to regulate it. The question was, What form would this government regulation take?

Government could opt for public radio, with popular participation. This approach would be as democratic as the society is. Public radio in the Soviet Union would have been totalitarian, but in, say, Canada or England, it would be partially democratic (insofar as those societies are democratic).

That debate was pursued all over the world—at least in the wealthier societies, which had the luxury of choice. Almost every country (maybe every one—I can't think of an exception) chose public radio, while the US chose private radio. It wasn't 100%; you were allowed to have small radio stations—say, a college radio station—that can reach a few blocks. But virtually all radio in the US was handed over to private power.

As McChesney points out, there was a considerable struggle about that. There were church groups and some labor unions and other public interest groups that felt that the US should go the way the rest of the world was going. But this is very much a business-run society, and they lost out.

Rather strikingly, business also won an ideological victory, claiming that handing radio over to private power constituted democracy, because it gave people choices in the marketplace. That's a very weird concept of democracy, since your power depends on the number of dollars you have, and your choices are limited to selecting among options that are highly structured by the real concentrations of power. But this was nevertheless widely accepted, even by liberals, as the democratic solution. By the mid- to late 1930s, the game was essentially over.

This struggle was replayed—in the rest of the world, at least—about a decade later, when television came along. In the US this wasn't a battle at all; TV was completely commercialized without any conflict. But again, in most other countries—or maybe every other country—TV was put in the public sector.

In the 1960s, television and radio became partly commercialized in other countries; the same concentration of private power that we find in the US was chipping away at the public-service function of radio and television. At the same time in the US, there was a slight opening to public radio and television.

The reasons for this have never been explored in any depth (as far as I know), but it appears that the private broadcasting companies recognized that it was a nuisance for them to have to satisfy the formal requirements of the Federal Communications Commission that they devote part of their programming to public-interest purposes. So CBS, say, had to have a big office with a lot of employees who every year would put together a collection of fraudulent claims about how they'd met this legislative condition. It was a pain in the neck.

At some point, they apparently decided that it would be easier to get the entire burden off their backs and permit a small and underfunded public broadcasting system. They could then claim that they didn't have to fulfill this service any longer. That was the origin of public radio and television—which is now largely corporate-funded in any event.

That's happening more and more. PBS [the Public Broadcasting Service] is sometimes called "the Petroleum Broadcasting Service."

That's just another reflection of the interests and power of a highly class-conscious business system that's always fighting an intense class war. These issues are coming up again with respect to the internet and the new interactive communications technologies. And we're going to find exactly the same conflict again. It's going on right now.

I don't see why we should have had any long-term hopes for something different. Commercially-run radio is going to have certain purposes—namely, the ones determined by people who own and control it.

As I mentioned earlier, they don't want decision-makers and participants; they want a passive, obedient population of consumers and political spectators—a community of people who are so atomized and isolated that they can't put together their limited resources and become an independent, powerful force that will chip away at concentrated power.

Does ownership always determine content?

In some far-reaching sense it does, because if content ever goes beyond the bounds owners will tolerate, they'll surely move in to limit it. But there's a fair amount of flexibility.

Investors don't go down to the television studio and make sure that the local talk-show host or reporter is doing what they want. There are other, subtler, more complex mechanisms that make it fairly certain that the people on the air will do what the owners and investors want. There's a whole, long, filtering process that makes sure that people only rise through the system to become managers, editors, etc., if they've internalized the values of the owners.

At that point, they can describe themselves as quite free. So you'll occasionally find some flaming independent-liberal type like Tom Wicker who writes, *Look, nobody tells me what to say. I say whatever I want. It's an absolutely free system.*

And, for *him,* that's true. After he'd demonstrated to the satisfaction of his bosses that he'd internalized their values, he was entirely free to write whatever he wanted.

Both PBS and NPR [National Public Radio] frequently come under attack for being left-wing.

That's an interesting sort of critique. In fact, PBS and NPR are elite institutions, reflecting by and large the points of view and interests of wealthy professionals who are very close to business circles, including corporate executives. But they happen to be liberal by certain criteria.

That is, if you took a poll among corporate executives on matters like, say, abortion rights, I presume their responses would be what's called liberal. I suspect the same would be true on lots of social issues, like civil rights and freedom of speech. They tend not to be fundamentalist, born-again Christians, for example, and they might tend to be more opposed to the death penalty than the general population. I'm sure you'll find plenty of private wealth and corporate power backing the American Civil Liberties Union.

Since those are aspects of the social order from which they gain, they tend to support them. By these criteria, the people who dominate the country tend to be liberal, and that reflects itself in an institution like PBS.

You've been on NPR just twice in 23 years, and on *The MacNeil-Lehrer News Hour* once in its almost 20 years. What if you'd been on *MacNeil-Lehrer* ten times? Would it make a difference?

Not a lot. By the way, I'm not quite sure of those numbers; my own memory isn't that precise. I've been on local PBS stations in particular towns.

I'm talking about the national network.

Then probably something roughly like those numbers is correct. But it wouldn't make a lot of difference.

In fact, in my view, if the managers of the propaganda system were more intelligent, they'd allow more leeway to real dissidents and critics. That would give the impression of broader debate and discussion and hence would have a legitimizing function, but it still wouldn't make much of a dent, given the overwhelming weight of propaganda on the other side. By the way, that propaganda system includes not just how issues are framed in news stories but also how they're presented in entertainment programming—that huge area of the media that's simply devoted to diverting people and making them more stupid and passive.

That's not to say I'm against opening up these media a bit, but I would think it would have a limited effect. What you need is something that presents every day, in a clear and comprehensive fashion, a different picture of the world, one that reflects the concerns and interests of ordinary people, and that takes something like the point of view with regard to democracy and participation that you find in people like Jefferson or Dewey.

Where that happens—and it has happened, even in modern societies—it has effects. In England, for example, you did have major mass media of this kind up until the 1960s, and it helped sustain and enliven a working-class culture. It had a big effect on British society.

What do you think about the internet?

I think that there are good things about it, but there are also aspects of it that concern and worry me. This is an intuitive response—I can't prove it—but my feeling is that, since people aren't Martians or robots, direct face-to-face contact is an extremely important part of human life. It helps develop self-understanding and the growth of a healthy personality.

You just have a different relationship to somebody when you're looking at them than you do when you're punching away at a keyboard and some symbols come back. I suspect that extending that form of abstract and remote relationship, instead of direct, personal contact, is going to have unpleasant effects on what people are like. It will diminish their humanity, I think.

Sports

In 1990, in one of our many interviews, we had a brief discussion about the role and function of sports in American society, part of which was subsequently excerpted in *Harper's*. I've probably gotten more comments about that than anything else I've ever recorded. You really pushed some buttons.

I got some funny reactions, a lot of irate reactions, as if I were somehow taking people's fun away from them. I have nothing against sports. I like to watch a good basketball game and that sort of thing. On the other hand, we have to recognize that the mass hysteria about spectator sports plays a significant role.

First of all, spectator sports make people more passive, because you're not doing them—you're watching somebody doing them. Secondly, they engender jingoist and chauvinist attitudes, sometimes to quite an extreme degree.

I saw something in the newspapers just a day or two ago about how high school teams are now so antagonistic and passionately committed to winning at all costs that they had to abandon the standard handshake before or after the game. These kids can't even do civil things like greeting one another because they're ready to kill one another.

It's spectator sports that engender those attitudes, particularly when they're designed to organize a community to be hysterically committed to their gladiators. That's very dangerous, and it has lots of deleterious effects.

I was reading something about the glories of the information superhighway not too long ago. I can't quote it exactly, but it was talking about how wonderful and empowering these new interactive technologies are going to be. Two basic examples were given.

For women, interactive technologies are going to offer highly improved methods of home shopping. So you'll be able to watch the tube and some model will appear with a product and you're supposed to think, *God, I've got to have that.* So you press a button and they deliver it to your door within a couple of hours. That's how interactive technology is supposed to liberate women.

For men, the example involved the Super Bowl. Every red-blooded American male is glued to it. Today, all they can do is watch it and cheer and drink beer, but the new interactive technology will let them actually participate in it. While the quarterback is in the huddle calling the next play, the people watching will be able to decide what the play should be.

If they think he should pass, or run, or punt, or whatever, they'll be able to punch that into their computer and their vote will be recorded. It won't have any effect on what the quarterback does, of course, but after the play the television channel will be able to put up the numbers—63% said he should have passed, 24% said he should have run, etc.

That's interactive technology for men. Now you're really participating in the world. Forget about all this business of deciding what ought to happen with healthcare—now you're doing something really important.

This scenario for interactive technology reflects an understanding of the stupefying effect spectator sports have in making people passive, atomized, obedient nonparticipants—nonquestioning, easily controlled and easily disciplined.

At the same time, athletes are lionized or—in the case of Tonya Harding, say—demonized.

If you can personalize events of the world—whether it's Hillary Clinton or Tonya Harding—you've succeeded in directing people away from what really matters and is important. The John F. Kennedy cult is a good example, with the effects it's had on the left.

Religious fundamentalism

In his book *When Time Shall Be No More,* historian Paul Boyer states that, "surveys show that from one-third to one-half of [all Americans] believe that the future can be interpreted from biblical prophecies." I find this absolutely stunning.

I haven't seen that particular number, but I've seen plenty of things like it. I saw a cross-cultural study a couple of years ago— I think it was published in England—that compared a whole range of societies in terms of beliefs of that kind. The US stood out— it was unique in the industrial world. In fact, the measures for the US were similar to pre-industrial societies.

Why is that?

That's an interesting question. This is a very fundamentalist society. It's like Iran in its degree of fanatic religious commitment. For example, I think about 75% of the US population has a literal belief in the devil.

There was a poll several years ago on evolution. People were asked their opinion on various theories of how the world of living creatures came to be what it is. The number of people who believed in Darwinian evolution was less than 10%. About half the population believed in a Church doctrine of divinely-guided evolution. Most of the rest presumably believed that the world was created a couple of thousand years ago.

These are very unusual results. Why the US should be off the spectrum on these issues has been discussed and debated for some time.

I remember reading something maybe ten or fifteen years ago by a political scientist who writes about these things, Walter Dean Burnham. He suggested that this may be a reflection of depoliticization—that is, the inability to participate in a meaningful fashion in the political arena may have a rather important psychic effect.

That's not impossible. People will find some ways of identifying themselves, becoming associated with others, taking part in something. They're going to do it some way or other. If they don't have the option to participate in labor unions, or in political organizations that actually function, they'll find other ways. Religious fundamentalism is a classic example.

We see that happening in other parts of the world right now. The rise of what's called Islamic fundamentalism is, to a significant extent, a result of the collapse of secular nationalist alternatives that were either discredited internally or destroyed.

In the nineteenth century, you even had some conscious efforts on the part of business leaders to promote fire-and-brimstone preachers who led people to look at society in a more passive way. The same thing happened in the early part of the industrial revolution in England. E.P. Thompson writes about it in his classic, *The Making of the English Working Class.*

In a State of the Union speech, Clinton said, "We can't renew our country unless more of us—I mean, all of us—are willing to join churches." What do you make of this?

I don't know exactly what was in his mind, but the ideology is very straightforward. If people devote themselves to activities that are out of the public arena, then we folks in power will be able to run things the way we want.

Don't tread on me

I'm not quite clear about how to formulate this question. It has to do with the nature of US society as exemplified in comments like *do your own thing, go it alone, don't tread on me, the pioneer spirit* — all that deeply individualistic stuff. What does that tell you about American society and culture?

It tells you that the propaganda system is working full-time, because there is no such ideology in the US. Business certainly doesn't believe it. All the way back to the origins of American society, business has insisted on a powerful, interventionist state to support its interests, and it still does.

There's nothing individualistic about corporations. They're big conglomerate institutions, essentially totalitarian in character. Within them, you're a cog in a big machine. There are few institutions in human society that have such strict hierarchy and top-down control as a business organization. It's hardly *don't tread on me* — you're being tread on all the time.

The point of the ideology is to prevent people who are outside the sectors of coordinated power from associating with each other and entering into decision-making in the political arena. The point is to leave the powerful sectors highly integrated and organized, while atomizing everyone else.

That aside, there is another factor. There's a streak of independence and individuality in American culture that I think is a very good thing. This *don't tread on me* feeling is in many respects a healthy one — up to the point where it keeps you from working together with other people.

So it's got a healthy side and a negative side. Naturally it's the negative side that's emphasized in the propaganda and indoctrination.

Toward greater inequality

In his column in the *New York Times,* Anthony Lewis wrote, "Since World War II, the world has experienced extraordinary growth." Meanwhile, at a meeting in Quito, Ecuador, Juan de Dias Parra, the head of the Latin American Association for Human Rights, said, "In Latin America today, there are 7 million more hungry people, 30 million more illiterate people, 10 million more families without homes, 40 million more unemployed persons than there were 20 years ago. There are 240 million human beings in Latin America without the necessities of life, and this when the region is richer and more stable than ever, according to the way the world sees it." How do you reconcile those two statements?

It just depends on which people you're talking about. The World Bank came out with a study on Latin America which warned that Latin America was facing chaos because of the extraordinarily high level of inequality, which is the highest in the world (and that's after a period of substantial growth). Even the things the World Bank cares about are threatened.

The inequality didn't just come from the heavens. There was a struggle over the course of Latin American development back in the mid-1940s, when the new world order of that day was being crafted.

The State Department documents on this are quite interesting. They said that Latin America was swept by what they called the "philosophy of the new nationalism," which called for increasing production for domestic needs and reducing inequality. The basic principle of this new nationalism was that the people of the country should be the prime beneficiary of the country's resources.

The US was sharply opposed to that and came out with an economic charter for the Americas that called for eliminating economic nationalism (as it's also called) in all of its forms and insisting that Latin American development be "complementary" to US development. That means we'll have the advanced industry and the technology and the peons in Latin America will produce export crops and do some simple operations that they can manage. But they won't develop economically the way we did.

Given the distribution of power, the US of course won. In countries like Brazil, we just took over—Brazil has been almost completely directed by American technocrats for about fifty years. Its enormous resources should make it one of the richest countries in the world, and it's had one of the highest growth rates. But thanks to our influence on Brazil's social and economic system, it's ranked around Albania and Paraguay in quality of life measures, infant mortality and so on.

It's true, as Lewis says, that there's been very substantial growth in the world. At the same time, there's incredible poverty and misery, and that's increased even more.

If you compare the percentage of world income held by the richest 20% and the poorest 20%, the gap has dramatically increased over the past thirty years. Comparing rich countries to poor countries, it's about doubled. Comparing rich people to poor people within countries, it's increased far more and is much sharper. That's the consequence of a particular kind of growth.

Do you think this trend of growth rates and poverty rates increasing simultaneously will continue?

Actually, growth rates have been slowing down a lot; in the past twenty years, they've been roughly half of what they were in the preceding twenty years. This tendency toward lower growth will probably continue.

One cause is the enormous increase in the amount of unregulated, speculative capital. The figures are really astonishing. John Eatwell, one of the leading specialists in finance at Cambridge University, estimates that, in 1970, about 90% of international capital was used for trade and long-term invest-ment—more or less productive things—and 10% for speculation. By 1990, those figures had reversed: 90% for speculation and 10% for trade and long-term investment.

Not only has there been radical change in the nature of unregulated financial capital, but the quantity has grown enormously. According to a recent World Bank estimate, $14 *trillion* is now moving around the world, about $1 trillion or so of which moves every *day*.

This huge amount of mostly speculative capital creates pressures for deflationary policies, because what speculative capital wants is low growth and low inflation. It's driving much of the world into a low-growth, low-wage equilibrium.

This is a tremendous attack against government efforts to stimulate the economy. Even in the richer societies, it's very difficult; in the poorer societies, it's hopeless. What happened with Clinton's trivial stimulus package was a good indication. It amounted to nothing—$19 billion, but it was shot down instantly.

In the fall of 1993, the *Financial Times* trumpeted, "the public sector is in retreat everywhere." Is that true?

It's largely true, but major parts of the public sector are alive and well—in particular those parts that cater to the interests of the wealthy and the powerful. They're declining somewhat, but they're still very lively, and they're not going to disappear.

These developments have been going on for about twenty years now. They had to do with major changes in the international economy that became more or less crystallized by the early 1970s.

For one thing, US economic hegemony over the world had pretty much ended by then, and Europe and Japan had reemerged as major economic and political powers. The costs of the Vietnam War were very significant for the US economy, and extremely beneficial for its rivals. That tended to shift the world balance.

In any event, by the early 1970s, the US felt that it could no longer sustain its traditional role as—essentially—international banker. (This role was codified in the Bretton Woods agreements at the end of the Second World War, in which currencies were regulated relative to one another, and in which the de facto international currency, the US dollar, was fixed to gold.)

Nixon dismantled the Bretton Woods system around 1970. That led to tremendous growth in unregulated financial capital. That growth was rapidly accelerated by the short-term rise in the price of commodities like oil, which led to a huge flow of petrodollars into the international system. Furthermore, the telecommunications revolution made it extremely easy to transfer capital—or, rather, the electronic equivalent of capital—from one place to another.

There's also been a very substantial growth in the internationalization of production. It's now a lot easier than it was to shift production to foreign countries—generally highly repressive ones—where you get much cheaper labor. So a corporate executive who lives in Greenwich, Connecticut and whose corporate and bank headquarters are in New York City can have a factory somewhere in the Third World. The actual banking operations can take place in various offshore regions where you don't have to worry about supervision—you can launder drug money or whatever you feel like doing. This has led to a totally different economy.

With the pressure on corporate profits that began in the early 1970s, a big attack was launched on the whole social contract that had developed through a century of struggle and that had been more or less codified around the end of the Second World War with the New Deal and the European social welfare states. The attack was led by the US and England, and by now has reached continental Europe.

It's led to a serious decline in unionization, which carries with it a decline in wages and other forms of protection, and to a very sharp polarization of the society, primarily in the US and Britain (but it's spreading).

Driving in to work this morning, I was listening to the BBC [the British Broadcasting Company, Britain's national broadcasting service]. They reported a new study that found that children living in workhouses a century ago had better nutritional standards than millions of poor children in Britain today.

That's one of the grand achievements of [former British Prime Minister Margaret] Thatcher's revolution. She succeeded in devastating British society and destroying large parts of British manufacturing capacity. England is now one of the poorest countries in Europe—not much above Spain and Portugal, and well below Italy.

The American achievement was rather similar. We're a much richer, more powerful country, so it isn't possible to achieve quite what Britain achieved. But the Reaganites succeeded in driving US wages down so far that we're now the second lowest of the major industrial countries, barely above Britain. Labor costs in Italy are about 20% higher than in the US, and in Germany they're maybe 60% higher.

Along with that goes a deterioration of the general social contract and a breakdown of the kind of public spending that benefits the less privileged. Needless to say, the kind of public spending that benefits the wealthy and the privileged—which is enormous—remains fairly stable.

"Free trade"

My local newspaper, the Boulder [Colorado] *Daily Camera,* which is part of the Knight-Ridder chain, ran a series of questions and answers about GATT. They answered the question, Who would benefit from a GATT agreement? by writing, "Consumers would be the big winners." Does that track with your understanding?

If they mean rich consumers—yes, they'll gain. But much of the population will see a decline in wages, both in rich countries and poor ones. Take a look at NAFTA, where the analyses have already been done. The day after NAFTA passed, the *New York Times* had its first article on its expected impact in the New York region. (Its conclusions apply to GATT too.)

It was a very upbeat article. They talked about how wonderful NAFTA was going to be. They said that finance and services will be particularly big winners. Banks, investment firms, PR firms, corporate law firms will do just great. Some manufacturers will also benefit—for example, publishing and the chemical industry, which is highly capital-intensive, with not many workers to worry about.

Then they said, Well, there'll be some losers too: women, Hispanics, other minorities, and semi-skilled workers—in other words, about two-thirds of the work force. But everyone else will do fine.

Just as anyone who was paying attention knew, the purpose of NAFTA was to create an even smaller sector of highly privileged people—investors, professionals, managerial classes. (Bear in mind that this is a rich country, so this privileged sector, although smaller, still isn't tiny.) It will work fine for them, and the general population will suffer.

The prediction for Mexico is exactly the same. The leading financial journal in Mexico, which is very pro-NAFTA, estimated that Mexico would lose about 25% of its manufacturing capacity in the first few years and about 15% of its manufacturing labor force. In addition, cheap US agricultural exports are expected to drive several million people off the land. That's going to mean a substantial increase in the unemployed workforce in Mexico, which of course will drive down wages.

On top of that, union organizing is essentially impossible. Corporations can operate internationally, but unions can't—so there's no way for the work force to fight back against the internationalization of production. The net effect is expected to be a decline in wealth and income for most people in Mexico and for most people in the US.

The strongest NAFTA advocates point that out in the small print. My colleague at MIT, Paul Krugman, is a specialist in international trade and, interestingly, one of the economists who's done some of the theoretical work showing why free trade doesn't work. He was nevertheless an enthusiastic advocate of NAFTA—which is, I should stress, not a free-trade agreement.

He agreed with the *Times* that unskilled workers—about 70% of the work force—would lose. The Clinton administration has various fantasies about retraining workers, but that would probably have very little impact. In any case, they're doing nothing about it.

The same thing is true of skilled white-collar workers. You can get software programmers in India who are very well trained at a fraction of the cost of Americans. Somebody involved in this business recently told me that Indian programmers are actually being brought to the US and put into what are kind of like slave-labor camps and kept at Indian salaries—a fraction of American salaries—doing software development. So that kind of work can be farmed out just as easily.

The search for profit, when it's unconstrained and free from public control, will naturally try to repress people's lives as much as possible. The executives wouldn't be doing their jobs otherwise.

What accounted for all the opposition to NAFTA?

The original expectation was that NAFTA would just sail through. Nobody would even know what it was. So it was signed in secret. It was put on a fast track in Congress, meaning essentially no discussion. There was virtually no media coverage. Who was going to know about a complex trade agreement?

That didn't work, and there are a number of reasons why it didn't. For one thing, the labor movement got organized for once and made an issue of it. Then there was this sort of maverick third-party candidate, Ross Perot, who managed to make it a public issue. And it turned out that as soon as the public learned anything about NAFTA, they were pretty much opposed.

I followed the media coverage on this, which was extremely interesting. Usually the media try to keep their class loyalties more or less in the background—they try to pretend they don't have them. But on this issue, the bars were down. They went berserk, and toward the end, when it looked like NAFTA might not pass, they just turned into raving maniacs.

But despite this enormous media barrage and the government attack and huge amounts of corporate lobbying (which totally dwarfed all the other lobbying, of course), the level of opposition remained pretty stable. Roughly 60% or so of those who had an opinion remained opposed.

The same sort of media barrage influenced the Gore-Perot television debate. I didn't watch it, but friends who did thought Perot just wiped Gore off the map. But the media proclaimed that Gore won a massive victory.

In polls the next day, people were asked what they thought about the debate. The percentage who thought that Perot had been smashed was far higher than the percentage who'd seen the debate, which means that most people were being told what to think by the media, not coming to their own conclusions.

Incidentally, what was planned for NAFTA worked for GATT—there was virtually no public opposition to it, or even awareness of it. It was rammed through in secret, as intended.

What about the position people like us find ourselves in of being "against," of being "anti-," reactive rather than pro-active?

NAFTA's a good case, because very few NAFTA critics were opposed to *any* agreement. Virtually everyone—the labor movement, the Congressional Office of Technology Assessment (a major report that was suppressed) and other critics (including me)—was saying there'd be nothing wrong with *a* North American Free Trade Agreement, but not *this* one. It should be different, and here are the ways in which it should be different—in some detail. Even Perot had constructive proposals. But all that was suppressed.

What's left is the picture that, say, Anthony Lewis portrayed in the *Times:* jingoist fanatics screaming about NAFTA. Incidentally, what's called the left played the same game. James Galbraith is an economist at the University of Texas. He had an article in a sort of left-liberal journal, *World Policy Review,* in which he discussed an article in which I said the opposite of what he attributed to me (of course—but that's typical).

Galbraith said there's this jingoist left—nationalist fanatics—who don't want Mexican workers to improve their lives. Then he went on about how the Mexicans are in favor of NAFTA. (True, if by "Mexicans" you mean Mexican industrialists and executives and corporate lawyers, not Mexican workers and peasants.)

All the way from people like James Galbraith and Anthony Lewis to way over on the right, you had this very useful fabrication—that critics of NAFTA were reactive and negative and jingoist and against progress and just wanted to go back to old-time protectionism. When you have essentially total control of the information system, it's rather easy to convey that image. But it simply isn't true.

Anthony Lewis also wrote, "The engine for [the world's] growth has been...vastly increased...international trade." Do you agree?

His use of the word "trade," while conventional, is misleading. The latest figures available (from about ten years ago—they're probably higher now) show that about 30% or 40% of what's called "world trade" is actually internal transfers within a corporation. I believe that about 70% of Japanese exports to the US are intrafirm transfers of this sort.

So, for example, Ford Motor Company will have components manufactured here in the US and then ship them for assembly to a plant in Mexico where the workers get much lower wages and where Ford doesn't have to worry about pollution, unions and all that nonsense. Then they ship the assembled part back here.

About half of what are called US exports to Mexico are intrafirm transfers of this sort. They don't enter the Mexican market, and there's no meaningful sense in which they're exports to Mexico. Still, that's called "trade."

The corporations that do this are huge totalitarian institutions, and they aren't governed by market principles—in fact, they promote severe market distortions. For example, a US corporation that has an outlet in Puerto Rico may decide to take its profits in Puerto Rico, because of tax rebates. It shifts its prices around, using what's called "transfer pricing," so it doesn't seem to be making a profit here.

There are estimates of the scale of governmental operations that interfere with trade, but I know of no estimates of internal corporate interferences with market processes. They're no doubt vast in scale, and are sure to be extended by the trade agreements.

GATT and NAFTA ought to be called "investor rights agreements," not "free trade agreements." One of their main purposes is to extend the ability of corporations to carry out market-distorting operations internally.

So when people like Anthony Lake talk about enlarging market democracy, he's enlarging something, but it's not markets and it's not democracy.

Mexico (and South Central LA)

I found the mainstream media coverage of Mexico during the NAFTA debate somewhat uneven. The *New York Times* has allowed in a number of articles that official corruption was—and is—widespread in Mexico. In fact, in one editorial, they virtually conceded that Salinas stole the 1988 presidential election. Why did that information come out?

I think it's impossible to repress. Furthermore, there were scattered reports in the *Times* of popular protest against NAFTA. Tim Golden, their reporter in Mexico, had a story a couple of weeks before the vote, probably in early November [1993], in which he said that lots of Mexican workers were concerned that their wages would decline after NAFTA. Then came the punch line.

He said that that undercuts the position of people like Ross Perot and others who think that NAFTA is going to harm American workers for the benefit of Mexican workers. In other words, the fact that they're *all* going to get screwed was presented as a critique of the people who were opposing NAFTA here!

There was very little discussion here of the large-scale popular protest in Mexico, which included, for example, the largest non-governmental trade union. (The main trade union is about as independent as the Soviet trade unions were, but there are some independent ones, and they were opposed to the agreement.)

The environmental movements and most of the other popular movements were opposed. The Mexican Bishops' Conference strongly endorsed the position the Latin American bishops took when they met at Santa Domingo [in the Dominican Republic] in December 1992.

That meeting in Santa Domingo was the first major conference of Latin American bishops since the ones at Puebla [Mexico] and Medellín [Colombia] back in the 1960s and 1970s. The Vatican tried to control it this time to make sure that they wouldn't come out with these perverse ideas about *liberation theology* and *the preferential option for the poor.* But despite a very firm Vatican hand, the bishops came out quite strongly against neoliberalism and structural adjustment and these free-market-for-the-poor policies. That wasn't reported here, to my knowledge.

There's been significant union-busting in Mexico.

Ford and VW are two big examples. A few years ago, Ford simply fired its entire Mexican work force and would only rehire, at much lower wages, those who agreed not to join a union. Ford was backed in this by the always-ruling PRI [the Institutional Revolutionary Party, which controlled Mexico from 1929 to 2000].

VW's case was pretty much the same. They fired workers who supported an independent union and only rehired, at lower wages, those who agreed not to support it.

A few weeks after the NAFTA vote in the US, workers at a GE and Honeywell plant in Mexico were fired for union activities. I don't know what the final outcome will be, but that's exactly the purpose of things like NAFTA.

In early January [1994], you were asked by an editor at the *Washington Post* to submit an article on the New Year's Day uprising in Chiapas [a state at the southern tip of Mexico, next to Guatemala]. Was this the first time the *Post* had asked you to write something?

It was the first time ever. I was kind of surprised, since I'm never asked to write for a national newspaper. So I wrote the article—it was for the *Sunday Outlook* section—but it didn't appear.

Was there an explanation?

No. It went to press, as far as I know. The editor who commissioned it called me, apparently after the deadline, to say that it looked OK to him but that it had simply been cancelled at some higher level. I don't know any more about it than that.

But I can guess. The article was about Chiapas, but it was also about NAFTA, and I think the *Washington Post* has been even more extreme than the *Times* in refusing to allow any discussion of that topic.

What happened in Chiapas [the Zapatista rebellion] doesn't come as very much of a surprise. At first, the government thought they'd just destroy the rebellion with tremendous violence, but then they backed off and decided to do it by more subtle violence, when nobody was looking. Part of the reason they backed off is surely their fear that there was just too much sympathy all over Mexico; if they were too up front about suppression, they'd cause themselves a lot of problems, all the way up to the US border.

The Mayan Indians in Chiapas are in many ways the most oppressed people in Mexico. Nevertheless, their problems are shared by a large majority of the Mexican population. This decade of neoliberal reforms has led to very little economic progress in Mexico but has sharply polarized the society. Labor's share in income has declined radically. The number of billionaires has shot up.

In that unpublished *Post* article, you wrote that the protest of the Indian peasants in Chiapas gives "only a bare glimpse of time bombs waiting to explode, not only in Mexico." What did you have in mind?

Take South Central Los Angeles, for example. In many respects, they are different societies, of course, but there are points of similarity to the Chiapas rebellion. South Central LA is a place where people once had jobs and lives, and those have been destroyed—in large part by the socio-economic processes we've been talking about.

For example, furniture factories went to Mexico, where they can pollute more cheaply. Military industry has somewhat declined. People used to have jobs in the steel industry, and they don't any more. So they rebelled [beginning on January 1, 1994]

The Chiapas rebellion was quite different. It was much more organized, and much more constructive. That's the difference between an utterly demoralized society like South Central Los Angeles and one that still retains some sort of integrity and community life.

When you look at consumption levels, doubtless the peasants in Chiapas are poorer than people in South Central LA. There are fewer television sets per capita. But by other, more significant criteria—like social cohesion—Chiapas is considerably more advanced. In the US, we've succeeded not only in polarizing communities but also in destroying their structures. That's why you have such rampant violence.

Haiti

Let's stay in Latin America and the Caribbean, which Henry Stimson called "our little region over here which has never bothered anyone." Jean-Bertrand Aristide was elected president of Haiti in what's been widely described as a free and democratic election. Would you comment on what's happened since?

When Aristide won in December 1990 [he took office in February 1991], it was a big surprise. He was swept into power by a network of popular grassroots organizations, what was called *Lavalas*—the flood—which outside observers just weren't aware of (since they don't pay attention to what happens among poor people). There had been very extensive and very successful organizing, and out of nowhere came this massive popular organization that managed to sweep their candidate into power.

The US was willing to support a democratic election, figuring that its candidate, a former World Bank official named Marc Bazin, would easily win. He had all the resources and support, and it looked like a shoo-in. He ended up getting 14% of the vote, and Aristide got about 67%.

The only question in the mind of anybody who knows a little history should have been, *How is the US going to get rid of Aristide?* The disaster became even worse in the first seven months of Aristide's administration. There were some really amazing developments.

Haiti is, of course, an extremely impoverished country, with awful conditions. Aristide was nevertheless beginning to get places. He was able to reduce corruption extensively, and to trim a highly bloated state bureaucracy. He won a lot of international praise for this, even from the international lending institutions, which were offering him loans and preferential terms because they liked what he was doing.

Furthermore, he cut back on drug trafficking. The flow of refugees to the US virtually stopped. Atrocities were reduced to way below what they had been or would become. There was a considerable degree of popular engagement in what was going on, although the contradictions were already beginning to show up, and there were constraints on what he could do.

All of this made Aristide even more unacceptable from the US point of view, and we tried to undermine him through what were called—naturally—"democracy-enhancing programs." The US, which had never cared at all about centralization of power in Haiti when its own favored dictators were in charge, all of a sudden began setting up alternative institutions that aimed at undermining executive power, supposedly in the interests of greater democracy. A number of these alleged human rights and labor groups became the governing authorities after the coup, which came on September 30, 1991.

In response to the coup, the Organization of American States declared an embargo of Haiti; the US joined it, but with obvious reluctance. The [first] Bush administration focused attention on Aristide's alleged atrocities and undemocratic activities, downplaying the major atrocities which took place right after the coup. The media went along with Bush's line, of course. While people were getting slaughtered in the streets of Port-au-Prince [Haiti's capital], the media concentrated on alleged human rights abuses under the Aristide government.

Refugees started fleeing again, because the situation was deteriorating so rapidly. The Bush administration blocked them—instituted a blockade, in effect—to send them back. Within a couple of months, the Bush administration had already undermined the embargo by allowing a minor exception—US-owned companies would be permitted to ignore it. The *New York Times* called that "fine-tuning" the embargo to improve the restoration of democracy!

Meanwhile, the US, which is known to be able to exert pressure when it feels like it, found no way to influence anyone else to observe the embargo, including the Dominican Republic next door. The whole thing was mostly a farce. Pretty soon Marc Bazin, the US candidate, was in power as prime minister, with the ruling generals behind him. That year—1992—US trade with Haiti was not very much below the norm, despite the so-called embargo (Commerce Department figures showed that, but I don't think the press ever reported it).

During the 1992 campaign, Clinton bitterly attacked the Bush administration for its inhuman policy of returning refugees to this torture chamber—which is, incidentally, a flat violation of the Universal Declaration of Human Rights, which we claim to uphold. Clinton claimed he was going to change all that, but his first act after being elected, even before he took office, was to impose even harsher measures to force fleeing refugees back into this hellhole.

Ever since then, it's simply been a matter of seeing what kind of finessing will be carried out to ensure that Haiti's popularly elected government doesn't come back into office. It doesn't have much longer to run [the next elections were scheduled for December 1995], so the US has more or less won that game.

Meanwhile, the terror and atrocities increase. The popular organizations are getting decimated. Although the so-called embargo is still in place, US trade continues and, in fact, went up about 50% under Clinton. Haiti, a starving island, is exporting food to the US—about 35 times as much under Clinton as it did under [the first] Bush.

Baseballs are coming along nicely. They're produced in US-owned factories where the women who make them get 10¢ an hour—if they meet their quota. Since meeting the quota is virtually impossible, they actually make something like 5¢ an hour.

Softballs from Haiti are advertised in the US as being unusually good because they're hand-dipped into some chemical that makes them hang together properly. The ads don't mention that the chemical the women hand-dip the balls into is toxic and that, as a result, the women don't last very long at this work.

In his exile, Aristide has been asked to make concessions to the military junta.

And to the right-wing business community.

That's kind of curious. For the victim—the aggrieved party—to make concessions to his victimizer.

It's perfectly understandable. The Aristide government had entirely the wrong base of support. The US has tried for a long time to get him to "broaden his government in the interests of democracy." This means throw out the two-thirds of the population that voted for him and bring in what are called "moderate" elements of the business community—the local owners or managers of those textile and baseball-producing plants, and those who are linked up with US agribusiness. When they're not in power, it's not democratic.

(The extremist elements of the business community think you ought to just slaughter everybody and cut them to pieces and hack off their faces and leave them in ditches. The moderates think you ought to have them working in your assembly plants for 14¢ an hour under indescribable conditions.)

Bring the moderates in and give them power and then we'll have a real democracy. Unfortunately, Aristide—being kind of backward and disruptive—has not been willing to go along with that.

Clinton's policy has gotten so cynical and outrageous that he's lost almost all major domestic support on it. Even the mainstream press is denouncing him at this point. So there will have to be some cosmetic changes made. But unless there's an awful lot of popular pressure, our policies will continue and pretty soon we'll have the "moderates" in power.

Let's say Aristide is "restored." Given the destruction of popular organizations and the devastation of civil society, what are his and the country's prospects?

Some of the closest observation of this has been done by Americas Watch. They gave an answer to that question that I thought was plausible. In early 1993, they said that things were reaching the point that even if Aristide were restored, the lively, vibrant civil society based on grassroots organizations that had brought him to power would have been so decimated that it's unlikely that he'd have the popular support to do anything anyway.

I don't know if that's true or not. Nobody knows, any more than anyone knew how powerful those groups were in the first place. Human beings have reserves of courage that are often hard to imagine. But I think that's the plan—to decimate the organizations, to intimidate people so much that it won't matter if you have democratic elections.

There was an interesting conference run by the Jesuits in El Salvador several months before the Salvadoran elections; its final report came out in January [1994]. They were talking about the buildup to the elections and the ongoing terror, which was substantial. They said that the long-term effect of terror—something they've had plenty of experience with—is to domesticate people's aspirations, to make them think there's no alternative, to drive out any hope. Once you've done that, you can have elections without too much fear.

If people are sufficiently intimidated, if the popular organizations are sufficiently destroyed, if the people have had it beaten into their heads that either they accept the rule of those with the guns or else they live and die in unrelieved misery, then your elections will all come out the way you want. And everybody will cheer.

Cuban refugees are considered political and are accepted immediately into the US, while Haitian refugees are considered economic refugees and are refused entry.

If you look at the records, many Haitians who are refused asylum in the US because they aren't considered to be political refugees are found a few days later hacked to pieces in the streets of Haiti.

There were a couple of interesting leaks from the INS [the Immigration and Naturalization Service]. One was from an INS officer who'd been working in our embassy in Port-au-Prince. In an interview with Dennis Bernstein of KPFA [a listener-supported radio station in Berkeley, California], he described in detail how they weren't even making the most perfunctory efforts to check the credentials of people who were applying for political asylum.

At about the same time, a document was leaked from the US interests section in Havana (which reviews applications for asylum in the US) in which they complain that they can't find genuine political asylum cases. The applicants they get can't really claim any serious persecution. At most they claim various kinds of harassment, which aren't enough to qualify them. So—there are the two cases, side by side.

I should mention that the US Justice Department has just made a slight change in US law which makes our violation of international law and the Universal Declaration of Human Rights even more grotesque. Now Haitian refugees who, by some miracle, reach US territorial waters can be shipped back. That's never been allowed before. I doubt that many other countries allow that.

Nicaragua

You recall the uproar in the 1980s about how the Sandinistas were abusing the Miskito Indians on Nicaragua's Atlantic coast. President Reagan, in his inimitable, understated style, said it was "a campaign of virtual genocide." UN Ambassador Jeane Kirkpatrick was a bit more restrained; she called it the "most massive human rights violation in Central America." What's happening now with the Miskitos?

Reagan and Kirkpatrick were talking about an incident in which, according to Americas Watch, several dozen Miskitos were killed and a lot of people were forcefully moved in a rather ugly way in the course of the Contra war. The US terrorist forces were moving into the area and this was the Sandinista's reaction.

It was certainly an atrocity, but it's not even visible compared to the ones Jeane Kirkpatrick was celebrating in the neighboring countries at the time—and in Nicaragua, where the overwhelming mass of the atrocities were committed by the so-called "freedom fighters."

What's happening to the Miskitos now? When I was in Nicaragua in October 1993, church sources—the Christian Evangelical Church, primarily, which works in the Atlantic coast—were reporting that 100,000 Miskitos were starving to death as a result of the policies we were imposing on Nicaragua. Not a word about it in the media here. (More recently, it did get some slight reporting.)

People here are worrying about the fact that one typical consequence of US victories in the Third World is that the countries where we win immediately become big centers for drug flow. There are good reasons for that—it's part of the market system we impose on them.

Nicaragua has become a major drug transshipment center. A lot of the drugs go through the Atlantic coast, now that Nicaragua's whole governmental system has collapsed. Drug transshipment areas usually breed major drug epidemics, and there's one among the Miskitos, primarily among the men who dive for lobsters and other shellfish.

Both in Nicaragua and Honduras, these Miskito Indian divers are compelled by economic circumstances to do very deep diving without equipment. Their brains get smashed and they quickly die. In order to try to maintain their work rate, the divers stuff themselves with cocaine. It helps them bear the pain.

There's concern about drugs here, so *that* story got into the press. But of course nobody cares much about the working conditions. After all, it's a standard free-market technique. You've got plenty of superfluous people, so you make them work under horrendous conditions; when they die, you just bring in others.

China

Let's talk about human rights in one of our major trading partners—China.

During the Asia Pacific summit in Seattle [in November 1993], Clinton announced that we'd be sending more high-tech equipment to China. This was in violation of a ban that was imposed to punish China for its involvement in nuclear and missile proliferation. The executive branch decided to "reinterpret" the ban, so we could send China nuclear generators, sophisticated satellites and supercomputers.

Right in the midst of that summit, a tiny little report appeared in the papers. In booming Kwangdong province, the economic miracle of China, 81 women were burned to death because they were locked into a factory. A couple of weeks later, 60 workers were killed in a Hong Kong-owned factory. China's Labor Ministry reported that 11,000 workers had been killed in industrial accidents just in the first eight months of 1993—twice as many as in the preceding year.

These sort of practices never enter the human rights debate, but there's been a big hullabaloo about the use of prison labor—front-page stories in the *Times*. What's the difference? Very simple. Because prison labor is state enterprise, it doesn't contribute to private profit. In fact, it undermines private profit, because it competes with private industry. But locking women into factories where they burn to death contributes to private profit.

So prison labor is a human-rights violation, but there's no right not to be burned to death. We have to maximize profit. From that principle, everything follows.

Russia

Radio listener: I'd like to ask about US support for Yeltsin vs. democracy in Russia.

Yeltsin was the tough, autocratic Communist Party boss of Sverdlovsk. He's filled his administration with the old party hacks who ran things for him under the earlier Soviet system. The West likes him a lot because he's ruthless and because he's willing to ram through what are called "reforms" (a nice-sounding word).

These "reforms" are designed to return the former Soviet Union to the Third World status it had for the five hundred years before the Bolshevik Revolution. The Cold War was largely about the demand that this huge region of the world once again become what it had been—an area of resources, markets and cheap labor for the West.

Yeltsin is leading the pack on pushing the "reforms." Therefore he's a "democrat." That's what we call a democrat anywhere in the world—someone who follows the Western business agenda.

Dead children and debt service

After you returned from a recent trip to Nicaragua, you told me it's becoming more difficult to tell the difference between economists and Nazi doctors. What did you mean by that?

There's a report from UNESCO (which I didn't see reported in the US media) that estimated the human cost of the "reforms" that aim to return Eastern Europe to its Third World status.

UNESCO estimates that about half a million deaths a year in Russia since 1989 are the direct result of the reforms, caused by the collapse of health services, the increase in disease, the increase in malnutrition and so on. Killing half a million people a year— that's a fairly substantial achievement for reformers.

The figures are similar, but not quite as bad, in the rest of Eastern Europe. In the Third World, the numbers are fantastic. For example, another UNESCO report estimated that about half a million children in Africa die every year simply from debt service. Not from the whole array of reforms—just from interest on their countries' debts.

It's estimated that about eleven million children die every year from easily curable diseases, most of which could be overcome by treatments that cost a couple of cents. But the economists tell us that to do this would be interference with the market system.

There's nothing new about this. It's very reminiscent of the British economists who, during the Irish potato famine in the mid-nineteenth century, dictated that Ireland must export food to Britain—which it did right through the famine—and that it shouldn't be given food aid because that would violate the sacred principles of political economy. These policies always happen to have the curious property of benefiting the wealthy and harming the poor.

How the Nazis won the war

In his book *Blowback,* Chris Simpson described Operation Paper Clip, which involved the importation of large numbers of known Nazi war criminals, rocket scientists, camp guards, etc.

There was also an operation involving the Vatican, the US State Department and British intelligence, which took some of the worst Nazi criminals and used them, at first in Europe. For example, Klaus Barbie, the butcher of Lyon [France], was taken over by US intelligence and put back to work.

Later, when this became an issue, some of his US supervisors didn't understand what the fuss was all about. After all, we'd moved in—we'd replaced the Germans. We needed a guy who would attack the left-wing resistance, and here was a specialist. That's what he'd been doing for the Nazis, so who better could we find to do exactly the same job for us?

When the Americans could no longer protect Barbie, they moved him over to the Vatican-run "ratline," where Croatian Nazi priests and others managed to spirit him off to Latin America. There he continued his career. He became a big drug lord and narcotrafficker, and was involved in a military coup in Bolivia—all with US support.

But Barbie was basically small potatoes. This was a big operation, involving many top Nazis. We managed to get Walter Rauff, the guy who created the gas chambers, off to Chile. Others went to fascist Spain.

General Reinhard Gehlen was the head of German military intelligence on the Eastern Front. That's where the real war crimes were. Now we're talking about Auschwitz and other death camps. Gehlen and his network of spies and terrorists were taken over quickly by American intelligence and returned to essentially the same roles.

If you look at the American army's counterinsurgency literature (a lot of which is now declassified), it begins with an analysis of the German experience in Europe, written with the cooperation of Nazi officers. Everything is described from the point of view of the Nazis: which techniques for controlling resistance worked, which ones didn't. With barely a change, that was transmuted into American counterinsurgency literature. (This is discussed at some length by Michael McClintock in *Instruments of Statecraft*, a very good book that I've never seen reviewed.)

The US left behind armies the Nazis had established in Eastern Europe, and continued to support them at least into the early 1950s. By then the Russians had penetrated American intelligence, so the air drops didn't work very well any more.

You've said that if a real post-World War II history were ever written, this would be the first chapter.

It would be a part of the first chapter. Recruiting Nazi war criminals and saving them is bad enough, but imitating their activities is worse. So the first chapter would primarily describe US—and some British—operations throughout the world that aimed to destroy the antifascist resistance and restore the traditional, essentially fascist, order to power. [This is also discussed on pp. 11, 15–18 and 33 above.]

In Korea (where we ran the operation alone), restoring the traditional order meant killing about 100,000 people just in the late 1940s, before the Korean War began. In Greece, it meant destroying the peasant and worker base of the anti-Nazi resistance and restoring Nazi collaborators to power.

When British and then American troops moved into southern Italy, they simply reinstated the fascist order—the industrialists. But the big problem came when the troops got to the north, which the Italian resistance had already liberated. The place was functioning—industry was running. We had to dismantle all of that and restore the old order.

Our big criticism of the resistance was that they were displacing the old owners in favor of workers' and community control. Britain and the US called this "arbitrary replacement" of the legitimate owners. The resistance was also giving jobs to more people than were strictly needed for the greatest economic efficiency (that is, for maximum profit-making). We called this "hiring excess workers."

In other words, the resistance was trying to democratize the workplace and to take care of the population. That was understandable, since many Italians were starving. But starving people were *their* problem—our problem was to eliminate the hiring of excess workers and the arbitrary dismissal of owners, which we did.

Next we worked on destroying the democratic process. The left was obviously going to win the elections; it had a lot of prestige from the resistance, and the traditional conservative order had been discredited. The US wouldn't tolerate that. At its first meeting, in 1947, the National Security Council decided to withhold food and use other sorts of pressure to undermine the election.

But what if the communists still won? In its first report, NSC 1, the council made plans for that contingency: the US would declare a national emergency, put the Sixth Fleet on alert in the Mediterranean and support paramilitary activities to overthrow the Italian government.

That's a pattern that's been relived over and over. If you look at France and Germany and Japan, you get pretty much the same story. Nicaragua is another case. You strangle them, you starve them, and then you have an election and everybody talks about how wonderful democracy is.

The person who opened up this topic (as he did many others) was Gabriel Kolko, in his classic book *Politics of War* in 1968. It was mostly ignored, but it's a terrific piece of work. A lot of the documents weren't around then, but his picture turns out to be quite accurate.

Chile

Richard Nixon's death generated much fanfare. Henry Kissinger said in his eulogy: "The world is a better place, a safer place, because of Richard Nixon." I'm sure he was thinking of Laos, Cambodia and Vietnam. But let's focus on one place that wasn't mentioned in all the media hoopla—Chile—and see how it's a "better, safer place." In early September 1970, Salvador Allende was elected president of Chile in a democratic election. What were his politics?

He was basically a social democrat, very much of the European type. He was calling for minor redistribution of wealth, to help the poor. (Chile was a very inegalitarian society.) Allende was a doctor, and one of the things he did was to institute a free milk program for half a million very poor, malnourished children. He called for nationalization of major industries like copper mining, and for a policy of international independence— meaning that Chile wouldn't simply subordinate itself to the US, but would take more of an independent path.

Was the election he won free and democratic?

Not entirely, because there were major efforts to disrupt it, mainly by the US. It wasn't the first time the US had done that. For example, our government intervened massively to prevent Allende from winning the preceding election, in 1964. In fact, when the Church Committee investigated years later, they discovered that the US spent more money per capita to get the candidate it favored elected in Chile in 1964 than was spent by both candidates (Johnson and Goldwater) in the 1964 election in the US!

Similar measures were undertaken in 1970 to try to prevent a free and democratic election. There was a huge amount of black propaganda about how if Allende won, mothers would be sending their children off to Russia to become slaves—stuff like that. The US also threatened to destroy the economy, which it could—and did—do.

Nevertheless, Allende won. A few days after his victory, Nixon called in CIA Director Richard Helms, Kissinger and others for a meeting on Chile. Can you describe what happened?

As Helms reported in his notes, there were two points of view. The "soft line" was, in Nixon's words, to "make the economy scream." The "hard line" was simply to aim for a military coup.

Our ambassador to Chile, Edward Korry, who was a Kennedy liberal type, was given the job of implementing the "soft line." Here's how he described his task: "to do all within our power to condemn Chile and the Chileans to utmost deprivation and poverty." That was the soft line.

There was a massive destabilization and disinformation campaign. The CIA planted stories in *El Mercurio* [Chile's most prominent paper] and fomented labor unrest and strikes.

They really pulled out the stops on this one. Later, when the military coup finally came [in September 1973] and the government was overthrown—and thousands of people were being imprisoned, tortured and slaughtered—the economic aid which had been cancelled immediately began to flow again. As a reward for the military junta's achievement in reversing Chilean democracy, the US gave massive support to the new government.

Our ambassador to Chile brought up the question of torture to Kissinger. Kissinger rebuked him sharply—saying something like, *Don't give me any of those political-science lectures. We don't care about torture—we care about important things.* Then he explained what the important things were.

Kissinger said he was concerned that the success of social democracy in Chile would be contagious. It would infect southern Europe—southern Italy, for example—and would lead to the possible success of what was then called Eurocommunism (meaning that Communist parties would hook up with social democratic parties in a united front).

Actually, the Kremlin was just as much opposed to Eurocommunism as Kissinger was, but this gives you a very clear picture of what the domino theory is all about. Even Kissinger, mad as he is, didn't believe that Chilean armies were going to descend on Rome. It wasn't going to be that kind of an influence. He was worried that successful economic development, where the economy produces benefits for the general population—not just profits for private corporations—would have a contagious effect.

In those comments, Kissinger revealed the basic story of US foreign policy for decades.

You see that pattern repeating itself in Nicaragua in the 1980s.

Everywhere. The same was true in Vietnam, in Cuba, in Guatemala, in Greece. That's always the worry—the threat of a good example.

Kissinger also said, again speaking about Chile, "I don't see why we should have to stand by and let a country go Communist due to the irresponsibility of its own people."

As the *Economist* put it, we should make sure that policy is insulated from politics. If people are irresponsible, they should just be cut out of the system.

In recent years, Chile's economic growth rate has been heralded in the press.

Chile's economy isn't doing badly, but it's based almost entirely on exports—fruit, copper and so on—and thus is very vulnerable to world markets.

There was a really funny pair of stories yesterday. The *New York Times* had one about how everyone in Chile is so happy and satisfied with the political system that nobody's paying much attention to the upcoming election.

But the *Financial Times* (which is the world's most influential business paper, and hardly radical) took exactly the opposite tack. They cited polls that showed that 75% of the population was very "disgruntled" with the political system (which allows no options).

There is indeed apathy about the election, but that's a reflection of the breakdown of Chile's social structure. Chile was a very vibrant, lively, democratic society for many, many years—into the early 1970s. Then, through a reign of fascist terror, it was essentially depoliticized. The breakdown of social relations is pretty striking. People work alone, and just try to fend for themselves. The retreat into individualism and personal gain is the basis for the political apathy.

Nathaniel Nash wrote the Chile story in the *Times*. He said that many Chileans have painful memories of Salvador Allende's fiery speeches, which led to the coup in which thousands of people were killed [including Allende]. Notice that they don't have painful memories of the torture, of the fascist terror—just of Allende's speeches as a popular candidate.

Cambodia

Would you talk a little about the notion of unworthy vs. worthy victims?

[Former *New York Times* and *Newsday* reporter and columnist] Sidney Schanberg wrote an op-ed piece in the *Boston Globe* in which he blasted Senator Kerry of Massachusetts for being two-faced because Kerry refused to concede that the Vietnamese have not been entirely forthcoming about American POWs. Nobody, according to Schanberg, is willing to tell the truth about this.

He says the government ought to finally have the honesty to say that it left Indochina without accounting for all the Americans. Of course, it wouldn't occur to him to suggest that the government should also be honest enough to say that we killed a couple of million people and destroyed three countries and left them in total wreckage and have been strangling them ever since.

It's particularly striking that this is Sidney Schanberg, a person of utter depravity. He's regarded as the great conscience of the press because of his courage in exposing the crimes of our official enemies—namely, Pol Pot [leader of Cambodia's Khmer Rouge rebel army]. He also happened to be the main US reporter in Phnom Penh [Cambodia's capital] in 1973. This was at the peak of the US bombardment of inner Cambodia, when hundreds of thousands of people (according to the best estimates) were being killed and the society was being wiped out.

Nobody knows very much about the bombing campaign and its effects because people like Sidney Schanberg refused to cover it. It wouldn't have been hard for him to cover it. He wouldn't have to go trekking off into the jungle—he could walk across the street from his fancy hotel in Phnom Penh and talk to any of the hundreds of thousands of refugees who'd been driven from the countryside into the city.

I went through all of his reporting—it's reviewed in detail in *Manufacturing Consent*, my book with Edward Herman [currently, editor of *Lies of Our Times*]. You'll find a few scattered sentences here and there about the bombing, but not a single interview with the refugees.

There is one American atrocity he did report (for about three days); *The Killing Fields,* the movie that's based on his story, opens by describing it. What's the one report? American planes hit the wrong village—a government village. That's an atrocity; *that* he covered. How about when they hit the *right* village? We don't care about that.

Incidentally, the United States' own record with POWs has been atrocious—not only in Vietnam, where it was monstrous, but in Korea, where it was even worse. And after WW II, we kept POWs illegally under confinement, as did the British.

World War II POWs

Other Losses, a Canadian book, alleges it was official US policy to withhold food from German prisoners in World War II. Many of them supposedly starved to death.

That's James Bacque's book. There's been a lot of controversy about the details, and I'm not sure what the facts of the matter are. On the other hand, there are things about which there's no controversy. Ed Herman and I wrote about them back in the late 1970s.

Basically, the Americans ran what were called "re-education camps" for German POWs (the name was ultimately changed to something equally Orwellian). These camps were hailed as a tremendous example of our humanitarianism, because we were teaching the prisoners democratic ways (in other words, we were indoctrinating them into accepting our beliefs).

The prisoners were treated very brutally, starved, etc. Since these camps were in gross violation of international conventions, they were kept secret. We were afraid that the Germans might retaliate and treat American prisoners the same way.

Furthermore, the camps continued after the war; I forget for how long, but I think the US kept German POWs until mid-1946. They were used for forced labor, beaten and killed. It was even worse in England. They kept their German POWs until mid-1948. It was all totally illegal.

Finally, there was public reaction in Britain. The person who started it off was Peggy Duff, a marvelous woman who died a couple of years ago. She was later one of the leading figures in the CND [the Campaign for Nuclear Disarmament] and the international peace movement during the 1960s and 1970s, but she started off her career with a protest against the treatment of German POWs.

Incidentally, why only German POWs? What about the Italians? Germany's a very efficient country, so they've published volumes of documents on what happened to their POWs. But Italy's sort of laid back, so there was no research on their POWs. We don't know anything about them, although they were surely treated much worse.

When I was a kid, there was a POW camp right next to my high school. There were conflicts among the students over the issue of taunting the prisoners. The students couldn't physically attack the prisoners, because they were behind a barrier, but they threw things at them and taunted them. There were a group of us who thought this was horrifying and objected to it, but there weren't many.

Consumption vs. well-being

The United States, with 5% of the world's population, consumes 40% of the world's resources. You don't have to be a genius to figure out what that's leading to.

For one thing, a lot of that consumption is artificially induced—it doesn't have to do with people's real wants and needs. People would probably be better off and happier if they didn't have a lot of those things.

If you measure economic health by profits, then such consumption is healthy. If you measure the consumption by what it means to people, it's very unhealthy, particularly in the long term.

A huge amount of business propaganda—that is, the output of the public relations and advertising industry—is simply an effort to create wants. This has been well understood for a long time; in fact, it goes back to the early days of the Industrial Revolution.

For another thing, those who have more money tend to consume more, for obvious reasons. So consumption is skewed towards luxuries for the wealthy rather than towards necessities for the poor. That's true within the US and on a global scale as well. The richer countries are the higher consumers by a large measure, and within the richer countries, the wealthy are higher consumers by a large measure.

Cooperative enterprises

There's a social experiment in Mondragón in the Basque region of Spain. Can you describe it?

Mondragón is basically a very large worker-owned cooperative with many different industries in it, including some fairly sophisticated manufacturing. It's economically quite successful, but since it's inserted into a capitalist economy, it's no more committed to sustainable growth than any other part of the capitalist economy is.

Internally, it's not worker-controlled—it's manager-controlled. So it's a mixture of what's sometimes called *industrial democracy*—which means ownership, at least in principle, by the work force—along with elements of hierarchic domination and control (as opposed to worker management).

I mentioned earlier that businesses are about as close to strict totalitarian structures as any human institutions are. Something like Mondragón is considerably less so.

The coming eco-catastrophe

Radio listener: What's happening in the growing economies in Southeast Asia, China, etc.? Is it going to be another example of capitalist exploitation, or can we expect to see some kind of change in their awareness?

Right now, it's catastrophic. In countries like Thailand or China, ecological catastrophes are looming. These are countries where growth is being fueled by multinational investors for whom the environment is what's called an "externality" (which means you don't pay any attention to it). So if you destroy the forests in Thailand, say, that's OK as long as you make a short-term profit out of it.

In China, the disasters which lie not too far ahead could be extraordinary—simply because of the country's size. The same is true throughout Southeast Asia.

But when the environmental pressures become such that the very survival of people is jeopardized, do you see any change in the actions?

Not unless people react. If power is left in the hands of transnational investors, the people will just die.

Nuclear power

At a conference in Washington DC, a woman in the audience got up and decried the fact that you're in favor of nuclear power. Are you?

No. I don't think anybody's in favor of nuclear power, even business, because it's too expensive. But what I am in favor of is being rational on the topic. That means recognizing that the question of nuclear power isn't a moral one—it's a technical one. You have to ask what the consequences of nuclear power are, versus the alternatives.

There's a range of other alternatives, including conservation, solar and so on. Each has its own advantages and disadvantages. But imagine that the only alternatives were hydrocarbons and nuclear power. If you had to have one or the other, you'd have to ask yourself which is more dangerous to the environment, to human life, to human society. It's not an entirely simple question.

For example, suppose that fusion were a feasible alternative. It could turn out to be nonpolluting. But there are also negative factors. Any form of nuclear power involves quite serious problems of radioactive waste disposal, and can also contribute to nuclear weapons proliferation. Fusion would require a high degree of centralization of state power too.

On the other hand, the hydrocarbon industry, which is highly polluting, also promotes centralization. The energy corporations are some of the biggest in the world, and the Pentagon system is constructed to a significant degree to maintain their power.

In other words, there are questions that have to be thought through. They're not simple.

The family

You've suggested that, to further democracy, people should be "seeking out authoritarian structures and challenging them, eliminating any form of absolute power and hierarchic power." How would that work in a family structure?

In any structure, including a family structure, there are various forms of authority. A patriarchal family may have very rigid authority, with the father setting rules that others adhere to, and in some cases even administering severe punishment if there's a violation of them.

There are other hierarchical relations among siblings, between the mother and father, gender relations, and so on. These all have to be questioned. Sometimes I think you'll find that there's a legitimate claim to authority—that is, the challenge to authority can sometimes be met. But the burden of proof is always on the authority.

So, for example, some form of control over children is justified. It's fair to prevent a child from putting his or her hand in the oven, say, or from running across the street in traffic. It's proper to place clear bounds on children. They want them—they want to understand where they are in the world.

However, all of these things have to be done with sensitivity and with self-awareness and with the recognition that any authoritarian role requires justification. It's never self-justifying.

When does a child get to the point where the parent doesn't need to provide authority?

I don't think there are formulas for this. For one thing, we don't have solid scientific knowledge and understanding of these things. A mixture of experience and intuition, plus a certain amount of study, yields a limited framework of understanding (about which people may certainly differ). And there are also plenty of individual differences.

So I don't think there's a simple answer to that question. The growth of autonomy and self-control, and expansion of the range of legitimate choices, and the ability to exercise them—that's growing up.

What you can do

Radio listener: Taking it down to the individual, personal level, I got a notice in my public service bill that said they're asking for a rate hike. I work, and I really don't have the time to sit down and write a letter of protest. This happens all the time, and not just with me. Most people don't have time to be active politically to change something. So those rate hikes go through without anybody ever really pointing out what's going on. I've often wondered why there isn't a limitation on the amount of profit any business can make (I know this probably isn't democratic).

I think it's highly democratic. There's nothing in the principle of democracy that says that power and wealth should be so highly concentrated that democracy becomes a sham.

But your first point is quite correct. If you're a working person, you just don't have time—alone—to take on the power company. That's exactly what organization is about. That's exactly what unions are for, and political parties that are based on working people.

If such a party were around, they'd be the ones speaking up for you and telling the truth about what's going on with the rate hike. Then they'd be denounced by the Anthony Lewises of the world for being antidemocratic—in other words, for representing popular interests rather than power interests.

Radio listener: I'm afraid there may be a saturation point of despair just from knowing the heaviness of the truth that you impart. I'd like to strongly lobby you to begin devoting maybe 10% or 15% of your appearances or books or articles towards tangible, detailed things that people can do to try to change the world. I've heard a few occasions where someone asks you that question and your response is, *Organize. Just do it.*

I try to keep it in the back of my mind and think about it, but I'm afraid that the answer is always the same. There is only one way to deal with these things. Being alone, you can't do anything. All you can do is deplore the situation.

But if you join with other people, you can make changes. Millions of things are possible, depending on where you want to put your efforts.

THE
COMMON
GOOD

FIRST PUBLISHED 1998

6 US PRINTINGS

6 FOREIGN EDITIONS

63,000 COPIES IN PRINT

TABLE OF CONTENTS

That dangerous radical Aristotle

Early in January 1997, you gave a talk at a conference in Washington DC. It was sponsored by several organizations, including the Progressive Caucus, a group of about fifty liberal and radical members of Congress. What did you think of the conference?

I was pretty encouraged by what I saw of it. There was a good, lively atmosphere, a lot of vitality. A dominant feeling there—with which I agree—was that a considerable majority of Americans are more or less in favor of New Deal-style liberalism. That's remarkable, since most Americans never hear *anybody* advocating that position.

Supposedly, the market has proved that the L-word is bad—that's what's drummed into everybody's head all the time. Yet many people in the Progressive Caucus who publicly stood for New Deal positions—like Sen. Paul Wellstone [D–Minn.], Rep. Jim McGovern [D–Mass.] and others—won their elections. The Progressive Caucus actually grew after the 1996 election.

Now I don't think New Deal liberalism is the end of the road... by any means. But its achievements, which are the result of a lot of popular struggle, are worth defending and expanding.

Your talk was entitled *The Common Good*.

That title was given to me, and since I'm a nice, obedient type, that's what I talked about. I started from the beginning, with Aristotle's *Politics*, which is the foundation of most subsequent political theory.

Aristotle took it for granted that a democracy should be fully participatory (with some notable exceptions, like women and slaves) and that it should aim for the common good. In order to achieve that, it has to ensure relative equality, "moderate and sufficient property" and "lasting prosperity" for everyone.

In other words, Aristotle felt that if you have extremes of poor and rich, you can't talk seriously about democracy. Any true democracy has to be what we call today a welfare state—actually, an extreme form of one, far beyond anything envisioned in this century.

(When I pointed this out at a press conference in Majorca, the headlines in the Spanish papers read something like, *If Aristotle were alive today, he'd be denounced as a dangerous radical.* That's probably true.)

The idea that great wealth and democracy can't exist side by side runs right up through the Enlightenment and classical liberalism, including major figures like de Tocqueville, Adam Smith, Jefferson and others. It was more or less assumed.

Aristotle also made the point that if you have, in a perfect democracy, a small number of very rich people and a large number of very poor people, the poor will use their democratic rights to take property away from the rich. Aristotle regarded that as unjust, and proposed two possible solutions: reducing poverty (which is what he recommended) or reducing democracy.

James Madison, who was no fool, noted the same problem, but unlike Aristotle, he aimed to reduce democracy rather than poverty. He believed that the primary goal of government is "to protect the minority of the opulent against the majority." As his colleague John Jay was fond of putting it, "The people who own the country ought to govern it."

Madison feared that a growing part of the population, suffering from the serious inequities of the society, would "secretly sigh for a more equal distribution of [life's] blessings." If they had democratic power, there'd be a danger they'd do something more than sigh. He discussed this quite explicitly at the Constitutional Convention, expressing his concern that the poor majority would use its power to bring about what we would now call land reform.

So he designed a system that made sure democracy couldn't function. He placed power in the hands of the "more capable set of men," those who hold "the wealth of the nation." Other citizens were to be marginalized and factionalized in various ways, which have taken a variety of forms over the years: fractured political constituencies, barriers against unified working-class action and cooperation, exploitation of ethnic and racial conflicts, etc.

(To be fair, Madison was *pre*capitalist and his "more capable set of men" were supposed to be "enlightened statesmen" and "benevolent philosophers," not investors and corporate executives trying to maximize their own wealth regardless of the effect that has on other people. When Alexander Hamilton and his followers began to turn the US into a capitalist state, Madison was pretty appalled. In my opinion, he'd be an anticapitalist if he were alive today—as would Jefferson and Adam Smith.)

It's extremely unlikely that what are now called "inevitable results of the market" would ever be tolerated in a truly democratic society. You can take Aristotle's path and make sure that almost everyone has "moderate and sufficient property"—in other words, is what he called "middle-class." Or you can take Madison's path and limit the functioning of democracy.

Throughout our history, political power has been, by and large, in the hands of those who own the country. There have been some limited variations on that theme, like the New Deal. FDR had to respond to the fact that the public was not going to tolerate the existing situation. He left power in the hands of the rich, but bound them to a kind of social contract. That was nothing new, and it will happen again.

━━━━━━━━━━━━━━━━━━━━━━━━━━━━━━━

Equality

Should we strive merely for equality of opportunity, or for equality of outcome, where everyone ends up in more or less the same economic condition?

Many thinkers, beginning with Aristotle, have held that equality of outcome should be a major goal of any just and free society. (They didn't mean identical outcomes, but at least relatively equal conditions.)

Acceptance of radical inequality of outcome is a sharp departure from the core of the humane liberal tradition as far back as it goes. In fact, Adam Smith's advocacy of markets was based on the assumption that under conditions of perfect liberty, free markets would lead to perfect equality of outcome, which he believed was a good thing.

Another grand figure of the pantheon, de Tocqueville, admired the relative equality he thought he saw in American society. (He exaggerated it considerably, but let's put aside the question of whether his perceptions were accurate.) He pointed out quite explicitly that if a "permanent inequality of conditions" ever developed, that would be the death of democracy.

Incidentally, in other parts of his work that aren't widely quoted, de Tocqueville condemned the "manufacturing aristocracy" that was growing up under his eyes in the US, which he called "one of the harshest" in history. He said that if it ever got power, we'd be in deep trouble. Jefferson and other Enlightenment figures had the same fear. Unfortunately, it happened far beyond their worst nightmares.

Ron Daniels, who's director of the Center for Constitutional Rights in New York, uses the metaphor of two runners in a race: One begins at the starting line and the other begins five feet from the finish line.

That's a good analogy, but I don't think it gets to the main point. It's true that there's nothing remotely like equality of opportunity in this country, but even if there were, the system would *still* be intolerable.

Suppose you have two runners who start at exactly the same point, have the same sneakers, and so on. One finishes first and gets everything he wants; the other finishes second and starves to death.

One of the mechanisms to address inequality is affirmative action. What do you think of it?

Many societies just take it for granted. In India, for example, a sort of affirmative action system called *reservations* was instituted back in the late 1940s, at the time of independence, in an effort to try to overcome very long-standing and deep-seated caste and gender differences.

Any such system is going to impose hardships on some people, in order (one hopes) to develop a more equitable and just society for the future. How it works as a practical matter can be tricky. I don't think there are any simple mechanical rules for it.

The attack on affirmative action is, to a large extent, an attempt to justify the oppressive, discriminatory patterns that existed in the past. On the other hand, affirmative action should certainly be designed so that it doesn't harm poor people who don't happen to be in the categories designated for support.

That can be done. There have been very effective applications of affirmative action—in the universities, the construction industry, the public service field and elsewhere. If you look in detail, you find plenty of things to criticize, but the main thrust of the program is humane and appropriate.

Libraries

Libraries were very important to your intellectual development when you were a kid, weren't they?

I used to haunt the main public library in downtown Philadelphia, which was extremely good. That's where I read all the offbeat anarchist and left-Marxist literature I'm always quoting. Those were days when people read, and used the libraries very extensively. Public services were richer in many ways back in the late 1930s and early 1940s.

I think that's one of the reasons why poor, even unemployed people living in slums seemed more hopeful back then. Maybe this is sentimentality, and it involves comparing a child's perceptions and an adult's, but I think it's true.

Libraries were one of the factors. They weren't just for educated people—a lot of people used them. That's much less true now.

I'll tell you why I asked. Recently I went back to visit the public library I used when I was a kid, on 78th and York in New York. I hadn't been there in thirty-five years, and it's now in one of the richest districts in the country.

I discovered they had very few political books. When the librarian explained that branch libraries carry mostly bestsellers, I told him I'd be happy to donate some of our books.

He expressed mild interest and suggested I fill out a form. When I went over to the desk to get one, I found out that it costs 30¢ to recommend a book you think the library should purchase!

It sounds similar to what you find in the publications industry in general, including bookstores. I travel a lot and often get stuck in some airport or other...because it's snowing in Chicago, say. I used to be able to find something I wanted to read in the airport bookstore—maybe a classic, maybe something current. Now it's almost impossible. (It's not just in the

US, by the way. I was stuck at the airport in Naples not long ago and the bookstore there was awful too.)

I think it's mostly just plain market pressures. Bestsellers move fast, and it costs money to keep books around that don't sell very quickly. Changes in the tax laws have exacerbated the problem, by making it more expensive for publishers to hold inventory, so books tend to get remaindered [sold off at cost and put out-of-print] much sooner.

I think political books are being harmed by this—if you go into the big chains, which pretty much dominate bookselling now, you certainly don't find many of them—but the same thing is true of most books. I don't think it's political censorship.

The right wing is promoting the idea of charging people to use the library.

That's part of the whole idea of redesigning society so that it just benefits the wealthy. Notice that they aren't calling for terminating the Pentagon. They're not crazy enough to believe it's defending us from the Martians or somebody, but they understand very clearly that it's a subsidy for the rich. So the Pentagon is fine, but libraries aren't.

Lexington, the Boston suburb where I live, is an upper-middle-class, professional town where people are willing and able to contribute to the library. I give money to it and use it, and benefit from the fact that it's quite good.

But I don't like the fact that zoning laws and inadequate public transportation virtually guarantee that only rich people can live in Lexington. In poorer neighborhoods, few people have enough money to contribute to the library, or time to use it, or knowledge of what to look for once they're there.

Let me tell you a dismal story. One of my daughters lived in a declining old mill town. It's not a horrible slum, but it's fading away. The town happens to have a rather nice public library— not a wonderful collection, but good things for children. It's nicely laid out, imaginatively designed, staffed by a couple of librarians.

I went with her kids on a Saturday afternoon, and nobody was there except a few children of local professional families. Where are the kids who ought to be there? I don't know, probably watching television, but going to the library just isn't the kind of thing they do.

It *was* the kind of thing you did if you were a working-class person fifty or sixty years ago. Emptying people's minds of the ability, or even the desire, to gain access to cultural resources—that's a tremendous victory for the system.

Freedom

The word *freedom* has become virtually synonymous with *capitalism,* as in the title of Milton Friedman's book, *Capitalism and Freedom.*

It's an old scam. Milton Friedman is smart enough to know that there's never been anything remotely resembling capitalism, and that if there were, it wouldn't survive for three seconds—mostly because business wouldn't let it. Corporations insist on powerful governments to protect them from market discipline, and their very existence is an attack on markets.

All this talk about capitalism and freedom has got to be a conscious fraud. As soon as you move into the real world, you see that nobody could actually believe that nonsense.

Dwayne Andreas, CEO of ADM [Archer Daniels Midland, a major NPR and PBS sponsor that calls itself "Supermarket to the World"] was quoted as saying, "There's not one grain of anything in the world that is sold in the free market. Not one! The only place you see a free market is in the speeches of politicians."

It must have been an internal memo or talk—that's not the kind of thing you tell the public. But in general it's true. As the United Nations Development Program put it, "survival in agricultural markets depends less on comparative advantage than on comparative access to subsidies."

Two technical economists in Holland found that *every single one* of the hundred largest transnational corporations on *Fortune* magazine's list has benefited from the industrial policy of its home country, and that at least twenty of them wouldn't have even survived if their governments hadn't taken them over or given them large subsidies when they were in trouble.

There was a front-page article in the *Boston Globe* that talked about our passing Japan in semiconductor production. It said that we've just seen "one of the great role reversals of the modern era—the transformation of Japan from behemoth to bungler....Japan's government-guided effort to dominate the chip industry, for example, was turned back. The US share of global chip production, which sank below Japan's in 1985, jumped back ahead of it in 1993 and has remained there." The article quoted Edward Lincoln, economic advisor to former US Ambassador to Japan Walter Mondale, as saying, "The lesson of the 1990s is that all nations obey the same economic laws."

What actually happened? During the 1980s, the Reagan-Bush administrations forced Japan to raise prices for chips and to guarantee US producers a share in Japanese markets. They also poured a lot of money into our own industry, through the military system and through Sematech, a government-industry consortium that was restricted to US companies. Because of this large-scale state intervention, the US did indeed regain control of the more sophisticated end of the microprocessor market.

Japan then announced it was starting up a new government-industry consortium for semiconductors in an effort to compete. (Some US corporations are to participate in Japan's projects in the new age that some business economists call "alliance capitalism.") Obviously, neither action had anything to do with the laws of the market.

The Mexican bailout is another example. The big investment firms in New York could have taken a beating if Mexico defaulted on its loans, or paid short-term loans in devalued pesos, as it was legally entitled to do. But they got the American public to guarantee their losses—as usual.

You can make as much money as you want, but if you get into trouble, it's the taxpayers' responsibility to fix things. Under capitalism, investment is supposed to be as risk-free as possible. No corporation wants free markets—what they want is power.

Another of the many areas where freedom and capitalism collide is what's laughably called *free trade*. About 40% of US trade is estimated to be internal to individual corporations. If a US auto manufacturer ships a part from Indiana to Illinois, that isn't called trade; if it ships the same part from Illinois to northern Mexico, it *is* called trade—it's considered an export when it leaves and an import when it comes back.

But that's nothing more than exploiting cheaper labor, avoiding environmental regulations and playing games about where you pay your taxes. This sort of activity also accounts for similar or even higher proportions of trade in other industrial countries. Furthermore, strategic alliances among firms play an increasing role in administration of the global economy.

So talk about "the growth in world trade" is largely a joke. What's growing is complicated interactions among transnational corporations—centrally managed institutions that really amount to private command economies.

The hypocrisy is pervasive. For example, free-trade boosters also demand intellectual property rights [copyrights, patents, etc.] that are highly protectionist. The World Trade Organization's version of patents (which today's rich countries would never have accepted while they were gaining their place in the sun) is not only extremely harmful to developing countries economically, but also undermines innovation—in fact, that's what they're *designed* to do. They call it "free trade," but what it really does is concentrate power.

The big transnationals want to reduce freedom by undermining the democratic functioning of the states in which they're based, while at the same time ensuring the government will be powerful enough to protect and support them. That's the essence of what I sometimes call "really existing market theory."

If you look through the whole history of modern economic development, you find that—virtually without exception—advocates of "free markets" want them applied to the poor and the middle-class but not to themselves. The government subsidizes corporations' costs, protects them from market risks and lets them keep the profits.

Can I smoke here in your office? If you deny me that, are you limiting my freedom?

I'm limiting your freedom but I'm increasing my rights. If you smoke in my office, it increases my chances of dying. Any effort to create a more human existence is going to inhibit somebody's freedom. If a kid crosses the street in front of me when I have a red light, that inhibits my freedom to run him over and get to work faster.

Public schools are another example. People who don't have children still have to pay school taxes, because we have a common feeling that it's good for our society if children are educated. Whether we personally have kids isn't relevant.

The most fanatic advocates of private despotism (who actually want to undermine freedom and democracy) naturally use nice words like *freedom*. What they really mean is that we have to have tyranny and a powerful state to ensure it. Just look at what they propose.

The Heritage Foundation, for instance, is full of talk about big philosophical issues, minimizing the state and so on, but they want to raise the Pentagon budget, because it's the major pipeline for public subsidy to high-tech industries. That's a hard line to defend, but as long as there isn't much in the way of intelligent public debate, they can get away with it.

The most extreme types, like Murray Rothbard, are at least honest. They'd like to eliminate highway taxes because they force you to pay for a road you may never drive on. As an alternative, they suggest that if you and I want to get somewhere, we should build a road there and then charge people tolls to go on it.

Just try generalizing that. Such a society couldn't survive, and even if it could, it would be so full of terror and hate that any human being would prefer to live in hell.

In any case, it's ridiculous to talk about freedom in a society dominated by huge corporations. What kind of freedom is there inside a corporation? They're totalitarian institutions— you take orders from above and maybe give them to people below you. There's about as much freedom as under Stalinism. Whatever rights workers have are guaranteed by the limited public authority that still exists.

When enormous, private, tyrannical institutions are granted the same rights as—or more rights than—human beings, freedom becomes something of a joke. The solution isn't to undermine freedom—it's to undermine the private tyrannies.

In Boulder [Colorado], where I live, an ordinance banning smoking in restaurants was put on the ballot. There was an enormous, well-funded campaign against it. Some city council members were threatened, and their actions were described as "fascist" and "Nazi-like." All in the name of freedom.

There's nothing new about that. In the past, the line was that Philip Morris has to be free to get twelve-year-old kids to smoke, and the kids' mothers are free to prevent them from smoking.

Of course, Philip Morris has greater resources, and therefore more persuasive power, than thousands of parents and hundreds of city councils, but that was supposed to be irrelevant.

There was a funny coincidence a while back. The *New York Times* ran an op-ed by a senior fellow of the Hoover Institute about the "profound philosophical differences" that separate liberals and conservatives. The liberals want to see social policy administered at the federal level, while "conservatives prefer to transfer power to the states, in the belief that policies should be made closer to the people."

The same day, the *Wall Street Journal* ran a story headlined: "What Fidelity Wants It Usually Gets, and It Wants Massachusetts Tax Cut." It opened by stating that "when Fidelity Investments talks, Massachusetts listens"—or else.

Massachusetts listens, the article explains, because Fidelity is one of the biggest firms in the state and can easily shift operations across the border to Rhode Island. That was exactly what it was threatening to do unless Massachusetts granted it "tax relief"—a subsidy, in effect, since "the people" pay more taxes to compensate for it. (New York recently had to do the same, when major financial firms threatened to move to New Jersey.) Massachusetts granted Fidelity the "relief."

A few months earlier, Raytheon had demanded tax and utility rate relief, perhaps to compensate for the fact that its shares had only about tripled in value in the past four years, while dividends per share rose 25% as well. The report on the business pages raised the (rhetorical) question whether Raytheon "is asking for tax dollars with one hand while passing money to shareholders with the other."

Again, Massachusetts listened to the threat to transfer out-of-state. Legislators had planned a big tax break for Massachusetts businesses generally, but restricted it to Raytheon and other "defense contractors."

It's an old story. Until the late 19th century, corporations were limited to functions explicitly determined by the state charters. That requirement effectively disappeared when New Jersey offered to drop it. Corporations began incorporating in New Jersey instead of New York, thus forcing New York to also drop the requirement and setting off a "race to the bottom."

The result was a substantial increase in the power of private tyrannies, providing them with new weapons to undermine liberty and human rights, and to administer markets in their own interest. The logic is the same when GM decides to invest in Poland, or when Daimler-Benz transfers production from Germany, where labor is highly paid, to Alabama, where it isn't.

By playing Alabama off against another competitor, North Carolina, Daimler-Benz received subsidies, protected markets and risk protection from "the people." (Smaller corporations can get into the act too, when states are forced to compete to bribe the powerful.)

Of course, it's far easier to play this game with states than countries. For Fidelity to move to Rhode Island, and for Raytheon to move to Tennessee, is no major problem—and Massachusetts knows it. Transferring operations overseas would be rather more difficult.

"Conservatives" are surely intelligent enough to understand that shifting decisions to the state level does not transfer power to "the people" but to those powerful enough to ask for subsidies with one hand and pocket them with the other. That's the "profound philosophical principle" that underlies the efforts of "conservatives" to shift power to the states.

There are still some defenses at the federal level, which is why it's been made the enemy (but not, of course, the parts that funnel money to large corporations—like the Pentagon, whose budget is going up, over the objections of more than 80% of the people).

According to a poll reported in the *Washington Post,* an enormous number of people think anything the federal government does is bad—except for the military, which we need (of course) to counter grave threats to US security. (Even so, people didn't want the military budget increased, as Clinton, Gingrich and the others proceeded to do.) *What could explain this?* the *Post* wondered.

Could it be fifty years of intense corporate propaganda, in the media and elsewhere, that has been trying to direct people's fear, anger and hatred against the government and make *private* power invisible to them? That isn't suggested as a reason. It's just a mystery why people have these strange ideas.

But there's no question they have them. When somebody wants to vent his anger at the fact that his life is falling apart, he's more likely to put a bomb in a federal building than in a corporate headquarters.

There are plenty of things wrong with government, but this propaganda opposes what's *right* with it—namely, that it's the one defense people have against private tyrannies.

To come back to the Boulder situation, is it an example of what you call "anti-politics"?

It's an example of opposition to democracy. It means that people shouldn't have a right to get together and democratically decide how they want to live.

You've frequently made the point that while corporate executives are getting everything they want on a silver platter, they're very leery of the far right, because they want to make sure their daughters continue to have access to abortion. But their daughters had access to abortions before Roe vs. Wade.

The executives don't want to have to do it secretly, and get involved in criminal activity. They want their wives and daughters to have normal freedoms and they want to live in a civilized society, not one in the grips of religious fundamentalism, where people around them think the world was created a couple of thousand years ago.

Another thing that worries them about this ultra-right tendency is that there's a populist streak in it. There's a lot of opposition to "bigness"—not just big government but big business too. The right wing doesn't see the point of things like funding for science, but business does, because it creates the technology and knowledge they'll exploit in the future.

Corporate executives also don't particularly like the idea of dismantling international institutions like the United Nations, or eliminating what's called foreign aid. They need those institutions, and they want them around. The jingoist, narrow-minded fanaticism that gave them deregulation, tort reform and the cutback of social services has another side to it, and they're definitely concerned about it.

The myth of hard times

When I called you the other day at home in Lexington, you were sitting in the dark, because the power had gone out.

I have a feeling we're going to be seeing more and more of that sort of thing. There simply hasn't been much investment in infrastructure. It's part of the drive for short-term profit: you let everything else go.

A lot of people are aware of it. We had a plumber in the other day, and he told us he had just bought himself a generator because he expects the power to be going off regularly.

Outsourcing is another aspect of it—it saves corporations' money today, but it destroys the potential work force. In the universities, they're hiring part-time junior faculty, who turn over fast. In research, there's a lot of pressure to do short-term, applied work, not the kind of basic, theoretical studies that were done in the 1950s and that laid the basis for the economy of today. The long-term effects of this are pretty obvious.

What do you think of this notion of scarcity—not enough jobs, not enough money, not enough opportunity?

Take a walk through any big city. Do you see anything that needs improvement?

There are huge amounts of work to be done, and lots of idle hands. People would be delighted to do the work, but the economic system is such a catastrophe it can't put them to work.

The country's awash in capital. Corporations have so much money they don't know what to do with it—it's coming out of their ears. There's no scarcity of funds—these aren't "lean and mean" times. That's just a fraud.

In 1996, President Clinton signed something called the Personal Responsibility and Work Opportunity Act, which eliminated the federal government's 61-year commitment to the poor. You've said that commitment was always very limited, and that it's declined sharply since about 1970.

When the assault began.

You've got to like the wording of that bill.

It says seven-year-old children have to take personal responsibility. It gives them opportunities they were deprived of before—like the opportunity to starve. It's just another assault against defenseless people, based on a very effective propaganda campaign to make people hate and fear the poor.

That's smart, because you don't want them looking at the *rich*, at what *Fortune* and *Business Week* call "dazzling" and "stupendous" profit growth, at the way the military system is pouring funds into advanced technology for the benefit of private industry. No, you want them to look at some imaginary black mother driving a Cadillac to pick up a welfare check so she can have more babies. *Why should I pay for that?* people ask.

The effectiveness of this campaign has been striking. Although most people think the government has a responsibility to ensure reasonable, minimal standards for poor people, they're also against welfare, which is what the government efforts to ensure reasonable, minimal standards for poor people are *called*. That's a propaganda achievement you have to admire.

There's another aspect of this that's much less discussed. One of the purposes of driving people away from welfare and into work is to lower wages by increasing the supply of workers.

The New York City government is now partially subsidizing workers driven out of the welfare system. The main effect has been to decrease unionized labor. Put a lot of unskilled labor into the workplace, make conditions so awful that people will take virtually any job, maybe throw in some public subsidy to keep them working, and you can drive down wages. It's a good way to make everybody suffer.

Ralph Nader calls the Republicans and the Democrats "Tweedledum and Tweedledee."

There's never been much of a difference between the two business parties, but over the years, what differences there were have been disappearing.

In my view, the last liberal president was Richard Nixon. Since him, there've been nothing but conservatives (or what are called "conservatives"). The kind of gesture to liberalism that was required from the New Deal on became less necessary as new weapons of class war developed in the early 1970s.

For the last twenty years, they've been used to bring about what the business press openly calls "capital's clear subjugation of labor." Under those circumstances, you can drop the liberal window-dressing.

Welfare capitalism was introduced in order to undercut democracy. If people are trying to take over some aspect of their lives and there doesn't seem any way to stop them, one standard historical response has been to say, *We rich folk will do it for you.* A classic example took place in Flint, Michigan, a town dominated by General Motors, around 1910.

There was a good deal of socialist labor organizing there, and plans had been developed to really take things over and provide more democratic public services. After some hesitation, the wealthy businessmen decided to go along with the progressive line. They said, *Everything you're saying is right, but we can do it a lot better, because we have all this money. You want a park? Fine. Vote for our candidate and he'll put in a park.*

Their resources undermined and eliminated the incipient democratic and popular structures. Their candidate won, and there was indeed welfare capitalism...until it wasn't needed any more, at which point it was dropped.

During the Depression, there was again a live union movement in Flint, and popular rights were again extended. But the business counterattack began right after the Second World War. It took a while this time, but by the 1950s, it was getting somewhere.

It slowed somewhat in the 1960s, when there was a lot more ferment—programs like the War on Poverty, things coming out of the civil rights movement—but by the early 1970s, it reached new heights, and it's been going pretty much full-steam ever since.

The typical picture painted by business propaganda since the Second World War—in everything from television comedies to scholarly books—has been: *We all live together in harmony. Joe Six-Pack, his loyal wife, the hard-working executive, the friendly banker—we're all one big happy family. We're all going*

223

to work together to defend ourselves against the bad guys out there—like union organizers and big government—who are trying to disrupt our harmony. That's always the picture presented: class harmony between the people with the hammers and the people getting beaten over the head with them.

There's a campaign to undermine public confidence in Social Security, by saying it's going broke and that when the baby boomers reach retirement age, there'll be no money for them.

Most of the talk about Social Security is pretty fraudulent. Take the question of privatizing it. Social Security funds can be invested in the stock market whether the system is public or private. But putting people in charge of their own assets breaks down the solidarity that comes from doing something together, and diminishes the sense that people have any responsibility for each other.

Social Security says, *Let's ensure that all of us have a minimal standard of living.* That puts a bad idea into people's heads—that we can all work together, get involved in the democratic process and make our own decisions. Much better to create a world in which people behave individually and the powerful win.

The goal is a society in which the basic social unit is you and your television set. If the kid next door is hungry, it's not your problem. If the retired couple next door invested their assets badly and are now starving, that's not your problem either.

I think that's what lies behind the Social Security propaganda. The other issues are technical and probably not very significant. A slightly more progressive tax system could keep Social Security functioning for the indefinite future.

So we're moving from the idea that an injury to one is an injury to all, to the idea that an injury to one is just an injury to one.

That's the ideal of a capitalist society—except for the rich. Boards of directors are allowed to work together, and so are banks and investors and corporations in alliances with one another and with powerful states. That's fine. It's just the poor who aren't supposed to cooperate.

Corporate welfare

In an op-ed in the *Boston Globe*, Bernie Sanders of Vermont, the only Independent member of Congress, wrote, "If we're serious about balancing the budget in a fair way, we must slash corporate welfare." You've said you're very uncomfortable with the term *corporate welfare*. Why?

I like Bernie Sanders, and that was a good column, but I think he starts off on the wrong foot. Why should we balance the budget? Do you know a business—or a household—that doesn't have any debt?

I don't think we should balance the budget at all. The whole idea is just another weapon against social programs and in favor of the rich—in this case, mostly financial institutions, bondholders and the like.

Putting that aside, I don't hesitate to use the term *corporate welfare* because corporate welfare doesn't exist, or because it isn't a serious problem, but because people typically use the term to refer to specific government programs—a subsidy for ethanol manufacturers, say—rather than the more pervasive and fundamental ways government helps business. That's a serious error.

If it hadn't been for massive government interference, our automobile, steel and semiconductor industries probably wouldn't even exist today. The aerospace industry is even more thoroughly socialized. When Lockheed—Gingrich's favorite—was in big trouble back in the early 1970s, it was saved from destruction by a $250 million loan subsidized by the federal government. The same with Penn Central, Chrysler, Continental Illinois Bank and many others.

Right after the 1996 elections (I assume the timing wasn't accidental), the Clinton administration decided to funnel what's expected to amount to $750 *billion* or more of public money into developing new jet fighters, which we don't need for military purposes. The contract is to be awarded not to the traditional fighter manufacturer, McDonnell Douglas, but to Lockheed Martin and/or Boeing, which hasn't produced a fighter plane for sixty years.

The reason is that Boeing sells commercial aircraft, our biggest civilian export. (The market for them is huge.) Commercial aircraft are often modified military aircraft, and adapt a lot of technology and design from them.

Boeing and McDonnell Douglas announced a merger, which was publicly subsidized to the tune of more than one billion dollars.

I'm sure the fact that McDonnell Douglas was knocked out of the competition for that fighter contract is part of the reason they're willing to be taken over by Boeing. In describing why Boeing was chosen over McDonnell Douglas, the Pentagon's undersecretary for acquisition and technology said, "We need to get hooks into the commercial research base to influence its growth." Defense Secretary William Perry explained that we've got to overcome earlier "barriers which limited timely access to rapidly evolving commercial technology."

"The Pentagon is ushering out the military-industrial complex and ushering in an industrial-military complex," *NY Times* reporter Adam Bryant added, "noting that it's "not just an idle reordering of adjectives" but reflects Pentagon efforts "to do more business with companies that have a diverse customer base."

An aerospace industry analyst at Merrill Lynch pointed out that "this effort to broaden the industrial base that supports the military has been going on for a couple of years, but the Pentagon's decision [about the new Joint Strike fighter] was a major milestone in this trend."

In fact, "this effort" has been has going on not for "a couple of years" but for half a century, and its roots lie much deeper, in the crucial role of the military in developing the basic elements of the "American system of manufacturing" (standardization and interchangeable parts) in the 19th century.

In other words, a major purpose of military production and procurement, along with research and development in government labs or publicly funded private industry (by the Department of Energy and other agencies, as well as the Pentagon) is to subsidize private corporations. The public is simply being deluded about how they're paying for high technology.

By now this stuff is described almost openly—usually on the business pages but sometimes even on the front page. That's one of the nice things about the end of the Cold War—the clouds lift a bit. More people now realize, at least to some extent, that the military system has been partially a scam, a cover for ensuring that advanced sectors of industry can continue to function at public expense. This is part of the underpinnings of the whole economic system, but it's off the agenda when most people talk about corporate welfare.

I'm not saying public financing shouldn't exist, by the way. I think it's a very good idea to fund research in the science and technology of the future. But there are two small problems: public funding shouldn't be funneled through private tyrannies (let alone the military system), and the public should decide what to invest in. I don't think we should live in a society where the rich and powerful determine how public money is spent, and nobody even knows about their decisions.

Ironically, the politicians who prate the most about minimizing government are exactly the ones most likely to expand its business-funding role. The Reagan administration poured money into advanced technology and was the most protectionist in postwar American history. Reagan probably didn't know what was going on, but the people around him virtually doubled various import restrictions. His Treasury secretary, James Baker, boasted that they'd raised tariffs higher than any post-WWII government.

Government subsidies to private industry are unusually large here, but they exist in all the industrial nations. The Swedish economy, for instance, rests heavily on big transnational corporations—weapons manufacturers, in particular. Sweden's military industry appears to have provided much of the technology that allowed Ericsson to carve out a large share of the mobile phone market.

Meanwhile, the Swedish welfare state is being cut back. It's still way better than ours, but it's being reduced—while the multinationals' profits increase.

Business wants the popular aspects of government, the ones that actually serve the population, beaten down, but it also wants a very powerful state, one that works for it and is removed from public control.

Do you think corporate welfare is a good wedge issue to get people involved in politics?

227

I'm not a great tactician, and maybe this is a good way to stir people up, but I think it would be better for them to think through the issues and figure out the truth. Then they'll stir themselves up.

Crime: in the suites vs. in the streets

The media pays a lot of attention to crime in the streets, which the FBI estimates costs about $4 billion a year. The *Multinational Monitor* estimates that white-collar crime—what Ralph Nader calls "crime in the suites"—costs about $200 billion a year. That generally gets ignored.

Although crime in the US is high by the standards of comparable societies, there's only one major domain in which it's really off the map—murders with guns. But that's because of the gun culture. The overall crime rate hasn't changed much for a long time. In fact, it's been decreasing recently.

The US is one of very few societies—maybe the only one—where crime is considered a political issue; in most parts of the world, it's looked at as a social problem. Politicians don't have to fight during elections about who's tougher on crime—they simply try to figure out how to deal with it.

Why does crime get all this attention here? I think it has more to do with social control than with crime itself. There's a very committed effort to convert the US into something resembling a Third World society, where a few people have enormous wealth and a lot of others have no security (for one reason, because their jobs might be sent to Mexico or some other place where employers don't have to worry about benefits, unions or the like).

Now that these workers are superfluous, what do you do with them? First of all, you have to make sure they don't notice that society is unfair and try to change that, and the best way to distract them is to get them to hate and fear one another. Every coercive society immediately hits on that idea, which has two other benefits: it reduces the number of superfluous people (by violence) and provides places to put the ones who survive (prisons).

The utterly fraudulent war on drugs was undertaken at a time when everyone knew that the use of every drug—even coffee—was falling among educated whites, and was staying sort of level among blacks. The police obviously find it much easier to make an arrest on the streets of a black ghetto than in a white suburb. By now, a very high percentage of incarceration is drug-related, and it mostly targets little guys, somebody who's caught peddling dope.

The big guys are largely ignored. The US Department of Commerce publishes regular data on foreign operations of US business (estimates only, with delays; the details are unknown). In late 1996 it reported that in 1993–95, about a quarter of direct foreign investment in the Western Hemisphere (apart from Canada) was in Bermuda.

The figures for majority-owned foreign affiliates of US corporations (other than banks) were about a quarter in Bermuda and another 15% in Panama, the British Caribbean islands and other tax havens. Most of the rest seems to be short-term speculative money—picking up assets in, say, Brazil.

Now, they're not building manufacturing plants in Bermuda. The most benign interpretation is that it's some form of tax evasion. Quite possibly it's narco-capital. The OECD [the Organization of Economic Cooperation and Development, a Paris-based group representing the 29 richest nations] estimates that more than half of all narco-money—something like $250 *billion*—goes through US banks each year. But, as far as I know, nobody's looking into this dirty money.

It's also been known for years that American industrial producers have been sending way more of the chemicals used in drug production to Latin America than there's any conceivable legal use for. This has occasionally led to executive orders requiring the manufacturers to monitor what chemicals they sell to whom, but I haven't seen any prosecutions on this.

Corporate crime isn't just ignored in the area of drugs. Take what happened with the S&Ls. Only a very small part of it was treated as crime; most of it was just picked up by the taxpayer with bailouts. Is that surprising? Why should rich and powerful people allow themselves to be prosecuted?

Russell Mokhiber of the *Corporate Crime Reporter* contrasts two statistics: 24,000 Americans are murdered each year, while 56,000 Americans die from job-related accidents and diseases.

That's another example of unpunished corporate crime. In the 1980s, the Reagan administration essentially informed the business world that it was not going to prosecute violations of OSHA [the Occupational Safety and Health Administration] regulations. As a result, the number of industrial accidents went up rather dramatically. *Business Week* reported that working days lost to injury almost doubled from 1983 to 1986, in part because "under Reagan and [Vice-President] Bush, [OSHA] was a hands-off agency."

The same is true of the environmental issues—toxic waste disposal, say. Sure, they're killing people, but is it criminal? Well, it *should* be.

Howard Zinn and I visited a brand-new maximum-security federal prison in Florence, Colorado. The lobby has high ceilings, tile floors, glass everywhere. Around the same time, I read that New York City schools are so overcrowded that students are meeting in cafeterias, gyms and locker rooms. I found that quite a juxtaposition.

They're certainly related. Both prisons and inner-city schools target a kind of superfluous population that there's no point educating because there's nothing for them to do. Because we're a civilized people, we put them in prison, rather than sending death squads out to murder them.

Drug-related crimes, usually pretty trivial ones, are mostly what's filling up the prisons. I haven't seen many bankers or executives of chemical corporations in prison. People in the rich suburbs commit plenty of crimes, but they're not going to prison at anything like the rate of the poor.

There's another factor too. Prison construction is by now a fairly substantial part of the economy. It's not yet on the scale of the Pentagon, but for some years now it's been growing fast enough to get the attention of big financial institutions like Merrill Lynch, who have been floating bonds for prison construction.

High-tech industry, which has been feeding off the Pentagon for research and development, is turning to the idea of administering prisons with supercomputers, surveillance technology, etc. In fact, I wouldn't be entirely surprised to see fewer people in prisons and more people imprisoned in their homes. It's probably within reach of the new technology to have surveillance devices that control people wherever they are. So if you pick up the telephone to make a call they don't like, alarms go off or you get a shock.

It saves the cost of building prisons. That hurts the construction industry, true, but it contributes to the high-tech sector, which is the more advanced, growing, dynamic part of the economy.

It sounds like an Orwellian *1984* scenario you're describing.

Call it Orwellian or whatever you like—I'd say it's just ordinary state capitalism. It's a natural evolution of a system that subsidizes industrial development and seeks to maximize short-term profit for the few at the cost of the many.

If you'd predicted, thirty or forty years ago, that there'd be smoke-free flights and restaurants, and that the tobacco companies would be under intense attack, no one would have believed you.

Through the 1980s, the use of all substances—drugs, smoking, coffee, etc.—declined, by and large, among the more educated and wealthier sectors of the population. Because the cigarette companies know they're going to end up losing that portion of their market, they've been expanding rapidly into foreign markets, which are forced open by US government power.

You still find plenty of poor, uneducated people smoking; in fact, tobacco has become such a lower-class drug that some legal historians are predicting that it will become illegal. Over the centuries, when some substance became associated with "the dangerous classes," it's often been outlawed. Prohibition of alcohol in this country was, in part, aimed at working-class people in New York City saloons and the like. The rich kept drinking as much as they wanted.

I'm not in favor of smoking being made illegal, by the way, any more than I'm in favor of making other class-related substances illegal. But it's a murderous habit that kills huge numbers of people and harms plenty of others, so the fact that it's come under some sort of control is a step forward.

In August 1996, Gary Webb wrote a three-part article in the *San Jose Mercury News,* which was expanded into a book called *Dark Alliance*. Webb alleged that the CIA had been making money selling crack cocaine in the black ghetto in Los Angeles, and in fact was re-sponsible for the explosion of that drug's popularity in the 1980s.

I've noticed that you tend to stay away from such stories—at least until you're asked about them during a question-and-answer period. You don't devote much energy to them.

I just look at them differently. The Webb story is fundamen-tally correct, but the fact that the CIA has been involved in drug-running has been well-known since Al McCoy's work 25 years ago [in books like *The Politics of Heroin: CIA Complicity in the Global Drug Trade*]. It started right after the Second World War. You can follow the trail through the French connection in Marseilles (a consequence of CIA efforts to undermine unions by reconstitut-ing the Mafia for strikebreaking and disruption), to the Golden Triangle in Laos and Burma, and on to Afghanistan, etc.

Bob Parry and Brian Barger exposed a lot of the story ten years ago. Their evidence was correct, but they were shut up very quickly. Webb's contribution was to trace some of the details and discover that cocaine got into the ghettos by a particular pathway.

When the CIA says they didn't know anything about it, I as-sume they're right. Why should they want to know anything about details like that? That it's going to end up in the ghettos isn't a plot—it's just going to happen in the natural course of events. It's not going to sneak into well-defended communities that can pro-tect themselves. It's going to break into devastated communities where people have to fight for survival, where kids aren't cared for because their parents are working to put food on the table.

So of course there's a connection between the CIA and drugs. The US was involved in massive international terrorism through -out Central America. It was mostly clandestine (which means people in powerful positions in government and the media knew about it, but it was enough below the surface that they could pretend they didn't). To get untraceable money and brutal thugs, our government naturally turned to narco-traffickers—like Nor-iega (he was our great friend, remember, until he became too in-dependent). None of this is a secret or a surprise.

Where I differ from a lot of other people is, I don't think the CIA has been involved as an independent agency: I think it does what it's told to do by the White House. It's used as an instrument of state policy, to carry out operations the government wants to be able to "plausibly deny."

The media

In *Manufacturing Consent,* the book you wrote with Ed Herman in 1988, you described five filters that news goes through before we see it. Would you revise that list? One of the filters, anticommunism, probably needs to be changed.

Temporarily, at least. I thought at the time it was put too narrowly. More broadly, it's the idea that grave enemies are about to attack us and we need to huddle under the protection of domestic power.

You need something to frighten people with, to prevent them from paying attention to what's really happening to them. You have to somehow engender fear and hatred, to channel the kind of rage—or even just discontent—that's being aroused by social and economic conditions.

By the early 1980s, it was clear that Communism wasn't going to remain usable as a threat for much longer, so when the Reagan administration came in, they immediately focussed on "international terrorism." Right from the start, they used Libya as a punching bag.

Then every time they had to rally support for aid to the Contras or something, they'd engineer a confrontation with Libya. It got so ludicrous that, at one point, the White House was sur - rounded with tanks to protect poor President Reagan from Libyan hit squads. It became an international joke.

By the late 1980s, Hispanic drug traffickers became the enemy; by now, they've been joined by immigrants, black criminals, welfare mothers and a whole host of other attackers on every side.

Toward the end of *Manufacturing Consent,* you conclude that "the societal purpose of the media is to...defend the economic, social and political agenda of privileged groups that dominate the domestic society and the state." Anything you'd want to add to that?

It's such a truism that it's almost unnecessary to put it into words. It would be amazing if it *weren't* true. Assuming virtually

233

nothing except that there's a free market—or something resembling one—virtually forces you to that conclusion.

In *Z* magazine, Ed Herman discussed the persistence of the idea that the media are liberal.

Ed's main point is perfectly valid—what really matters is the desires of the people who own and control the media. But I may slightly disagree with him about whether they're liberal. In my view, national media like the *Washington Post* and the *New York Times* probably meet the current definition of the word *liberal.* Sometimes they even run things I approve of.

For instance, to my amazement, the *New York Times* actually had an editorial in favor of greater workers' rights in Indonesia (as opposed to the right-wing view that it's OK to strangle Indonesian workers if you can make more money that way). The *Times* also has columnists—Bob Herbert is one example—that I don't think you would have seen there forty years ago, and they often write very good stuff.

But in general, the mainstream media all make certain basic assumptions, like the necessity of maintaining a welfare state for the rich. Within that framework, there's some room for differences of opinion, and it's entirely possible that the major media are toward the liberal end of that range. In fact, in a well-designed propaganda system, that's exactly where they should be.

The smart way to keep people passive and obedient is to strictly limit the spectrum of acceptable opinion, but allow very lively debate within that spectrum—even encourage the more critical and dissident views. That gives people the sense that there's free thinking going on, while all the time the presuppositions of the system are being reinforced by the limits put on the range of the debate.

So you're allowed to discuss whether the Mideast "peace process" should be implemented immediately or should be delayed, or whether Israel is sacrificing too much or just the right amount. But you're not allowed to discuss the fact—and it certainly is a fact—that this so-called "peace process" wiped out a 25-year, internationally-supported diplomatic effort recognizing the national rights of both contending parties, and rammed home the US position that denies these rights to the Palestinians.

Let's clarify what it really means to say the media are liberal. Suppose 80% of all journalists vote Democratic. Does that mean they're liberal in any meaningful sense of the word, or just that they're at the left end of an extremely narrow, center-right spectrum? (Most of my writing has been a criticism of the liberal end of the media, the ones who set the leftmost boundary for acceptable opinion.)

Take it a step further. Suppose it turns out that 80% of all journalists are flaming radicals who'd really rather be writing for *Z*. Would that show that the media themselves are radical? Only if you assume that the media are open to the free expression of ideas (by their reporters, in this case).

But that's exactly the thesis under debate, and you can't establish it by presupposing it. The empirical evidence that this thesis is false is overwhelming, and there has been no serious attempt to address it. Instead, it's just *assumed* that the media are open. It's possible to get away with that kind of thinking if power is sufficiently concentrated and educated sections of the population are sufficiently obedient.

The University of Illinois Press has published a US edition of *Taking the Risk out of Democracy* by the noted Australian scholar Alex Carey. One of the chapters is entitled *Grassroots and Treetops Propaganda*. What does Carey mean by that?

Treetops propaganda is the kind of thing that Ed Herman and I are mostly commenting on. It's the elite media, aimed at educated sectors of the population that are more involved in decision-making and setting a general framework and agenda for others to adhere to. Grassroots propaganda is aimed at the vulgar masses, to keep them distracted and out of our hair, and to make sure they don't interfere in the public arena, where they don't belong.

Do you find it ironic that one of the major works on US propaganda is written by an Australian?

Not at all. Alex Carey was an old friend; in fact, we dedicated *Manufacturing Consent* to him. He really pioneered the study of corporate propaganda, of which the media is just one aspect. He was working on a big book on the subject, but he died before it was completed.

Although corporate propaganda is a major force in contemporary history, it's very little studied, because people aren't supposed to know that major corporations are deeply dedicated to controlling the public mind, and have been for a long time. Carey quotes the business press as saying that the public mind is the greatest "hazard facing industrialists."

We're supposed to believe that the press is liberal, dangerous, adversarial, out-of-control. That itself is an extremely good example of corporate propaganda.

More than 700 people died in a Chicago heat wave in the summer of 1995. They were mostly old people living in poor neighborhoods who couldn't afford air conditioning. I think the headlines should have read, *Market Kills 700*.

You're absolutely right—honest media would have reported how the workings of the market system added more deaths to the toll. Every story in the paper could be recast with a more honest and humane point of view, one not reflecting the interests of the powerful. But expecting them to do that on their own initiative is like expecting General Motors to give away its profits to poor people in the slums.

Anthony Lewis, someone you often identified as the outer liberal fringe allowed in the *Times,* celebrated the Pentagon Papers on their 25th anniversary as a great example of media heroism and courage. He wrote that "we were a much tamer press before 1971."

There's been a bit of a change. The 1960s opened up society in many ways, from personal attitudes to dress codes to beliefs. That affected everything, including corporations and the corporate media—which now are, in many respects, less automatically disciplined than they were back in the sixties.

There was a column around the same time by Randolph Ryan. He's someone who came out of the 1960s and did extremely good reporting on Central America for the *Boston Globe* in the 1980s. The 1960s culture also affected the *Globe*'s editor, Tom Winship—whose son was a draft resister, incidentally. What was happening influenced his thinking and improved the newspaper in lots of ways. So sure, the 1960s had a big effect. But the publication of the *Pentagon Papers* in 1971 wasn't really part of it.

In 1968, after the Tet offensive [a massive assault by the southern resistance (called the "Viet Cong" by the US) with the support of North Vietnamese troops, during the Vietnamese holiday of Tet], corporate America basically decided that the war wasn't worth it. They came to believe that we'd essentially achieved what we needed to, and that continuing was just too costly. So they told Johnson to enter into some form of negotiations and to start withdrawing American troops.

It wasn't until about a year and a half later that the media here began to respond to the opening that corporate America had left for them by voicing very timid criticisms of the war. As I recall, the first newspaper to call for American withdrawal from Vietnam was the *Boston Globe*.

It was around then that Lewis started saying that the war began with "blundering efforts to do good" but that by 1969(!) it had become clear that it was "a disastrous mistake" and that the US "could not impose a solution except at a price too costly to itself." (By the same token, *Pravda* was probably saying, around 1980 or 1981, *The war in Afghanistan began with blundering efforts to do good, but now it's clear that it's a disastrous mistake and too costly for Russia.*)

Of course, Vietnam wasn't a "disastrous mistake"—it was murderous aggression. When the *Times* starts writing *that*, we'll know something has changed.

Most of the important parts of the Pentagon Papers never appeared in the *Times* and haven't been discussed in the mainstream literature either. The parts the *Times did* publish weren't all that revealing. Although they contained some new information, for the most part they simply confirmed what was already available in the public record. The *Times'* willingness to publish them, three years after the main centers of American power had decided the war should be ended, wasn't exactly an act of enormous heroism.

Because the government is giving less funding to public radio and TV, they're being forced more and more to turn to corporate funding.

Public radio and TV have always been very marginal enterprises. As Bob McChesney describes, there was a struggle back in the 1920s and 1930s over whether radio should be in the public arena or handed over to private power. You know which side won. When television came along, there wasn't even much of a debate—it was just given to business.

Both times this was done in the name of democracy! It tells you what a strange intellectual culture this is—we take the media out of the hands of the public, give them to private tyrannies, and call it *democracy.*

Over time, this attitude has solidified. The 1996 Telecommunications Act was the biggest giveaway of public assets in history. Even token payments weren't required.

McChesney makes the interesting and important point that this wasn't treated as a social and political issue—you read about it in the business pages, not on the front page. The issue of *whether* we should give away these public resources to private power wasn't discussed—just *how* we should give them away. That was a tremendous propaganda victory.

Public radio and television are permitted around the fringes, partly because the commercial media were criticized for not fulfilling the public-interest duties required of them by law. So they said, *Let the public stations take care of that. Let* them *run Hamlet.* Now, even that marginal function is being narrowed.

This doesn't necessarily mean the death of public radio and television, by the way. Back in the Middle Ages, the arts were supported almost entirely by benevolent autocrats like the Medicis; maybe today's benevolent autocrats will do the same. After all, they're the ones who support the operas and symphonies.

McChesney also notes that most broadcast innovation has taken place in public radio and television, not commercial. FM radio was public until it started making money, then it became private. The internet is a dramatic example today—it was designed, funded and run in the public sector as long as you couldn't make money on it, but as soon as it showed a potential for profitability, it was handed over to megacorporations.

Two Academy Award-winning documentaries, *Deadly Deception* (about General Electric) and *The Panama Deception,* and a film about you, *Manufacturing Consent,* were hardly shown on public TV.

Things used to be even worse. I spent a couple of weeks in Indochina in early 1970. At that point I was pretty well known in the Boston area, which is home to NPR's flagship affiliate, WGBH. With great reluctance, WGBH's big liberal leader, Louis M. Lyons, agreed to interview me—extremely hostilely—for a few minutes. That was probably the only time I was on local public radio back then.

I'm not a great admirer of today's media, but I think they're way better and more open than they were thirty or forty years ago. People who went through the 1960s are now in the media and are writing—at least partially—from more humane points of view.

What would the media look like in a genuinely democratic society?

They'd be under public control. Their design, the material they present, access to them, would all be the result of public participation—at least to the extent that people want to be involved, and I think they would.

Some of the media in this country were once more democratic. Not to be too exotic, let's go back to the 1950s, when eight hundred labor newspapers, reaching twenty or thirty million people a week, were devoted to struggling against the commercial press, which was "damning labor at every opportunity," as they put it, and "selling [the] virtues of big business"—driving the mythology into people's heads.

Bob McChesney says that in the early 1940s, there were about a thousand labor-beat reporters. Today there are seven.

Every newspaper has a business section, which responds to the interests of a small part of the population—the part that happens to control the newspaper, oddly enough. But I've never seen a labor section in a newspaper. When labor news is run at all, it's in the business section, and is looked at from that point of view. This simply reflects, in a very transparent way, who's in power.

Lots of people criticize the ongoing tabloidization of the news. The program directors respond by saying, *We're giving the public what it wants. No one's forcing them to turn on the TV and watch our programs.* What do you think about that?

First of all, I don't agree that that's what the public wants. To take just one example, I think people in New York would have been interested in learning that NAFTA was expected to

harm "women, blacks and Hispanics [and] semi-skilled production workers" (70% of all workers are categorized as "semi-
skilled")—as a very careful reader of the *Times* could discover
the day *after* Congress passed NAFTA.

Even then, the facts were concealed in an upbeat story about
the likely winners: "the region's banking, telecommunications
and service firms, from management consultants and public
relations to law and marketing...banks and Wall Street securi-
ties firms," the capital-intensive chemical industry, publishing
(including media corporations), etc.

But that aside, what people want is in part socially created—
it depends on what sort of experiences they've had in their lives,
and what sort of opportunities. Change the structure and
they'll choose different things.

I visited a working-class slum in Brazil where people gather
during prime television time to watch locally produced films on
a large outdoor screen. They prefer them to the soap operas
and other junk on commercial TV, but they can only have that
preference because they were offered the choice.

When people in the US are surveyed, it turns out that what
they want—overwhelmingly—is commercial-free television. Do
you see commercial-free television? Of course not. In US television, big corporations sell audiences to other businesses, and
they're not interested in providing us with other options.

In an article titled *The Strange Disappearance of Civic America,*
Robert Putnam named TV as the culprit.

Putnam is a sociologist at Harvard who's quite mainstream.
He found about a 50% decline since the 1960s in *any* form of
interaction—visiting a neighbor, going to PTA meetings, joining
a bowling league. One reason children watch so much TV is
that parent-child interaction has dropped 40% or so from the
1960s to today—at least in part because both parents have to
work fifty hours a week to put food on the table. There's little
day care and few support systems available, so what are you
left with? TV baby-sitting.

But it's a little thin to blame TV itself. It isn't a force of nature—it's the core of the marketing culture, and it's designed
to have certain effects. It's not trying to empower you. You

don't find messages on TV about how to join a union and do something about the conditions of your life. Over and over again, it rams into your head messages designed to destroy your mind and separate you from other people. That eventually has an effect.

What's happening with TV is part of something much broader. Elites always regard democracy as a major threat, something to be defended against. It's been well understood for a long time that the best defense against democracy is to distract people. That's why 19th-century businessmen sponsored evangelical religion, people talking in tongues, etc.

Kids are watching TV forty hours a week. It's a form of pacification.

It is a kind of pacification program.

More money, fewer voters

Clinton said the 1996 elections were a vindication of "the vital center," which he locates somewhere between "overheated liberalism and chilly conservatism." What was your reading of these elections?

Was there any choice other than "the vital center"? Clinton and Dole behaved slightly differently, and had somewhat different constituencies, but both were moderate Republicans, old-time government insiders and more or less interchangeable representatives of the business community.

I think the election was a vote *against* the vital center. Both candidates were unpopular and very few people expected anything from either one of them. Voter turnout was 49%—as low as it's ever been—and I think that reflected the general feeling that the political system isn't functioning.

I thought the turnout was the lowest since 1924.

1924 is misleading, because it was the first year women were allowed to vote. A smaller percentage of the electorate voted simply because a lot of women weren't used to voting and didn't do it the first time around. If you take that into account, 1996 may have been the lowest voter turnout ever.

241

The 1996 campaign also cost the most ever—$1.6 billion that we know about. More and more money is being spent, and fewer and fewer people are voting.

As one of the television commentators pointed out, these weren't conventions—they were coronations. It's just another step towards eliminating whatever functioning elements remain in formal democracy, and is all part of the general business attack on freedom, markets and democracy.

Compare Haiti, the poorest country in the hemisphere. The creation of a vibrant, lively, independent civil society there during the last few years has been remarkable, and was the basis for a remarkable triumph of democracy (which was extinguished very quickly and brutally with US help, and in a way that bars its revival).

If there were an independent intelligentsia in the US, they'd be falling off their chairs laughing at the idea that we have something to teach Haiti about democracy. Civil society is collapsing here. We have to go *there* to learn something about democracy.

Another commentator compared elections to auctions, with the prize going to the highest bidder.

They've never been much different from that, but yes, they're getting worse. On the other hand, if the public responds—if, for example, union organizing increases and grassroots organizations develop—things will change. The first change will be the political establishment saying, *Okay, we'll be more benevolent autocrats*. If they're pressured beyond that, we can get significant social change.

Most people realize that the political parties don't care about them. Public disaffection is enormous, but it's mostly directed against government. That's because business propaganda, which dominates the media, directs it that way. There may also be a lot of disaffection with business, but we don't really know, since that kind of question isn't asked much in the polls.

What's your take on campaign finance reform?

It's not a bad thing, but it's not going to have much effect. There are too many ways to cheat. It's like pretending to try to stop drug importation. There are so many ways to bring drugs in that there's no stopping them.

The real problem isn't campaign financing—it's the overwhelming power corporate tyrannies wield. Campaign finance reform isn't going to change that.

Is corporate power invincible?

Let me run a couple of quotes by you. The first is from Robert Reich, Clinton's former secretary of labor: "The jury is still out on whether the traditional union is necessary for the new workplace." The second is from Clinton's former commerce secretary, the late Ron Brown: "Unions are OK where they are, and where they're not, it's not clear yet what sort of organization should represent workers."

I think that's not surprising, coming from what amounted to a moderate Republican administration. Why let working people have ways to defend themselves against private power?

Maybe something else is needed in the high-tech workplace—"flexibility," which is a fancy way of saying that when you go to sleep at night, you don't know if you'll have a job in the morning (but you *do* know you won't have benefits). "Flexibility" is terrific for profits, but it destroys human beings.

There was a famous quote—at it least it should be famous—by a Brazilian general (around 1970, I think). Speaking of the Brazilian "economic miracle," he said, *The economy is doing fine—it's just the people that aren't*. That pretty much says it all.

Something about this puzzles me. It's in corporations' interest to make sure consumers have enough money to buy their products. This was the logic behind Henry Ford's raising his workers' pay to $5 a day, so that they could afford to buy the cars they were building.

It's in your interest to make profit, but there are other ways to do it than by selling a large quantity of goods to a mass market that's partially made up of your own workers. Maybe it's more in your interest to use extremely cheap, oppressed labor to produce fewer goods for relatively wealthy people, while at the same time making money through financial speculation.

When the managers of transnational corporations are asked about the very low wages they pay their workers in the Third World, they say, *These people didn't have a job before, we're giving them work, they're learning a trade,* and so on. How would you respond to that?

If they're serious about that, they would use some of their profits to support better working conditions in Indonesia. How often do they do that? They're not short of money—just read the *Fortune* 500 reports every year.

By the way, I'm not criticizing corporate executives individually. If one of them tried to use corporate funds to improve working conditions in Indonesia, he'd be out on his ear in three seconds. In fact, it would probably be illegal.

A corporate executive's responsibility is to his stockholders—to maximize profit, market share and power. If he can do that by paying starvation wages to women who'll die in a couple of years because their working conditions are so horrible, he's just doing his job. It's the *job* that should be questioned.

Aren't corporate managers quick to adjust and make small concessions, like letting people go the bathroom twice a day instead of once?

Absolutely. The same was true of kings and princes—they made plenty of concessions when they weren't able to control their subjects. The same was true of slave owners.

Small concessions are all to the good. People in the Third World may suffer a little less, and people here may see that activism can work, which will inspire them to push farther. Both are good outcomes. Eventually you get to the point where you start asking, *Why should we be asking them to make concessions? Why are they in power in the first place? What do we need the king for?*

I was recently in Trinidad, which is under "structural adjustment." While talking to some laborers, I asked them how they got to their job site. They said they had to take a taxi. I asked, *Isn't there any bus service?* and they told me that the route from the poor part of Port of Spain where they lived had been eliminated, and they now had to pay a substantial part of their earnings on private taxis.

It's happening everywhere. Transferring costs from the rich to the poor is the standard device of improving "efficiency."

I drove to work this morning. The roads are full of potholes, and there were big traffic jams, but it's hard to use public transportation, because it takes too long and is, in fact, more expensive than driving.

Depriving people of an alternative to driving forces them to buy more cars and more gas. Potholes increase car repairs and purchases. More driving increases pollution, and dealing with the health effects of that pollution costs even more money.

All the discomfort of all these people increases the gross national product (allowing celebration of the great economy) and is highly efficient from the point of view of the corporations who own the place. The costs that are transferred to the public, like the taxi fares those poor workers in Trinidad have to pay, aren't measured.

Los Angeles had a very extensive public transportation network that was simply bought up and destroyed.

Yes, and the same was true around here. Earlier in this century, you could get all around New England via electric railways.

Why do we have a society where everyone has to drive a car, live out in the suburbs, go to big shopping malls? In the 1950s, the government began a huge highway construction program called the National Defense Highway System. They had to put in the word "Defense" to justify the huge sums they were pouring into it, but in effect, it was a way of shifting from public transportation like railroads to a system that would use more automobiles, trucks, gasoline and tires (or airplanes).

It was part of one of the biggest social engineering projects in history, and it was initiated by a true conspiracy. General Motors, Firestone Tire and Standard Oil of California (Chevron) simply bought up and destroyed the public transportation system in Los Angeles, in order to force people to use their products.

The issue went to court, the corporations were fined a few thousand dollars, and then the government took over the whole process. The same happened elsewhere. State and local governments also joined in, and a wide range of business power. It's had enormous effects, and it certainly didn't happen by market principles.

It's still happening. One new plan in Boston is to dismantle parts of the public transportation system and privatize them—to make them more "efficient" (they claim) by letting private tyrannies run them. It's obvious what they'll do. If you're the head of a corporation that runs the transportation system and your responsibility is to make sure your stockholders make money, what would you do? Cut off unprofitable routes, get rid of unions, etc.

There's quite a bit of activism against sweatshops that transnationals like The Gap, Disney, Nike, Reebok, etc. profit from. Do you think these campaigns are getting to systemic issues?

I think they're really good campaigns. To ask whether they're getting to systemic issues is, I think, misleading—the kind of question that undermined a lot of traditional Marxist politics.

Systemic questions grow out of people learning more and more about how the world works, step-by-step. If you become aware that people in Haiti are being paid a couple of cents an hour to make money for rich people here, that ultimately—and maybe a lot sooner than ultimately—leads to questions about the structure of power in general.

The current economic system appears to be triumphant, but you've said that it's going to self-destruct—that that's inherent in its logic. Do you still feel that way?

I actually said something different. The current system has elements in it that look like they're going to self-destruct. But it's unclear whether the whole world is going to turn into something like a Third World country where wealth is highly concentrated, resources are used to protect the wealthy, and the general public finds itself somewhere between unpleasantness and actual misery.

I don't think that kind of world can survive very long, but I can't prove it. It's kind of an experiment. Nobody knows the answer, because nobody understands these things well enough.

Opinion polls show how much people dislike this system. When *Business Week* surveyed public attitudes towards business, they were pretty startled by the results. 95% of the people—that's a number you almost never see in a poll—said corporations have a responsibility to reduce profit for the benefit of their workers and the communities they do business in. 70% thought businesses have too much power, and roughly the same number thought business has gained more by deregulation and similar measures than the general population has.

Other studies taken around the same time show that over 80% of the population think that working people don't have enough say in what goes on, that the economic system is inherently unfair, and that the government basically isn't functioning, because it's working for the rich.

The poll questions still fall way short of what working people in eastern Massachusetts (and elsewhere) were asking for about 150 years ago. They weren't saying: *Be a little more benevolent. Give us few scraps.* They were saying: *You have no right to rule. We should own the factories. The people who work in the mills ought to own them.*

Many people today just want business to be a bit nicer, for there to be a little less corporate welfare and a little more welfare capitalism. But others would like to see more radical changes; we don't know how many, because the polls don't ask about radical alternatives, and they aren't readily available for people to think about.

People are tremendously cynical about institutions. A lot of this cynicism takes very antisocial and irrational forms, and the amount of propaganda and manipulation is so enormous that most people don't see alternatives, but the attitudes that might lead to acceptance—even enthusiastic acceptance—of alternatives are just below the surface.

You can see it in their actions—both destructive, like selling drugs in the streets, and constructive, like the strikes in South Korea. What South Korean workers consider totally intolerable is the idea that private power should have the right to replace strikers with permanent replacement workers. And they're right—that's against international labor standards.

There *is* a country that's been censured by the International Labor Organization for carrying out those practices—the US. That tells us something about who's civilized and who isn't.

People concerned about corporate power and its excesses are urged to invest in "socially responsible businesses." What do you think of that?

I have no criticism of that idea, but people shouldn't have any illusions about it. It's like preferring benevolent autocrats to murderous ones. Sometimes you get a benevolent ruler, but he can always stop being benevolent whenever he feels like it. Sure, I'd rather have an autocrat who doesn't go around torturing children, but it's the autocracy itself that needs to be eliminated.

Richard Grossman, Ward Morehouse and others have been advocating the revocation of corporate charters [the documents that create corporations and allow them to conduct business]. I'm wondering how realistic this is. This would have to happen in state legislatures, which are almost entirely under the control of big business.

I certainly think people should begin to question the legitimacy of corporate institutions. In their current form, they're a rather recent phenomenon; their rights were created, mostly by the judicial system, in the late 1800s and were dramatically expanded early in this century.

In my view, corporations are illegitimate institutions of tyrannical power, with intellectual roots not unlike those of fascism and Bolshevism. (There was a time when that kind of analysis wasn't uncommon—for example, in the work of political economist Robert Brady over fifty years ago. It has very deep roots in working-class movements, Enlightenment thought and classical liberalism.)

There are, as you point out, legal mechanisms for dissolving corporations, since they all have to have state charters. But let's not delude ourselves—these are massive changes. Just suggesting charter revocation as a tactic doesn't make any sense—it can only be considered after legislatures reflect the public interest instead of business interests, and that will require very substantial education and organization, and construction of alternative institutions to run the economy more democratically.

But we can—and should—certainly begin pointing out that corporations are fundamentally illegitimate, and that they don't have to exist at all in their modern form. Just as other oppressive institutions—slavery, say, or royalty—have been changed or eliminated, so corporate power can be changed or eliminated. What are the limits? There aren't any. Everything is ultimately under public control.

Is globalization inevitable?

Germany has unemployment levels not seen there since 1933. Companies like Siemens and Bosch are closing down their German factories and moving overseas. You've commented on Daimler-Benz's operations in Alabama and BMW's in South Carolina.

German industry has been treating the US as a Third World country for several years. Wages are low here, benefits are poor and the states compete against each other to bribe foreign companies to relocate. German unions have been trying to join with American ones to work on this problem, which hurts them both.

I suspect that the collapse of the Soviet empire has a lot to do with this. As was predictable, its main significance has been to return most of Eastern Europe to what it had been for five hundred years before—the original Third World. Areas that were part of the West—like the Czech Republic and western Poland—will end up resembling Western Europe, but most of Eastern Europe was submerged in deep Third World poverty, and they're going back to a kind of service role.

A while back, the *Financial Times* ran an article under the headline "Green Shoots in Communism's Ruins." The green shoots were Western European industrialists' ability to pay Eastern European laborers much less than they pay "pampered western workers" with their "luxurious lifestyles" (as *Business Week* put it in another article).

Now they can get workers who are well-educated, because Communism did do a good job with that—even white and blue-eyed, though no one says that openly. They're also pretty healthy—maybe not for long, because the healthcare systems are declining—but for a while, at least. And there's reasonable infrastructure.

Western companies typically insist on plenty of state protection, so when General Motors or VW invests in an auto plant in Poland or the Czech Republic, they insist on substantial market share, subsidies, protection, etc.—just as they do when they move into a Third World country or the US.

George Soros, the billionaire financier, has written several articles expressing his view that the spread of brutal global capitalism has replaced communism as the main threat to democratic societies.

It's not a new point. Working people 150 years ago were struggling against the rise of a system they saw as a great threat to their freedom, their rights and their culture. They were, of course, correct, and Soros is correct insofar as he reiterates that view.

On the other hand, he also makes the common assumption that the market system is spreading, which just isn't true. What's spreading is a kind of corporate mercantilism that's supported by—and crucially relies on—large-scale state power. Soros made his money by financial speculations that become possible when telecommunications innovations and the government's destruction of the Bretton Woods system (which regulated currencies and capital flow) allowed for very rapid transfers of capital. That isn't global capitalism.

As we sit here, the World Economic Forum is being held in Davos, Switzerland. It's a six-day meeting of political and corporate elites, with people like Bill Gates, John Welch of GE, Benjamin Netanyahu, Newt Gingrich and so on.

The companies represented at this forum do something like $4.5 trillion worth of business a year. Do you think it's a significant event that we should pay attention to?

Sure, we should pay attention to it, but I frankly wouldn't expect anything to come out of it that's not pretty obvious. Whether or not there's anything serious being discussed there, what reaches us will be mostly vacuous rhetoric.

We should also pay attention to the Trilateral Commission, but when you read its reports, they're rather predictable. The only really interesting thing I've ever seen from them was their first book—not because they were saying anything new, but because they were saying it so openly.

It's unusual to see an almost hysterical fear of democracy and a call for repressive measures to combat it expressed so explicitly. I suspect that's why the book was taken off the market as soon as it got to be noticed. I don't think it was meant to be read beyond select circles.

The Trilateral Commission, the Council on Foreign Relations and the like reflect a kind of consensus among business power, government power and intellectuals who aren't too far out of line. (They try to bring in other elements too; for instance, John Sweeney, president of the AFL-CIO, was at the Davos conference. They'd very much like to co-opt labor leadership, as they've done in the past.) There's plenty of evidence about what their views and goals are, and *why* they're their views and goals.

So you don't see any dark conspiracies at work in these organizations.

Having a forum in Switzerland would certainly be a pretty dumb way to plan a conspiracy.

I don't deny that there sometimes are conspiracies, by the way. In 1956, Britain, France and Israel planned an invasion of Egypt in secret. You can call that a conspiracy if you like, but it was really just a strategic alliance among huge power centers.

Admiral William Owens [former vice-chair of the Joint Chiefs of Staff] and Joseph Nye [former Clinton Defense Department official who's now dean of the Kennedy School at Harvard] predict that the 21st century will be "the US century" because the US dominates world media, the internet and telecommunications.

They also say that the US has an unrecognized "force multiplier" in its international diplomacy and actions, which comes from worldwide recognition of American democracy and free markets. They cite telecommunications and information technology, both textbook examples of how the public has been deluded into subsidizing private power.

The public assumes the risks and the costs, and is told it's defending itself against foreign enemies. *That's* supposed to be an illustration of democracy and markets. The delusion is so ingrained that nobody even comments on it.

Through Hollywood films and videos, TV and satellites, American culture is coming to dominate global culture.

When India began opening up its economy and American corporations were able to really start moving in, the first domain they took over was advertising. Very quickly, Indian advertising agencies became subsidiaries of big foreign ones, mostly based in the US.

The public relations industry has always aimed "to regiment the public mind every bit as much as an army regiments the bodies of its soldiers"—in the case of India, to create a system of expectations and preferences that will lead them to prefer foreign commodities to domestic ones.

There's been some resistance to this in India—massive demonstrations around Kentucky Fried Chicken, for example.

That's true in many places, even within Europe. There are moves towards creating a common European popular culture, common media and so on, making society more homogenous and controlled, but there are also moves in exactly the opposite direction—towards regionalization and the reviving of individual cultures and languages. These two movements are going on side-by-side, all over the world.

The US has created a global culture, but it's also created resistance to it. It's no more an inevitable process than any of the others.

In the last couple of years you've visited Australia, India, South America. What have you learned from your travels?

It's not hard to find out what's going on just sitting here in Boston.

But then you're just dealing with words on paper.

You're right—the colors become a lot more vivid when you actually see it. It's one thing to read the figures about poverty in India and another thing to actually see the slums in Bombay and see people living in hideous, indescribable poverty...and these are people who *have* jobs—they're manufacturing fancy leather clothes that sell on Madison Avenue and in shops in London and Paris.

It's a similar story throughout the world. But if you walk through downtown Boston, you'll also see appalling poverty. I've seen things in New York that are as horrifying as anything I've seen in the Third World.

Comparable to the *favelas* [shantytown slums] in Brazil?

It's hard to say "comparable." The poverty and suffering in Haiti or Rio de Janeiro or Bombay is well beyond what we have here—although we're moving in that direction. (As you know, black males in Harlem have roughly the same mortality rate as men in Bangladesh.)

But psychological effects are also crucially significant—how bad conditions seem depends on what else is around. If you're much poorer than other people in your society, that harms your health in detectable ways, even by gross measures like life expectancy.

So I'd say that there are parts of New York or Boston that are similar to what you find in the Third World. A Stone Age person could be very happy without a computer or a TV, and no doubt the people in the *favelas* live better than Stone Age people by a lot of measures—although they probably aren't as well-nourished or healthy.

But going back to your earlier point, seeing things firsthand gives them a vividness and significance you don't get by reading, and you also discover a lot of things that are never written about—like the way popular struggles are dealing with problems.

How can we organize against globalization and the growing power of transnational corporations?

It depends what time range you're thinking of. You read constantly that globalization is somehow inevitable. In the *New York Times,* Thomas Friedman mocks people who say there are ways to stop it.

According to him, it's not hawks and doves any more—there's a new dichotomy in the ideological system, between *integrationists,* who want to accelerate globalization, and *anti-integrationists,* who want to slow it down or modulate it. Within each group, there are those who believe in a safety net and those who believe people should be out on their own. That creates four categories.

He uses the Zapatistas as an example of the anti-integrationist pro-safety-net position, and Ross Perot as an example of the anti-integrationist anti-safety-net position, and dismisses them both as crazy. That leaves the two "sensible" positions, which are illustrated by Clinton (integrationist pro-safety-net) and Gingrich (integrationist anti-safety-net).

To test Friedman's analysis, let's look at Gingrich. To see if he represents maximization of free markets and undermining of safety nets, let's ask if he opposed the Reagan administration when it carried out the most protectionist policies since the 1930s? Did he object when Lockheed, his favorite cash cow, got big public subsidies for its merger with Martin Marietta? Did he resist the closing off of American markets to Japan, so our automotive, steel and semiconductor industries could reconstruct?

As these questions make clear, Gingrich is not an integrationist. He simply wants globalization when it's good for the people he's paid to represent, and not when it isn't.

What about safety nets? If he's opposed to welfare dependency, then he should certainly be opposed to providing federal subsidies to his constituents. But, in fact, he's a champion at bringing them home to his district.

So it's easy to see that Friedman's picture is mostly mythology. The fact that he can get away with it is the only interesting part of the story. The same is true of his belief that globalization is like a law of nature.

For one thing, in terms of gross measures like trade and investment flow (relative to the economy), globalization is more or less just getting back to where it was early in the century. (This is well known and has been pointed out in quite mainstream circles.)

There are also new factors. Capital flows are extremely fast and huge in scale. That's the result of two things: the telecommunications revolution (which is largely just another gift of publicly developed technology to private businesses), and the decision, during the Nixon administration, to break down the Bretton Woods system. But there's nothing *inevitable* about either—especially not the particular forms they've taken.

Also remember that huge corporations depend very extensively on their own states. Every single one of the companies on the *Fortune 100* list of the largest transnational corporations has benefited from interventionist industrial policies on the part of the countries in which they're based, and more than 20 wouldn't have even survived if it weren't for public bailouts.

About two-thirds of the international financial transactions take place within and between Europe, the US and Japan. In each of those places, parliamentary institutions are more or less functioning, and in none of them is there any danger of a military coup. That means it's possible to control, modify and even eliminate the supposedly uncontrollable forces driving us toward a globalized economy, even without substantial institutional change.

The myth of Third World debt

All over the world, but especially in the US, many workers vote against their own interests—assuming they vote at all.

I'm not sure that's true. Neither major party here represents workers' interests, but suppose there were candidates who did, and that US workers trusted them and were confident they'd try to do exactly what the workers wanted. There still might be a good reason not to vote for them.

When poor people in Central America vote for their own interests, the result is terror—organized and directed by the superpower of the hemisphere, and supervised on the local level by the upper classes of that country. Many countries are so weak that they can't really solve their internal problems in the face of US power; they can't even control their own wealthy. Their rich have virtually no social obligations—they don't pay taxes and don't keep their money in the country.

Unless these problems are dealt with, poor people will sometimes choose to vote for oppressors, rather than suffer the violence of the rich (which can take the form of terror and torture, or can simply be a matter of sending the country's capital somewhere else).

Is capital flight a serious problem?

Not so much in the US, though even here the threat is able to constrain government planning (Clinton in 1993 is a well-known case). But look at virtually any country south of the Rio Grande. Take Brazil.

As happened almost everywhere in the Third World, Brazil's generals, their cronies and the super-rich borrowed huge amounts of money and sent much of it abroad. The need to pay off that debt is a stranglehold that prevents Brazil from doing anything to solve its problems; it's what limits social spending and equitable, sustainable development.

But if I borrow money and send it to a Swiss bank, and then can't pay my creditors, is that your problem or mine? The people in the slums didn't borrow the money, nor did the landless workers. In my view, it's no more the debt of 90% of the people of Brazil than it is the man in the moon's.

Discussions about a debt *moratorium* are not really the main point. If the wealthy of Brazil hadn't been out of control, Brazil wouldn't have the debt in the first place. Let the people who *borrowed* the money pay it back. It's nobody else's problem.

I discussed these matters all over Brazil—with poor people, at the national bishops' conference, with elite television reporters and high officials. They didn't consider it very surprising. In educated circles here, you could hardly get the basic issues taken seriously. One of the very striking differences you notice as soon as you get out of the First World is that minds are much less open here. We live in a highly indoctrinated society.

Breaking out of doctrinal shackles isn't easy. When you have as much wealth and power as we do, you can be blind and self-righteous; you don't have to think about anything. In the Third World, even wealthy and powerful people tend to have much more open minds.

Why hasn't foreign debt held back the developing countries of East Asia?

Japan, South Korea and Taiwan not only controlled labor and the poor, but also capital and the rich. Their debt went for internal investment, not export of capital.

Japan didn't allow export of capital until its economy had already reconstructed. South Korea didn't either, until forced to remove capital controls and regulation of private borrowing, largely under US pressure, in very recent years. (It's widely recognized that this forced liberalization was a significant factor in South Korea's 1997 liquidity crisis.)

Latin America has the worst income inequality in the world, and East Asia has perhaps the least. Latin America's typical imports are luxury goods for the wealthy; East Asia's have been mostly related to capital investment and technology transfer. Countries like Brazil and Argentina are potentially rich and powerful, but unless they can somehow gain control over their wealthy, they're always going to be in trouble.

Of course, you can't really talk about these countries as a whole. There are different groups within them, and for some of these groups, the current situation is great—just as there were people in India who thought the British Empire was fine. They were linked to it, enriched themselves through it, and loved it.

It's possible to live in the poorest countries and be in very privileged surroundings all the time. Go to, say, Egypt, take a limousine from the fancy airport to your five-star hotel by the Nile, go to the right restaurants, and you'll barely be aware that there are poor people in Cairo.

You might see some out the car windows when you're driving along, but you don't notice them particularly. It's the same in New York—you can somehow ignore the fact that there are homeless people sleeping in the streets and hungry children a couple of blocks away.

Mexico, Cuba and Guatemala

William Greider's book *One World, Ready or Not* describes the appalling economic conditions in Mexico. He says the country is very explosive, politically and socially.

That's absolutely correct. Throughout the 1980s, wages fell (it depends on how you measure them, but they were roughly cut in half, and they weren't high before that). Starvation increased, but so did the number of billionaires (mostly friends of the political leaders who picked up public assets for a few pennies on the dollar). Things finally collapsed in December 1994, and Mexico went into the worst recession of its history. Wages, already poor, declined radically.

A journalist I know at a Mexican daily called to interview me after the collapse. He reminded me of some interview of mine from a couple of months earlier where I'd said that the whole economy was going to fall apart.

I don't know much about Mexico or economics, but it was pretty obvious. Very short-term speculative funds were pouring in, and the speculative bubble had no basis. The economy was actually declining. Everybody could see this, including the economists at the international financial institutions, who (according to some specialists) kept it quiet because they didn't want to trigger the collapse.

Mexico was the star pupil. It did everything right, and religiously followed the World Bank and IMF prescriptions. It was called another great economic miracle, and it probably was... for the rich. But for most of the Mexican people, it's been a complete disaster.

What do you hear from the Zapatistas?

Negotiations have been stalled for a couple of years, but I think it's clear what the government's strategy is: continue negotiations which won't get anywhere and ultimately, when the Zapatistas lose their capacity to arouse international interest, when people get tired of signing petitions—then the government will move in with force and wipe the Zapatistas out. That's my suspicion, anyway.

I think the only reason they didn't wipe them out right away is because the Zapatistas had so much popular support throughout Mexico and the world (which they managed to garner with a good deal of imagination). The fact that they've been able to remain in opposition for several years is pretty remarkable in itself.

But as it stands, it doesn't seem to me that they have any sort of a winning strategy. I don't say that as a criticism—I can't think of one either. Unless international support becomes really significant, I don't see how their position can be maintained.

What's happening with Cuba? A lot of people were bewildered when David Rockefeller [grandson of John D. and former chairman of the Chase Manhattan Bank] gave a party for Fidel Castro in New York in October 1995.

Cuba itself isn't of tremendous importance to the American economy. If it didn't exist, the effect wouldn't be noticeable. But the idea that other competitors are making inroads in this traditionally American market doesn't appeal to David Rockefeller and his friends. If investors elsewhere are going to break the American embargo, business here is going to call for it to end.

The same thing happened with Vietnam. US business was perfectly happy to punish Vietnam for failing to totally capitulate to US power. They would have kept their stranglehold on forever, dreaming up one fraudulent reason after another, except that by the mid-1980s Japan and other countries were starting to disregard the US embargo and move into the area, which has an educated population and low labor costs.

You followed the Jennifer Harbury case in Guatemala.

I wrote the introduction to her book, *Bridge of Courage*. She's a very courageous woman, and is still fighting. Sister Dianna Ortiz is another. It takes a lot of guts to do what these women have done.

Does the Guatemala peace treaty of December 1996 signal the end to this three-decade-old bloodbath?

I'm glad it's being signed, because it's a step forward. But it's also the very ugly outcome of one of the biggest state terror operations of this century, which started in 1954 when the US took part in overthrowing the only democratic government Guatemala ever had.

Let's hope the treaties may put an end to the real horrors. State terror has successfully intimidated people, devastated serious opposition, and made a government of right-wing business interests not only seem acceptable to many people, but even desirable.

Brazil, Argentina and Chile

What kind of contact did you have with the media in Brazil, Argentina and Chile?

I immediately had a lot of contact with the mass media. That happens almost everywhere except in the US.

State television and radio?

Commercial stations too. The mass media are a lot more open there.

What about independent media?

There's an independent left journal published in São Paulo. It's in Portuguese, so I have only a superficial sense of what's in it, but the material looks extremely interesting. The journal is very well-designed and well-printed, as professional as *Harper's* or the *Atlantic*. We don't have anything like it here.

There are also more popular efforts. My wife and I spent an evening in one of the biggest of Rio's suburbs, Nova Iguaçu, where several million people—a mixture of poor, working-class, unemployed and landless peasants—live. (Unlike here, the rich live in the center of most Latin American cities, and the poor in the suburbs.) We were warned that we shouldn't go to Nova Iguaçu—too dangerous—but the people there were perfectly friendly.

We went with people from an NGO [nongovernmental (nonprofit) organization]—progressive artists, professionals and intellectuals who want to provide the population with an alternative to having their minds destroyed by commercial television. Their idea was to drive a truck with a huge screen into some public area and show documentaries dealing with real problems.

They spent a fair amount of time with the leaders of popular organizations in the community, figuring out how to make their points accessible, and how to put some humor in. I haven't seen the films, but apparently they were very well-done. But when they showed them in the poor neighborhoods, they completely bombed. People came by to check them out, watched for a while and left.

When the NGO did wrap-up sessions to try to figure out what happened, they discovered something very interesting: the leaders in the community spoke a different dialect, full of intellectual words and Marxist rhetoric, than the people they lived among. The process that made them leaders had also drawn them out of the mainstream of the community.

So the NGO went back, and this time they avoided the community leaders and tried to get members of the community—sixteen-year-old kids and the like—interested in writing the scripts and making the films. It wasn't easy, but it worked.

By the time we visited, which was a couple of years later, the NGO simply brought in the truck and the big screen. The people in the community—mostly young, but not entirely—wrote, shot and acted in the films themselves. They got a little technical assistance from the urban professionals, but essentially nothing else.

There was a big screen in the middle of a public area with little bars around. Lots and lots of people from the community were there—children and old people, racially mixed. It was in prime television time, nine o'clock in the evening. The people watching were obviously very much engaged in what was happening.

The dialog was in Portuguese, so I couldn't understand a lot of it, but I got enough to see that they were dealing with quite serious issues—although with humor and clowns mixed in. There was a skit on racism. (In theory, there isn't supposed to be any in Brazil.)

A black person would go to an office and ask for a job, then a white person would do the same, and of course they were treated totally differently. Everybody in the audience was laughing and making comments. There was a segment on AIDS, and something about the debt.

Right after the films ended, one of the actresses—who was quite good and looked about seventeen (at most)—started walking around the audience with a microphone, interviewing people about what they'd just seen. Their comments and criticisms were filmed live, eliciting more reactions.

This is very impressive community-based media of a sort that I've never seen before, accomplished in spite of the initial failure I described. It was in an extremely poor area. It was an experience I'm sure I would never have read about in a book.

We saw something similar in Buenos Aires. Some friends from the university took my wife and me to a shantytown where they work as activists. It's a very poor community in a very rich city; most of its inhabitants are Guaraní, indigenous people who migrated there from Paraguay.

School facilities there are awful, and any kid who causes even a small problem is just kicked out. An enormous number of the kids never make it through school. So some mothers set up what they call a cultural center, where they try to teach these kids reading and arithmetic, basic skills and a little artwork, and also try to protect them from drug gangs. (It's very typical in such communities for women to do most of the organizing.)

261

Somehow they managed to find a small, abandoned concrete building and put a roof on it. It's kind of pathetic—about the size of this office. The provisions are so meager that even a pencil is a significant gift.

They also put out a journal. Written by the people in the shantytown, including some teenagers, it's full of information relevant to the community—what's going on, what the problems are.

Several of the women are becoming educated; a few are close to college degrees in professions like nursing. But they all say they'll never get out of the shantytown, no matter how many degrees they have. They haven't got a chance when they go for a job interview because they don't have the right clothes, the right look.

These activists are dedicated and they work hard, trying to save the children. They get some assistance from outside people, like those university friends of ours. The church also helps some. (This varies from community to community, depending on who the local priests are.)

They don't get any help from the government, I assume?

The Argentine government is in the grips of a neoliberal frenzy, obeying the orders of international financial institutions like the World Bank and the IMF. (*Neoliberalism* is basically nothing more than the traditional imperial formula: free markets for you, plenty of protection for me. The rich themselves would never accept these policies, but they're happy to impose them on the poor.)

So Argentina is "minimizing the state"—cutting down public expenditures, the way our government is doing, but much more extremely. Of course, when you minimize the state, you maximize something else—and it isn't popular control. What gets maximized is private power, domestic and foreign.

I met with a very lively anarchist movement in Buenos Aires, and with other anarchist groups as far away as northeastern Brazil, where nobody even knew they existed. We had a lot of discussions about these matters. They recognize that they have to try to use the state—even though they regard it as totally illegitimate.

The reason is perfectly obvious: When you eliminate the one institutional structure in which people can participate to some extent—namely the government—you're simply handing over power to unaccountable private tyrannies that are much worse. So you have to make use of the state, all the time recognizing that you ultimately want to eliminate it.

Some of the rural workers in Brazil have an interesting slogan. They say their immediate task is "expanding the floor of the cage." They understand that they're trapped inside a cage, but realize that protecting it when it's under attack from even worse predators on the outside, and extending the limits of what the cage will allow, are both essential preliminaries to dismantling it. If they attack the cage directly when they're so vulnerable, they'll get murdered.

That's something anyone ought to be able to understand who can keep two ideas in their head at once, but some people here in the US tend to be so rigid and doctrinaire that they don't understand the point. But unless the left here is willing to tolerate that level of complexity, we're not going to be of any use to people who are suffering and need our help—or, for that matter, to ourselves.

In Brazil and Argentina, you can discuss these issues even with people in the highest political echelons, and with elite journalists and intellectuals. They may not agree with you, but at least they understand what you're talking about.

There are now organizations of landless peasants in Brazil.

Brazil has an enormous agrarian problem. Land ownership is highly concentrated, incredibly unequal, and an enormous amount of land is unused, typically because it's being held as a hedge against inflation or for investment purposes.

A very big and important organization, the Landless Workers' Movement, has taken over a lot of land. It has close links to the people in the *favelas,* who were mostly driven off their land too.

Brazil's army is very brutal, even more so since the coup of 1964. There's lots of killing and violence, one striking example being the murder of a couple of dozen peasants who took over some land in one of the northern regions. When I was in Brazil, informal judicial proceedings were being held about these murders, because the formal judicial system hadn't done anything about them.

You met with people in the Workers' Party.

It was very interesting. Brazil's Workers' Party is the largest labor-based party in the world. It has its problems, but it's an impressive organization with a radical democratic and socialist thrust, a lot of popular support and lots of potential. It's doing many important and exciting things.

Lula [Luis Inácio Lula da Silva, 1944– , founder and leader of the Workers' Party and president of Brazil, 2003–2010] is extremely impressive. If Brazil's presidential elections were even remotely fair, he would have won them. (It's not so much that votes were stolen but that the media resources were so overwhelmingly on the other side that there wasn't a serious election.)

Many workers have also become organized into rural unions, which are very rarely discussed. There's some degree of coop-eration between the landless workers and groups in the *fave-las*. Both are linked in some fashion to the Workers' Party, but the people I asked couldn't say exactly how. Everyone agrees that most of the landless workers support the Workers' Party, and vote for it, but organizationally they're separate.

What were your impressions of Chile?

I wasn't there long enough to get much of an impression, but it's very clearly a country under military rule. We call it a democracy, but the military sets very narrow bounds on what can happen. You can see it in people's attitudes—they know there are limits they can't go beyond, and in private they tell you that, with many personal examples.

The Mideast

About 1980, you, Eqbal Ahmad [Pakistani scholar and activist, and professor at Hampshire College] and Edward Said [noted author, Palestinian activist and professor at Columbia] had a meeting with some top PLO officials. You've said you found this meeting rather revealing.

Revealing, but not surprising. It confirmed some very critical comments I'd made about the PLO in left journals a few years earlier, and which there was a big dispute over. The meeting was an attempt to make the PLO leadership, which happened to be visiting New York, aware of the views of a number of people who were very sympathetic to the Palestinians but quite critical of the PLO.

The PLO leadership wasn't interested. It's the only Third World movement I've ever had anything to do with that made no effort to build any kind of solidarity movement here, or to gain sympathy in the US for its goals.

It was extremely hard to get anything critical of Israel published, let alone distributed. The PLO could easily have helped, simply by buying books and sending them to libraries, but they were completely unwilling to do anything. They had huge amounts of money—they were brokering big deals between Kuwait and Hungary and who knows who else—but it was a very corrupt organization.

They insisted on portraying themselves as flaming revolutionaries, waving guns...which of course is going to alienate everyone. If they'd portrayed themselves as what they actually were—conservative nationalists who wanted to make money and maybe elect their own mayors—it would have increased the support in the United States for a Palestinian state from about 2 to 1 to about 20 to 1.

I think they believed that politics isn't about what the general population thinks or does, but about deals you make in back rooms with powerful people. (Incidentally, I heard much harsher criticisms of the PLO from activists and leaders in the Occupied Territories when I was there a few years later.)

If, as you've said, Israel is the local cop on the beat in the Mideast, why did the US go to such lengths to keep it out of the Gulf War?

Because if Israel had become directly involved, it would have been impossible for the US to keep the passive support of the major oil-producing countries in the region, and that's all Washington was really concerned with. Certainly they didn't need Israel's support to fight a war against a virtually defenseless Third World country. After the war, the US reestablished its domination of the region very strongly and told everybody, "What we say, goes" (as George [H. W.] Bush put it).

Eqbal Ahmad is rather pessimistic about Israel's long-term future. He says that sooner or later the relative weakness of Arab states will change.

I don't think it makes a lot of sense to try to make predictions about the long-term future. You can imagine a future in which the US is an embattled island, barely able to hold its own against the emerging powers of Asia that surround it. But as far as I can see, the US has about as much control and domination of the Middle East as any outside force could hope to maintain.

Our outpost there, Israel, is by far the main military, technological, industrial and even financial center. The huge oil resources of the region (which are still going to be needed for another couple of generations) are mostly in the hands of family dictatorships, brutal tyrannies that are highly dependent on the US and subordinated to its interests.

It's quite possible that the system will break down in the long term—but if you're talking about, say, two centuries from now, the US isn't even going to care about Mideast oil by then. For the kind of time frame within which policy planning makes any sense—which isn't long—things are working out as well as US planners could possibly have imagined. If it turns out, at some far distant time, that Israel is no longer necessary for US purposes, our support for Israel will end.

You've held that view for a very long time. You don't see any reason to change it?

None at all; in fact, I think we've had more and more evidence of it. For example, when a tiny disagreement came up between Israel and the US about how openly settlement of the West Bank should be pursued, [The first] President Bush didn't hesitate to make thinly veiled antisemitic remarks in front of a public audience. The Israeli lobby backed off and the US did what it wanted.

This is from Edward Said: "The crisis in Palestinian ranks deepens almost daily. Security talks between Israel and the PLO are advertised as a 'breakthrough' one day, stalled and deadlocked the next. Deadlines agreed upon come and go with no other timetable proposed, while Israel increases...the building of settlement residences [and] the punitive measures keeping Palestinians from leaving the territories and entering Jerusalem." He wrote this years ago, but it reads like today's news.

It does. The "peace process" goes up and down because the US-Israeli principles that define it have never offered anything meaningful to the Palestinians. The basic structure of US and Israeli policy has been clear for a long time. The principles are, strictly speaking, "rejectionist"—that is, they reject the rights of one of the two contestants in the former Palestine.

In the US, the term "rejectionist" is used in a racist sense, applying only to those who reject the rights of Jews. If we can bring ourselves to adopt nonracist usage, we will describe the US as the leader of the rejectionist camp.

In December 1989, when the Bush-Baker administration was supposed to be very hostile to Israel, the State Department came out with the Baker Plan. It called for a "dialog" in which only Palestinians acceptable to Israel and the US could participate. Discussion would be limited to implementation of Israel's official Shamir-Peres plan, which stipulated that:

- there can be no "additional Palestinian state" (other than Jordan, they meant)

- Israel should have effective control of as much of the Occupied Territories as it wants (however much that turns out to be)

- it's possible to hold "free elections" in territories that are under Israeli military supervision and with most of the educated elite in prison.

That was official US policy, under an administration that was supposed to be anti-Israel. (It was never accurately reported here. I wrote about it at the time.) The US was finally able to achieve these goals after the Gulf War, when the rest of the world backed off.

Large sections of the West Bank and Gaza are still occupied by the Israeli army.

The Oslo II Interim Agreement of September 1995 left Israel in control of about 70% of the West Bank, and in effective control of about another 26%. It put the urban centers of Palestinian towns in the hands of the Palestinian Authority, which is subordinate to Israel. (It's as if the New York police didn't have to patrol the worst slums—the local authorities did that for them—while the people in power took everything they wanted.)

I think Israel has way too much territory for its own potential needs or interests, and thus will probably be willing to relinquish some. If Israel is smart, it will work towards something like the Allon plan of 1968, which gave it control of the resources, water and usable territory (about 40% of the West Bank, the Gaza Strip and other areas), while relinquishing responsibility for the population.

Since then, the Gaza Strip has been more of a burden than something you'd want to hang onto. I think Israel will keep the so-called Gush Katif, down in the south; along with other parts they control, that probably amounts to 30% of all Gaza. (This is all for a couple of thousand Jewish settlers who use most of the resources, particularly water.) Israel will probably build strings of tourist hotels and keep up agricultural exports.

They'd be out of their minds to want to control Gaza City. They'd much rather leave it to the Palestinian Authority, along with the other urban centers and maybe 100 or so dots scattered around the West Bank and Gaza, with impassable roads connecting them.

There's a big superhighway system, but that's for the use of Israeli settlers and visitors. You can travel through the West Bank on these superhighways and barely know that Palestinians exist; you might see a remote village somewhere and maybe somebody selling something on the roadside.

It's like the Bantustans in South Africa, except that—as Norman Finkelstein has pointed out—the South African government gave much more support to the Bantustans than Israel is giving to those scattered regions.

In the epilogue to the latest edition of your book *World Orders, Old and New,* you say that Israel will eventually give some kind of state status to the Palestinians.

Israel and the US would be really stupid if they don't call whatever they decide to leave to Palestinian jurisdiction a state, just as South Africa insisted on calling the Bantustans "states," even though virtually no other country would do so. This new Palestinian "state" will get international recognition, however, because the US makes the rules.

What about the issue of Hebron and the agreement of January 1997?

It left the settlers in place, which is exactly what everyone should have expected them to do. There's no way for Israel to maintain control of the overwhelmingly Arab areas; they'd much rather have Palestinian police and joint Israeli-Palestinian patrols do that.

In the Israeli press, Clinton has been called "the last Zionist."

That was several years ago, in response to positions he took that were more extreme than almost anyone in mainstream Israeli politics.

Netanyahu got a five-minute ovation when he told the US Congress that Jerusalem will be the eternal, united capital of Israel, prompting him to remark, "If only I could get the Knesset [Israel's parliament] to vote like this."

Since 1967, US opinion—including liberal opinion—has pretty much been aligned with the more extremist elements in Israel. For example, the takeover of Arab East Jerusalem has been really ugly (I give a lot of details in World Orders and elsewhere). What's now called Jerusalem is an area much bigger than anything that has ever been called Jerusalem in the past; in fact, it's a substantial part of the whole West Bank.

World opinion has repeatedly condemned this annexation as illegal. The US publicly agreed with this position, but meanwhile gave Israel authorization to do what it liked.

Much of the land annexation and Israeli settlements in Arab East Jerusalem is funded by money from the US.

Some of it's from American citizens, who probably are doing it tax-free (at least in part), which means that the rest of us are paying for it. Part of it comes from the US government, which again means that US taxpayers are financing it.

Theoretically, the US reduces its loan guarantees so as to exclude any funds spent for settling the West Bank, but the amount that's restricted is way below what's actually spent. Is-raelis know this is a joke—it's all over the Israeli press.

Furthermore, funds from the Jewish National Fund and sev-eral other so-called charitable organizations in the US also sup-port settlements in various ways (in part indirectly, by funding development programs in Israel for Jewish citizens only, so that government funds can be transferred to subsidize settlers and infrastructure). That's again at taxpayer expense (since contri-butions to these charities are tax-deductible). All together, it amounts to quite a lot of money.

Many of the most militant settlers in the West Bank and Gaza are from the US. Does the American Jewish community foster this kind of militancy?

The American Jewish community is split, but a large number of the right-wing Jewish terrorists and extremists in Israel come from America. The Israelis don't like it—they don't want terrorists in their own society.

It's gotten to the point where there were even proposals—not entirely in jest—to control immigration from the US. Even very mainstream Israelis were saying, *Look, they're just sending us the crazies they don't know how to take care of. We don't want them.*

But I don't think this is unique to the American Jewish com-munity. For whatever reason, diaspora communities tend to be, by and large, more extremist, chauvinistic and fanatic than people in the home country. That's true of just about every US immigrant society I can think of.

Support for the Israeli-US position in the Middle East has been largely uniform among American intellectuals, except for yourself, Edward Said and a handful of others. What do you attribute that to?

Things shifted very dramatically in 1967. The love affair be-tween American intellectuals and Israel grew out of Israel's smashing military victory over all the Arab world. That was at a time when the US wasn't succeeding in its effort to demolish and control Indochina. There were all sorts of jokes about how we should send Moshe Dayan over there to show us how to do it.

There was also a lot of internal turmoil here, which worried elite opinion, including liberal opinion, a lot. Israel showed how to deal with the lower orders—really kick them in the face—and that won them a lot of points among American intellectuals.

There was an op-ed in the *New York Times* by an Israeli journalist, Ari Shavit, who also happens to be a veteran of Israel's 1978 invasion of Lebanon. In criticizing Israel's April 1996 attack on Lebanon, he wrote, "We killed [several hundred Lebanese] believing with absolute certitude that now, with the White House, the Senate and much of the American media in our hands, the lives of others do not count as much as our own." You had access to the Hebrew original of this. Did the *Times* make any changes?

There were a number of interesting changes. For example, Shavit didn't say "the American media"—he specified the *New York Times*. And he mentioned, as other institutions giving them confidence, AIPAC, the Anti-Defamation League, the Holocaust Museum [in Washington DC] and Yad Vashem [the Holocaust Memorial in Jerusalem].

This vulgar exploitation of the Holocaust is used to justify oppressive control over others. That's what Shavit was talking about—Israelis who think they can kill anybody, because they think that they have the *New York Times*, Yad Vashem and the Holocaust Museum behind them.

East Timor

José Ramos-Horta and Bishop Carlos Belo of East Timor, who both have labored against enormous odds, were honored with the 1996 Nobel Peace Prize. Any observations on that?

That was great—a wonderful thing. José Ramos-Horta has been a personal friend for twenty years. I haven't seen his official speech yet, but I ran into him in São Paulo and he was saying publicly that the prize should have been given to Xanana Gusmao, the leader of the resistance against Indonesian aggression, who's been in an Indonesian jail since 1992 [and later became president and prime minister of an independent East Timor].

The recognition of the struggle is very important—or it will be, if we can turn it into something. The mainstream media will suppress it as quickly as possible; they'll give it some polite applause and then try to forget about it. If that happens, it will be our fault—nobody else's.

Right now, Clinton is planning to send arms to Indonesia. He'll get away with that unless there's a real public outcry. The Nobel Peace Prize offers a golden opportunity for people who care about the fate of a few hundred thousand people. But it's not going to happen by itself.

Some of the major issues have never even made it into the American press. For instance, Timor's rich oil resources were part of the reason the US and Australia supported the Indonesian invasion in 1975. These resources are now being plundered under a disgraceful Australian-Indonesian treaty, with US oil companies involved. This issue has yet to be discussed, except really out at the fringes. We can do something about that.

Didn't you once go to the *New York Times* editorial offices with someone from East Timor?

At that time, they'd been refusing to interview Timorese refugees in Lisbon and Australia, claiming—like the rest of the mainstream media—that they had no access to them. I was asked to pay for plane tickets for some Timorese refugees in Lisbon to fly to New York. But the *Times* still wouldn't talk to them.

On another occasion, I managed to get the *Times* to interview a Portuguese priest, Father Leoneto do Rego, who had been living in the mountains with the Timorese resistance and had been driven out during the nearly genocidal campaign of 1978. That's when Carter increased the flow of weapons to Indonesia. The only reason they didn't murder Father Leoneto was because he was Portuguese.

He was a very interesting man and a very credible witness, a classmate of the cardinal of Boston, pretty hard to disregard—but nobody would talk to him. Finally I got the *Times* to interview him.

The article that resulted, by Kathleen Teltsch, was an utter disgrace. It said almost nothing about what was happening: there was one line that said something like, *Things aren't nice in Timor.* I suspect the badness of that article must have been what induced the *Times'* editors to run their first serious editorial on the issue.

272

Meanwhile, I was trying to get the *Boston Globe* to cover the story. They were just publishing State Department handouts and apologetics from Indonesian generals. They offered to let me write an op-ed, but I said, *No, I don't want to write an op-ed. I want one of your reporters to look into it.*

I finally got them to agree to look at the facts, but they didn't take them too seriously. Instead of putting an international reporter on the story, they gave it to a local reporter, Robert Levey. Fortunately, he was extremely good.

We helped him with some leads, and he picked up the ball and ran with it. Somebody in the State Department leaked him a transcript of the actual *New York Times* interview with Father Leoneto, which was very powerful and said extremely important things. His article was the best story on East Timor that had appeared in the American press.

All of this was in 1979 and early 1980. Before that, suppression of the East Timor issue had been total in the US press, and I mean *total;* when the atrocities peaked in 1978, there were literally *no* stories.

(It's not that nobody knew about East Timor. It was covered extensively back in 1974–75, when the Portuguese empire was collapsing—although the articles then were mostly apologetics and propaganda.)

The first article after the invasion that the *Reader's Guide to Periodical Literature* lists as specifically dealing with East Timor is one of my own; it was published in January 1979 in *Inquiry*, a right-wing libertarian journal I sometimes wrote for in those days. The article was based on testimony I'd given at the UN on the suppression of the issue by the Western—primarily the US—press. Arnold Kohen had discussed Timor in an earlier article about Indonesia he'd written in the *Nation*, and that was it for the journals.

Incidentally, here's a case where a very small number of people—the most important by far being Arnold Kohen—managed to save tens of thousands of lives, as a result of getting an issue into the public arena. The Red Cross was allowed in, and although the terror continued, it lessened.

It's also a case where the internet made a difference. The East Timor Action Network was a very small and scattered group until Charlie Scheiner and others used the internet to bring information to people who otherwise couldn't get it.

Friends in Australia had been sending me articles from the Australian press, but how many people have that luxury? Now everybody could get information very fast. The movement grew and became significant enough to have an impact.

India

Didn't Adam Smith criticize the British crown for giving the East India Company [chartered in 1600 by Queen Elizabeth I] a monopoly in India?

Yes, he did. He was very critical of what the British were doing there; he said the "savage injustice of the Europeans" was destroying Bengal [in the northeast part of the country]. One example was the activities of the British East India Company. It forced farmers to destroy food crops and plant opium instead, which the Company then sold in China.

India had substantial industry in the 1700s, before the British crushed it. As late as the 1820s, the British were going to India to learn how to make steel. Bombay made locomotives that competed with those made in England.

India's steel industry might have developed—it just wasn't allowed to. Very heavy protectionism enabled England to develop while India was basically ruralized. There was virtually no growth in India under British rule.

India grew its own cotton, but Indian fabric was virtually barred from the British market because it undercut British textiles. The justification was, *Asian wages are so cheap we can't compete—we have to protect our markets.*

Adam Smith questioned that, and a recent dissertation in economic history at Harvard suggests he may have been right. According to this research, real wages may have been *higher* in India than in England, and Indian workers may also have had better benefits and more control over their work.

Fortunately for the US, things were different here. During the railroad boom of the 1800s, we were able to develop a steel industry because we imposed very high protectionist barriers to keep out British steel, which was better and cheaper. We did the same thing in order to develop our textile industry fifty years before.

Adam Smith pointed out that British merchants and manufacturers used state power to make sure that their interests were "most peculiarly attended to," however grievous the impact on others—including not only people in the Third World, but also in England. The "principal architects of policy" got very rich, but the guys working in the satanic mills and in the British Navy surely did not.

Smith's analysis is truisms, but it's now considered extreme un-American radicalism, or something like that. The same pattern shows up today, when the US farms out export industries to El Salvador and Indonesia. A few people get richer and a lot of people don't—they may even get poorer—and our military power helps things stay that way.

In his book *Representations of the Intellectual,* Edward Said writes, "One of the shabbiest of all intellectual gambits is to pontificate about abuses in someone else's society and to excuse exactly the same practices in one's own." As examples, he cites de Tocqueville, who was critical of certain things in the US but cast a blind eye towards them in the French colony of Algeria, and John Stuart Mill, who had great ideas about democratic freedoms in England that he wasn't willing to apply to India.

Very far from it. Like his father, the famous liberal James Mill, John Stuart Mill was an official of the East India Company. In 1859, he wrote an absolutely appalling article about whether England should intervene in the ugly, dirty affairs of Europe.

A lot of people were saying, *It's none of our business. Let those backward people take care of themselves.* Mill objected, on the grounds that England had such a magnificent record of humane behavior that it would simply be unfair to the poor people of the world if England didn't intervene on their behalf. (You can see the same attitude in the US today, of course.)

The timing of Mill's article was interesting. It was written not long after the Indian Mutiny of 1857, which was suppressed with *extreme* brutality. The facts were well-known in England, but that didn't affect Mill's view of England as an angelic power that ought to help other countries out by intervening in their affairs.

You've just been to India for the first time in 25 years. What were the highlights of your visit?

I was there for just nine days, in six cities, so I don't have very deep impressions. It's a fascinating country, very diverse. Lots of resources, both human and material, are being wasted in a horrifying fashion.

There's extraordinary wealth and opulence, and incredible poverty (as there was under the British). The slums of Bombay are just appalling, and some rural areas are probably worse. India is still devastated by the effects of British colonialism, but many exciting things are going on as well.

India's constitution provides for village self-government, but that's apparently only been implemented in two states, West Bengal and Kerala [in southwestern India]. Both states are pretty poor, but because both have had Communist governments (West Bengal still does) and continue to have extensive social programs, neither foreign nor domestic investors seem to want to put money into them.

Despite that, Kerala is well ahead of other Indian states in health, welfare, literacy and women's rights. For instance, fertility rates have declined dramatically, and that's almost always a reflection of women's rights. I was there only briefly, but I could easily see the difference.

West Bengal is a much more complex area. Calcutta is a wreck—although not more so than other Indian cities, as far as I could see. (Based on what I'd read, I expected it to be worse than it seemed to be.)

The Bengali countryside is quite interesting. There's a history of peasant struggle in West Bengal, and it was apparently very violent in the 1970s. Indira Gandhi tried to put it down with a great deal of brute force, but it survived. They've gotten rid of most landlord control—maybe all of it.

I went to a part of West Bengal fifty miles or so from Calcutta. I was a guest of the government, accompanied by an Indian friend, an economist who works on rural development, and a government minister (who happened to have a PhD in economics from MIT). The villagers didn't know we were coming until about 24 hours before, so there was no particular preparation.

I've seen village development programs around the world, and this one was impressive. It's relatively egalitarian and appears to be really self-governing. We met with the village committee and a group of villagers, and they could answer every question we asked, which is unusual.

In other programs I've visited, people usually can't tell you what the budget is, what's planned for agricultural diversification next year, and so on. Here they knew all that stuff immediately, and spoke with confidence and understanding.

The composition of the committee was interesting. It was strikingly obvious that caste and tribal distinctions (tribal are usually worse) have been pretty much overcome. The governing committee was half women, one of them tribal. The guy who was more or less in charge of the committee was a peasant who had a little piece of land. Some of the people who spoke up were landless laborers who'd been given small plots.

They had an extensive land reform program and the literacy level has gone up. We went to a school that had a library of maybe thirty books, of which they were very proud.

Simple tube wells have been designed (with government support) that can be sunk by a group of families. Women, who've been trained to install and maintain them, seemed to be in charge. They took a tube well out for us and put it back in— also with lots of obvious pride.

We passed a place with a bunch of cans of milk out front, and I asked to stop in. It turned out to be a dairy cooperative set up by women. They said it wasn't particularly profitable, but they wanted to be self-employed and work together. These are all very important things, and unusual.

Unlike Kerala, Bengal was devastated by the British.

It was, but it was also very culturally advanced. For example, in the early 1800s, Bengal produced more books per capita than any place in the world. At that time, Dhaka [now the capital of Bangladesh] was so developed it was compared to London.

The Bengali literary tradition is extremely rich. Only the educated and wealthy took part in it (although even in the 19th century, caste differences were reported to be declining).

Kerala also has quite an interesting history. Although the British ruled it, they more or less left it alone. Apparently the local ruler initiated populist programs in order to gain popular support in a battle he was waging against feudal landlords.

The British were relaxed enough about Kerala to let these programs proceed, and after independence, they were picked up by the Communist government. By now, they're deeply imbedded, part of the way of life in Kerala, and when the Congress Party wins an election, it doesn't try to dismantle them.

One of the legacies of British colonialism is Kashmir. Did you have any discussions about that?

Most people I met said the Kashmiri separatists are terrorists. Some civil libertarians in India are pushing the issue courageously, and people do listen to them. But my impression (from six cities in nine days) is that it's not something a lot of Indians want to talk about honestly and openly.

Has the Indian government adopted neoliberal economics?

There's a tremendous amount of discussion, in the press and everywhere, about neoliberalism and structural adjustment. That's the main topic everybody wants to talk about.

They discuss it as if it's something new, but it's pretty much what India has been subjected to for three hundred years. When that's pointed out to them, they tend to recognize it, because they know their own history. That knowledge contributes to popular resistance to neoliberalism, which is why India hasn't accepted the harshest forms of it.

How far neoliberalism will get in India is an open question. For example, the government is trying to "liberalize" the media—which means, basically, sell them off to the likes of Rupert Murdoch. The media in India are mostly owned by the rich (as they are virtually everywhere), but they're resisting the attempt to turn them into subsidiaries of a half dozen international megacorporations.

Although they're pretty right-wing, they'd rather run their own system of control internally than be taken over by outsiders. They've managed to maintain some sort of cultural autonomy...at least so far. There's some diversity in the Indian media—more than here—and that's very significant. It's much better to have your own right-wing media than Murdoch's.

As mentioned earlier, the same isn't true of India's small advertising industry—it's been mostly bought up by big, mostly American (maybe all American) multinationals. What they push—of course—is foreign products. That undermines domestic production and is harmful to the Indian economy, but many privileged people like it. Somebody always benefits from these programs.

Intellectual property rights are also a big issue. The new international patent rules are very strict and may well destroy the Indian pharmaceutical industry, which has kept drugs quite cheap. The Indian companies are likely to become subsidiaries of foreign firms, and prices will go up. (The Indian parliament actually voted the proposed patent rules down, but the government is apparently going to try to institute them anyway.)

There used to be only *process patents*, which permit people to figure out smarter ways to make products. The World Trade Organization has introduced *product patents;* they allow companies to patent not only a process, but also the product that's the result of the process. Product patents discourage innovation, are very inefficient and undermine markets, but that's irrelevant—they empower the rich and help big multinationals exercise control over the future of pharmaceuticals and biotechnology.

Countries like the US, England and Japan would never have tolerated anything remotely like product patents, or foreign control of their press, during their development. But they're now imposing this sort of "market discipline" on the Third World, as they did throughout the colonial period. That's one reason India is India, and not the US.

Another example is recruitment of scientists. Foreign firms pay salaries way beyond what Indian researchers are used to, and set up research institutes with facilities Indian scientists can't dream of getting anywhere else. As a result, foreign firms can skim off the best scientists.

The scientists may be happy, and the companies are happy. But it's not necessarily good for India, which once had some of the most advanced agricultural research in the world.

An Indian farmer used to have a place he could go to and say, *There's some funny pest in my fields. Can you take a look at it?* But now that's being bought up by foreign firms, and will therefore be oriented towards export crops for specialized markets, and subsidized foreign imports that will undercut domestic production.

There's nothing new about this. It's part of a long history of "experiments" carried out by the powerful of the world. The first major one in India was what the British called the Permanent Settlement of 1793, which rearranged all the land holdings in Bengal.

When the British Parliament looked into this thirty or forty years later, they conceded that it was a disaster for the Bengalis. But they also pointed out that it enriched the British, and created a landlord class in Bengal, subordinated to British interests, that could control the population.

We've already discussed a recent example of such experiments, in Mexico. Such experiments regularly seem to fail for the experimental animals, but succeed for the *designers* of the experiment. It's an oddly consistent pattern. If you can find an exception to that pattern over the last couple of hundred years, I'd be interested in hearing about it. I'd also be interested in knowing who in the mainstream talks about it, since I haven't been able to find anyone.

Getting free from the colonial powers generated a tremendous burst of energy in India, as did presenting a neutralist challenge to US domination.

That challenge is pretty much gone—from Indian policy, at least, if not from the general population.

The US was very much opposed to Indian independence and also, of course, to Nehru's attempts at nonalignment. Any Indian with a streak of independence was bitterly hated and condemned by US policymakers. Eisenhower called Nehru a "schizophrenic" who suffered from an "inferiority complex" and had a "terrible resentment [of] domination by whites" (really surprising, given how the British treated India).

The US basically brought the Cold War to South Asia by arming Pakistan, which was part of our system of control of the

Middle East. It ended up with India and Pakistan fighting several wars with each other, sometimes with American arms.

US policymakers were also worried about Indonesia. In 1948, George Kennan, one of the chief architects of US policy, described Indonesia as "the most crucial issue of the moment in our struggle with the Kremlin." (The USSR wasn't really the issue, of course—that was just code for "independent Third World development.")

He was very much afraid that a Communist Indonesia would be an "infection [that] "would sweep westward through all of South Asia"—not by conquest, of course, but by example. That concern wasn't really overcome until the mass slaughter in Indonesia in 1965, which the US government, the press and other commentators were all exhilarated about.

They had the same fear about China—not that it was going to conquer South Asia, but that it was developing in ways that might be a model for other Asian countries. US policymakers remained ambivalent toward India. They had to support it as an alternative model to China, but they hated to do it, because India was following a somewhat independent line and had established close relations with the Soviet Union.

The US gave some aid to India, which was supposed to be the democratic alternative to China. But it was given grudgingly, and the US wouldn't permit India to develop its own energy resources; instead, they had to import oil, which was much more expensive. India's petroleum resources are apparently significant, but they still haven't been developed.

The results of US ambivalence towards India have sometimes been pretty ugly. Right after independence, in the early 1950s, India had a very serious famine, in which millions of people died. US internal records show that we had a huge food surplus, but Truman refused to send any, because we didn't like Nehru's independence. When we finally did send some, it was under stringent conditions. (There's a good book on this by the historian Dennis Merrill.)

What was your overall impression of India?

The questions being debated in India—whether to use import restrictions, or to adopt neoliberal policies—can't really be answered in general. Like debt, import restrictions aren't good or bad in themselves—it depends on what you use them for. In Japan, Taiwan and South Korea, where they were used to build up a domestic industrial base and market (as in Britain and the US in earlier years), they proved to be a good idea (for the home country, at least). But if you use them to protect an inefficient system and the super-rich who profit from it, they're bad.

Here's a personal anecdote that illustrates things that are very real, but that you can't measure. After a talk in Hyderabad, some friends were driving me to the airport. When we were about two miles away, the traffic completely froze up. Every inch of the road was covered by a bicycle, a rickshaw, an oxcart, a car or whatever. The people were sort of quiet; nobody was making a fuss.

After about twenty minutes, we realized that the only way to get to the airport on time was to walk. So my friends and I started threading our way through this immense traffic jam.

Finally we got to a big highway that was blocked off. There are lots of cops and security forces everywhere in India, but here there were *tons*. My friends talked them into letting us cross the road, which we weren't supposed to do, and we finally made it to the airport (which was semifunctional because it was cut off from the city).

Why was the highway closed down? There were signs next to it saying *VVIP*, which I was told means *Very Very Important Person*. Because some "VVIP"—we later found out it was the prime minister—was expected at some indefinite time in the future, the city was closed down.

That's bad enough—what's worse is the fact that people tolerated it. (Just imagine the same thing happening here in Boston, say.) Feudalistic attitudes run very deep in India, and they're going to be hard to uproot.

That's what was so striking about the village in West Bengal. Poor, landless workers, including lots of women, were active and engaged. You can't put numbers on that kind of change, but it makes a huge difference. That's real popular resistance and activism, like the democratic institutions that developed in Haiti before Aristide's election (and that still exist there) and what happened in Central America in the 1970s and 1980s.

(In Haiti, democracy elicited instant US hostility and a murderous military coup, tacitly supported by the US; in Central America, a US-run terrorist war. In both places, the US permitted democratic forms after establishing conditions that prevented them from functioning—amidst much self-congratulation about the nobility of our leaders.)

What has to be overcome in India is enormous. The inefficiency is unbelievable. While I was there, the Bank of India came out with an estimate that about a third of the economy is "black"—mainly rich people who don't pay their taxes. Economists there told me one-third is an underestimate. A country can't function that way.

As elsewhere, the real question for India is, can they control their own wealthy? If they can figure out a way to do that, there are lots of policies that might work.

International organizations

In *World Orders, Old and New,* you say that the UN has become virtually an agency for US power.

The UN mostly does what the US—meaning US business—wants done. A lot of its peacekeeping operations are aimed at maintaining the level of "stability" corporations need in order to do business. It's dirty work and they're happy to have the UN do it.

If that's so, how do you explain the hostility toward [former UN Secretary General] Boutros Boutros-Ghali?

In the first place, there was an element of racism there—even though the next choice, Kofi Annan, was also from Africa. When George [H. W.] Bush talked about "Bou-Bou Ghali," nobody batted an eyelash, although I doubt very much that a presidential candidate in the US would survive very long if he referred to the former prime minister of Israel as, say, "Itzy-Schmitzy Rabin."

There's a lot of opposition to the UN on the extreme right. Some of it's tied in with fantasies about black helicopters and loss of sovereignty to world government. But some of it's simply a case of avoiding blame.

Take the atrocities carried out in Somalia, where the US quietly concedes that thousands of Somali civilians—perhaps up to ten thousand—were killed by US forces. If somebody threatened US forces, they'd call in helicopter gunships. That doesn't sound so heroic, so the resulting catastrophes became the fault of the UN.

Similarly, the US evaded the burdens and difficulties of the conflict in the former Yugoslavia until things were more or less settled, then moved in and took over (effectively imposing a kind of partition between Greater Croatia and Greater Serbia). That way, the US could blame everything that went wrong on the UN. Very convenient.

It's easy to focus anti-UN hostility on the secretary-general. Let's kick him in the pants, and kick the rest of the world in the pants too. Why should we bother with what other countries think about us anyway?

Do you think the very critical UN report on the Israeli attack on the UN compound in Qana, Lebanon may have been a factor in undermining support for Boutros-Ghali?

It might have been a small factor, but who paid any attention to it? It was so marginalized that I frankly doubt it had much effect. Amnesty International came out with a study that strongly corroborated the UN report. That also disappeared very quickly; I'm not even sure it was reported on at all.

These sorts of things can be brushed off very quickly when they're inconvenient for power and career interests. Both reports are quite shocking, and confirmed by veteran journalists on the scene (notably Robert Fisk). But it's the wrong story.

The basic reason there's hostility to international institutions here is that they don't always do exactly what the US orders them to do. The World Court is a perfect example. The US government isn't going to accept being condemned by it—as it was in 1986, for "unlawful use of force" against Nicaragua. The Court ordered the US to desist and pay substantial reparations, and ruled explicitly that no aid to the Contras could be considered "humanitarian." We don't have to waste time noting how the US, the press and educated opinion reacted to this.

The International Labor Organization is another example. Not only does it stand up for workers' rights, but it condemned the US for violating international labor standards. So it's dismissed, and the US refuses to pay the roughly $100 million owed to it.

The US has little use for the UN Development Program or the Food and Agriculture Organization, since they're mostly concerned with developing countries. UNCTAD (the UN Conference on Trade and Development) has, to some extent, advocated the interests of developing countries and has been an expert critical voice opposing certain Washington policies, so it's been undermined and tamed as well.

As soon as UNESCO called for opening up the world information system, it was out of luck. The US forced it to abandon its evil ways, and significantly modified its role.

The attack on these organizations is all part of reconstructing the world in the interests of the most powerful and the most wealthy. There's lots wrong with the UN, but it's still a somewhat democratic institution. Why tolerate that?

The US attitude was expressed rather neatly by Madeleine Albright in a remark which, as far as I know, wasn't reported. She was trying to get the Security Council to accept one of our punitive actions toward Iraq; none of the other countries wanted to go along with it, since they recognized that it was really just a part of US domestic politics. So she told them that the US will act "multilaterally when we can and unilaterally as we must." So would anyone else, if they had the power.

The US owes the UN over $1 billion—more than any other country.

Of course. Why should we spend money on anybody but the rich?

The World Trade Organization is the successor to GATT. Has the US been fairly happy with the WTO?

Not entirely. The US has been brought up more than once for violation of WTO principles, and was also condemned by the GATT council earlier. But in general, the US is more or less favorable to the WTO, whose mixture of liberalization and protectionism is pretty much tailored to the needs of powerful transnational corporations and financial institutions.

The Uruguay Round treaty that led to the WTO was called a free-trade agreement, but it's really more of an investor-rights agreement. The US wants to use WTO rules in areas it expects to dominate, and is certainly in a position to cancel any rule it doesn't like.

For example, a while back the US forced Mexico to cut back exports of its tomatoes. It's a violation of NAFTA and WTO rules and will cost Mexican producers close to a billion dollars a year. The official reason was that Mexican producers were selling tomatoes at a cost American producers can't match.

If the WTO rules in favor of the European Union's request to condemn the Helms-Burton Act [which strengthened the US embargo against Cuba] as an illegal interference with world trade, the US will just go on acting unilaterally. If you're powerful enough, you can do whatever you want.

What do you think of the expansion of NATO?

I don't think there's a simple answer to that—it depends how the economic and political structure of Eastern Europe and Western Asia evolves.

As mentioned above, when the Cold War ended I expected that the former Soviet empire would pretty much revert to what it had been before. The areas that had been part of the industrial West—the Czech Republic, western Poland, Hungary—would essentially be reintegrated into the West, and the other parts, which had been Third World before the Soviet Union, would return to that status, with substantial poverty, corruption, crime and so on. Partial extension of the NATO system to industrial—or partially industrial—countries like the Czech Republic, Poland and Hungary would help formalize all this.

But there will be conflicts. Europe and the US have differing expectations and goals for the region, and there are also differences within Europe. Russia isn't a trivial force either; it can't be disregarded and doesn't like being excluded. There are more complex power plays, like the jockeying that's going on around the oil fields in Central Asia, where the *people* involved won't have much of a voice in the process.

In the case of NATO, there are other factors, like the special interests of military industry, which is looking forward to a huge market with NATO expansion and standardization of weapons (which are mainly produced by the US). That translates into another substantial taxpayer subsidy to high-tech industry, with the usual inefficiencies of our system of industrial policy and "state socialism for the rich."

Are left and right meaningful terms?

Historically, the left has been somewhat ambivalent about political power. The right has no such inhibitions—they *want* political power.

I don't much like the terms *left* and *right*. What's called the left includes Leninism, which I consider ultra-right in many respects. The Leninists were certainly very interested in political power—in fact, more so than anyone.

Leninism has nothing to do with the values of the left—in fact, it's radically opposed to them. That was recognized at the time by mainstream left Marxists like Anton Pannekoek, Paul Mattick and Karl Korsch. Even Trotsky had predicted that the Leninists would turn to dictatorial rule (before he decided to join them).

Rosa Luxemburg warned of the same things (in a more or less friendly way, because she didn't want to harm the movement). So did Bertrand Russell. And, of course, most of the anarchists did.

Conventional terms of political discourse like *left* and *right* have been almost evacuated of meaning. They're so distorted and irrelevant it's almost better to throw them out.

Take Witness for Peace, which has been a very important organization since the 1980s. People from an imperial country actually went down and lived in Third World villages, in the hope that a white face might protect the inhabitants from state terrorists organized by their own country. That's never happened before.

Was that left or right? It certainly represents the traditional ideals of the left, like justice, freedom, solidarity, sympathy. On the other hand, a lot of it came out of the conservative Christian community. I don't know where to put Witness for Peace on any political spectrum. It's just human beings acting decently.

What's currently lambasted as "political correctness" is supposed to be left. But in many places I go—including campuses that are extremely conservative, where there is hardly any political activity—very delicate judgments are made about just what it's OK to say with regard to minuscule questions of gender, race, color, etc. Is that left or right? I don't know.

Part of what the propaganda system does is deprive terms of meaning. It probably starts at some relatively conscious level and then just gets into your bones. Sometimes it's done quite deliberately.

One dramatic case in recent years is the disappearance of the word *profits*. Profits don't exist anymore—just jobs. So when Clinton came back from Indonesia with a $40-billion contract for Exxon, the media all talked about jobs for Americans. But profits for Exxon? Perish the thought. (Exxon's stock shot up, but that's just because investors were so delighted about the new jobs.)

That's conscious evacuation of meaning, and even the left falls into it, talking about how Congressmen vote for the Pentagon because they want jobs for their district. Are *jobs* what Congressmen are worried about, not profits and public subsidies for firms?

In a lead story, the *New York Times Week in Review* made an amazing discovery: the new kind of "populism"—as practiced by Steve Forbes, Pat Buchanan and the like—is different from the old kind of populism. The old kind opposes big corporations and plutocrats; the new kind *is* big corporations and plutocrats. That you can have a character like Steve Forbes on the national scene without people cracking up with laughter shows how intense the propaganda is.

The narcissism of small differences

In his book *The Twilight of Common Dreams,* Todd Gitlin says the left is polarized by identity politics, which he calls the "narcissism of small differences." He writes, "The right has been building…but the left has been…cultivating difference rather than commonality."

The left does tend to get caught up in sectarianism, but I think he's describing something that's happening in the country in general, not just in what might be called realistically "the left." The activism of the 1960s had a very civilizing effect—it brought to the fore all sorts of oppression and discrimination that had been suppressed.

The killing off of the native populations—which had been pretty much ignored even in scholarship—was put on the agenda for the first time. Environmental issues (which basically have to do with the rights of future generations), respect for other cultures, the feminist movement—these had all existed in some form earlier, but they really took off in the 1970s and spread throughout the whole country. The Central America solidarity movement wouldn't have existed in the form it did if not for what happened in the 1960s.

Concerns about oppression, authority and rights can sometimes take the unhealthy forms that Gitlin is criticizing, but they needn't, and commonly didn't.

Louis Farrakhan and the Million Man March seemed to be the epitome of identity politics, since the participants were defined not only by race but by gender. What did you think of that?

I think it's a more complicated phenomenon. There were also elements of self-help, rebuilding viable communities and lives, taking responsibility for what you do. These are all good things.

But Farrakhan's economic program is small-scale capitalism.

I didn't see anything much in the way of an economic program, but when you're crushed, even small-scale capitalism can be a step forward. It shouldn't be the end of the road, obviously, but it can be a step.

I think this movement is much more nuanced than some of the commentary has assumed. It has opportunities to go a lot of different ways, and how it comes out depends on what people do with it.

There's a reason why it's men—look at what's happened to black men in the last twenty years. There's been a virtual war against minorities and the poor. It included plenty of scapegoating, like Reagan's anecdotes about black welfare mothers with Cadillacs, and the Willie Horton concoctions. The fraudulent war on drugs, which has little to do with drugs or crime, is another part of it.

Michael Tonry points out that those who crafted the programs had to know they were going right after young blacks. Every indicator pointed in that direction. Tonry further points out that, in the law, conscious foreknowledge is evidence of intent of criminal action.

I think he's right about that. The so-called "war on drugs" was in no small measure a criminal attempt to criminalize the black male population and, more generally, segments of the population that are sometimes called "disposable people" in our Latin American dependencies, because they don't contribute to profit-making.

You're aware of Farrakhan's comments on…

I don't have anything to say about Farrakhan—I'm talking about the phenomenon. Probably he's just an opportunist trying to get power—that's what leaders usually are. But I don't know what's in his mind, and I don't presume to judge what he's up to. I'm too far out of it.

Christopher Hitchens, who writes for the *Nation* and *Vanity Fair,* recalls that the first time he heard the slogan "the personal is political," he felt a deep sense of impending doom. To him, the slogan sounds escapist and narcissistic, implying that nothing will be required of you except being able to talk about yourself and your own oppression. He was talking about the growth of identity movements.

I agree with him. It certainly opened itself up to that, and it's been exploited that way, sometimes in ugly—and often in comical—ways. But it doesn't only have that aspect. It can also mean that people have the right to adopt their individual ways of living if they want, without oppression or discrimination.

Postmodernism

A respected NYU physics professor, Allen Sokal, got an article published in *Social Text,* which has been described as the leading cultural-studies journal in the country. To point out the decline in intellectual rigor in certain parts of American academia, he intentionally filled the article with errors. What do you make of that?

His article was cleverly done. He quoted—accurately—from advanced physics journals, then juxtaposed quotes from postmodern critiques of science, including *Social Text*, as if the for-

290

mer somehow supported the latter. No one with any familiarity with the material could read the article without laughter.

Sokal's point was that postmodern critiques of science are based on ignorance—they're flights of fancy that lack minimal critical standards. There's something healthy about this sort of criticism, but his article is also going to be used as a weapon against attitudes and work that have merit.

It was immediately interpreted by the *New York Times* and the *Wall Street Journal* as just one more demonstration that some sort of left-fascist political correctness movement has taken over academic life—when what's really going on is a significant right-wing assault against academic freedom and intellectual independence.

Well, we live in this world, unfortunately. What we do is going to be used by powerful people and institutions for their purposes, not for ours.

Postmodernists claim to represent some kind of a subversive critique. Have you been able to detect that?

Very little of it. I'm not a big expert on postmodern literature; I don't read it much, because I find most of it pretty unilluminating, often complicated truisms or worse. But within it there are certainly things that are worth saying and doing. It's very valuable to study the social, institutional and cultural assumptions within which scientific work is done, but the best work of that sort isn't by postmodernists (at least as far as I can understand their work).

For instance, fascinating work has been done in the last thirty or forty years on what Isaac Newton, the great hero of science, actually thought he was doing. His theory of gravity was very disturbing to him and to everyone else at the time. Because gravity works at a distance, Newton agreed with other leading scientists of his day that it was an "occult force," and spent most of the rest of his life trying to come to terms with that unacceptable conclusion.

In the final edition of his great work, the *Principia*, he said that the world consists of three things: active force, passive matter and some semi-spiritual force (which, for various reasons, he identified with electricity) that functions as an intermediary between the two. Newton was an expert on church history (physics was a very small part of his interests) and the framework for his theory of an intermediary force was the fourth-century Arian heresy, which said that Jesus is semi-divine, not divine, and acts as an intermediary between God and man.

After Newton's death, his papers were given over to the physicists at Cambridge University. They were appalled by what they found in them, so they simply gave them back to his family, which held onto them and never published them.

Around the 1930s, they started selling the material off; [the British economist John Maynard] Keynes was one of those who recognized their enormous value. After WWII, some of this stuff started surfacing at antique dealers, and scholars began to gather it together and do important analytical work.

Now *that's* serious cultural-sociological analysis of some of the greatest moments of science, and there's plenty more like it. You can bring it right up to the present. People do scientific work within a framework of thought, and their work is affected by cultural factors, by power systems, by all sorts of things. Nobody denies that.

What the postmodernists claim to be fighting is *foundationalism,* the idea that science is divorced from society and culture and provides foundations for certain, absolute truth. Nobody has believed that since the 1700s.

From what I've looked at, I find postmodernism very dense, jargon-laden and hard to read.

I do too. A lot of it has the appearance of a kind of careerism, an escape from engagement.

But they claim to be socially engaged.

In the 1930s, left intellectuals were involved in worker education and writing books like *Mathematics for the Millions*. They considered it an ordinary, minimal responsibility of privileged people to help others who'd been deprived of formal education to enter into high culture.

Today's counterparts of these 1930s left intellectuals are telling people, *You don't have to know anything. It's all junk, a power play, a white male conspiracy. Forget about rationality and science.* In other words, put those tools in the hands of your enemies. Let them monopolize everything that works and makes sense.

Plenty of very honorable left intellectuals think this tendency is liberatory, but I think they're wrong. A lot of personal correspondence on related topics between me and my close, valued old friend Marc Raskin has been published in a book of his. There were similar interchanges in *Z Papers* in 1992–93, both with Marc and a lot of other people with whom I basically feel in sympathy, but with whom I differ very sharply on this issue.

Excommunicated by the illuminati

You've long been excommunicated, if I can use that word, not only from the mass media but also from the "illuminati" circles of the Upper West Side [of Manhattan] and their publications, like the *New York Review of Books* [often referred to simply as the *New York Review*].

It has nothing to do with me.

What happened?

The *New York Review* started in 1964. From about 1967 to about 1971, as political engagement grew among young intellectuals, it was open to dissident analysis and commentary from people like Peter Dale Scott, Franz Schurmann, Paul Lauter, Florence Howe and myself.

Then, within a few years, we all disappeared from its pages. I think what happened is that the editors wanted to keep ahead of the game. They knew their audience and couldn't fail to see that the young intellectuals who constituted a large part of it were changing.

It ended for me personally in late January 1973. Nixon and Kissinger's "peace treaty" with Hanoi had just been announced. The *New York Times* published a big supplement that included the text of the treaty and a long interview with Kissinger in which he went through the treaty paragraph by paragraph. *The war is over*, he said, *everything is just fantastic.*

I was suspicious. Something similar had happened about three months earlier, in October 1972, when Radio Hanoi had announced a peace agreement the US had been keeping secret. It was the last week of Nixon's re-election campaign. Kissinger went on television and said, *Peace is at hand.* Then he went through the peace agreement, rejected every single thing in it, and made it very clear that the US was going to continue bombing.

The press only picked up Kissinger's first line, *Peace is at hand.* Wonderful. It's all over. Vote for Nixon. What he was actually saying was, *We're not going to pay any attention to this, because we don't want this agreement, and we're going to keep bombing until we get something better.*

Then came the Christmas bombings, which didn't work. The US lost a lot of B-52s and faced big protests all around the world. So they stopped the bombings and accepted the October proposals they'd previously rejected. (That's not what the press said, but that's essentially what happened.)

The January farce was the same. Kissinger and the White House made it clear and explicit that they were rejecting every basic principle of the treaty they were compelled to sign, so that they could go on with the war, seeking to gain what they could.

I was pretty angry. I happened to have a talk scheduled for a peace group at Columbia that evening. I called Robert Silvers, a friend who was the editor of the *New York Review,* and asked him if we could meet for dinner. We spent an hour or two going through the texts in the *Times* special supplement. It was easy enough to see what they meant.

I said, *Look, I'd like to write about this. I think it's the most important thing I'll ever write, because you know as well as I do the press is going to lie flat out about it. The destruction and killing will go on, and then, when the whole thing collapses because of the US initiatives, they're going to blame the Vietnamese* (which is exactly what happened).

He said, *Don't worry—you don't need to write an article. I'll make sure your point of view gets in.* It was supposed to be in an article written by Frances FitzGerald, but it wasn't; she didn't understand or didn't agree with the point.

I published articles about this right away, but in *Ramparts* and in *Social Policy*. That was essentially the end of any association with the *New York Review*. We understood each other.

Why are you in the *Nation* so infrequently?

It's complicated. I don't recall any contact with them until about the late 1970s, I guess. At that point I wrote some book reviews for them. Occasionally they'd invite me to take part in a symposium, but mostly we were sort of at arm's length. We didn't really see things the same way.

In the late 1980s, I interviewed Victor Navasky [former editor, publisher and editorial director of the *Nation* magazine]. He said he was uncomfortable with your views on the Middle East.

Victor, who I like, once called me to say that people kept asking him why I wasn't in the magazine. He explained to them that it was because I kept sending him huge articles that were way too long. In fact, the only article I'd ever submitted to the *Nation* was about two pages long.

It was right after the bombing of Beirut ended in mid-August 1982. There was a flurry of talk about how there was going to be peace and everything was going to be wonderful.

My article, based mainly on the Israeli press, said this was nonsense, that the US and Israel intended to continue fighting, and that there were going to be further atrocities. (I didn't know then, of course, that the massacres at the Sabra and Shatila Palestinian refugee camps were going to take place a few weeks later, but that was the sort of thing I was anticipating.)

I sent the article to the *Nation* and never heard a word from them. It's the only one I've ever submitted. Actually, that's the reason I wrote my book *The Fateful Triangle*. I was so infuriated at my inability to get one word even into the left press about Sabra and Shatila that I figured I'd better write a book about it. (I wrote it mostly at night, because I had no other free time.)

Somebody asked me to ask you about critiques of your work, so let's talk about an article by Richard Wolin in *Dissent,* a very serious and scholarly journal. He wrote that your book *World Orders, Old and New* is a "heavy-handed, fact-filled, citation-laden jeremiad," that you're "ideologically obsessed," that your views coincide with those of the far right, and that you have a "long-standing contempt for Israel."

If that's the most cogent criticism you can find, there's nothing much to talk about. I wasn't going to respond to that article, but some friends associated with *Dissent* asked me to, so I did, putting aside the flow of insults and keeping to the few identifiable points.

Wolin's main complaint was that I'm always saying the US is a "totalitarian" and "fascist" country. It just so happened that articles appeared in London and Greece at about the same time I got that issue of *Dissent*. Both raised the question I'm commonly asked overseas: *Why do I keep talking about the US as the freest country in the world?* That's what people in other countries hear. What Wolin hears is my calling the US a totalitarian, fascist state.

He also says I use Orwellisms. That refers to my quoting a few sentences of an unpublished article of Orwell's that was supposed to be an introduction to *Animal Farm*. Orwell pointed out that in a very free society (he was talking about England), there are all sorts of ways to keep unpopular ideas from being expressed.

One factor is that the press is owned by wealthy men who have every reason not to have certain ideas expressed. He identified education as another factor. When you go through Oxford or Cambridge, you learn that there are certain things it "wouldn't do to say." If you don't learn that, you're not in the system.

What can you say about being criticized for having a fact-filled, citation-laden book? They've got you coming and going. If you don't cite facts...

It's not just me—any critic on the left is going to have to face that. If you don't footnote every word, you're not giving sources—you're lying. If you *do* footnote every word, you're a ridiculous pedant. There are lots of devices in relatively free societies to achieve the goals that Orwell described.

Signs of progress (and not)

Over the last twenty or thirty years, new attitudes about gay rights, smoking, drinking, guns, animal rights, vegetarianism, etc. have come into the mainstream. But there hasn't really been a strong transformation in other areas.

It's a much more civilized society than it was thirty years ago. Plenty of crazy stuff goes on, but in general, there's an overall improvement in the level of tolerance and understanding in this country, a much broader recognition of the rights of other people, of diversity, of the need to recognize oppressive acts that you yourself have been involved in.

There's no more dramatic illustration of that than the attitude towards the original sin of American society—the elimination of the native population. The founding fathers sometimes condemned it, usually long after their own involvement, but from then to the 1960s, it was hardly mentioned.

When I grew up, we played cowboys and Indians (and I was supposed to be some kind of young radical). My children certainly wouldn't have played like that, and obviously my grandchildren don't.

Looking at the timing, I suspect that a lot of the hysteria about political correctness was whipped up out of frustration over the fact that it wasn't going to be possible, in 1992, to have the kind of 500th anniversary celebration of Columbus' landing in the New World you could have had thirty years earlier. There's much more understanding today of what actually took place.

I'm not saying things are great now, but they are much better, in virtually every area. In the 1700s, the way people treated each other was an unbelievable horror. A century ago, workers' rights in the US were violently repressed.

Even fifty years ago, things were pretty bad. Repression of blacks in the South was obscene. Options for women were highly restricted. There was plenty of upper-class antisemitism too.

Harvard had almost no Jewish faculty when I got there about 1950. When my wife and I were looking for a house in the suburbs, we were told by realtors that "we wouldn't be happy" in certain areas we liked. Blacks were of course treated far worse.

The 1890s—the "Gay Nineties"—weren't so gay for the workers in western Pennsylvania. They lived under a brutal tyranny instituted by the great pacifist Andrew Carnegie and the troops he called out in Homestead (and elsewhere).

It wasn't until the 1930s, forty years later, that people were even willing to talk about what happened. People who grew up around there tell me that their parents (or grandparents) were afraid to talk about it to the end of their lives.

In 1919 or so, almost thirty years after Homestead, there was a steel strike in western Pennsylvania. [The union activist] Mother Jones [1830–1930], who was then about 90, came to give a talk. Before she could speak, the police dragged her off and threw her into jail. That's pretty rough.

In the 1920s—the "Roaring Twenties"—business control seemed total, and the means used to achieve it could hardly "proceed in anything remotely resembling a democracy," as political scientist Thomas Ferguson pointed out. He was referring to state repression, violence, destruction of unions and harsh management controls.

Yale labor historian David Montgomery, extensively reviewing the same period, wrote that modern America was "created over its workers' protests [with] fierce struggle [in a] most undemocratic America." The 1920s aren't very long ago.

In the early 1960s, the South was a terror state; it's not at all like that now. The beginnings of some kind of commitment to decent medical care for the entire population only go back to the 1960s. Concern for environmental protection doesn't really begin until the 1970s.

Right now we're trying to defend a minimal healthcare system; thirty years ago there wasn't a minimal healthcare system to defend. That's progress.

All those changes took place because of constant, dedicated struggle, which is hard and can look very depressing for long periods. Of course you can always find ways in which these new attitudes have been distorted and turned into techniques of oppression, careerism, self-aggrandizement and so on. But the overall change is toward greater humanity.

Unfortunately, this trend hasn't touched the central areas of power. In fact, it can be tolerated, even supported, by major institutions, as long as it doesn't get to the heart of the matter—their power and domination over the society, which has actually *increased*. If these new attitudes really started affecting the distribution of power, you'd have some serious struggles.

Disney is a good example of the kind of accommodation you're describing. It exploits Third World labor in Haiti and elsewhere, but domestically it has very liberal policies on gay rights and healthcare.

It's perfectly consistent for the kind of corporate oligarchy we have to say that we shouldn't discriminate among people. They're all equal—equally lacking in the right to control their own fate, all capable of being passive, apathetic, obedient consumers and workers. The people on top will have greater rights, of course, but they'll be *equally* greater rights—regardless of whether they're black, white, green, gay, heterosexual, men, women, whatever.

You arrived very late for a talk you gave in Vancouver. What were the circumstances?

The event was organized by the British Columbia labor movement. My talk was scheduled for about 7 pm. I should have made it in ample time, but every imaginable thing went wrong with the airlines, and I didn't get there until about 10:30 or 11:00.

To my amazement there were still (what looked like) 800 or 900 people there—they'd been watching documentaries and having discussions. I didn't bother with the talk—it was too late for that—so we just started off with a discussion. It was quite lively, and went on for a couple of hours.

Toward the end of the question-and-answer period, someone asked you about the power of the system and how to change it. You said it's "a very weak system. It looks powerful but could easily be changed." Where do you see the weaknesses?

I see them at every level. We've discussed them earlier, but here's a summary:

- People don't like the system. As mentioned earlier, 95% of Americans think corporations should lower their profits to benefit their workers and the communities they do business in, 70% think businesses have too much power and more than 80% think that working people don't have enough say in what goes on, that the economic system is inherently unfair, and that the government basically isn't functioning, because it's working for the rich.

- Corporations—the major power system in the West—are chartered by states, and legal mechanisms exist to take away their charters and place them under worker or community control. That would require a democratically organized public, and it hasn't been done for a century. But the rights of corporations were mostly given to them by courts and lawyers, not by legislation, and that power system could erode very quickly.

    Of course, the system, once in place, cannot simply be dismantled by legal tinkering. Alternatives have to be constructed within the existing economy, and within the minds of working people and communities. The questions that arise go to the core of socioeconomic organization, the nature of decision-making and control, and the fundamentals of human rights. They are far from trivial.

- Since government is to some extent under public control—at least potentially—it can also be modified.

- About two-thirds of all financial transactions in the globalized economy take place in areas dominated by the US, Japan and Germany. These are all areas where—in principle, at least—mechanisms already exist that allow the public to control what happens.

People need organizations and movements to gravitate to.

If people become aware of constructive alternatives, along with even the beginnings of mechanisms to realize those alternatives, positive change could have a lot of support. The current tendencies, many of which are pretty harmful, don't seem to be all that substantial, and there's nothing inevitable about them. That doesn't mean constructive change *will* happen, but the opportunity for it is definitely there.

═══════════════════════

Resistance

Who knows where the next Rosa Parks [the African-American woman whose refusal to sit in the back of the bus ignited the Montgomery bus boycott in 1955] will sit down and spark a movement?

Rosa Parks is a very courageous and honorable person, but she didn't come out of nowhere. There had been an extensive background of education, organizing and struggle, and she was more or less chosen to do what she did. It's that kind of background that we should be seeking to develop.

Union membership in the US is very low, but it's even lower in France. Yet the support for French general strikes—which shut down cities and, at one point, the whole country—was extraordinarily high. What accounts for that difference?

One factor is the power of business propaganda in the US, which has succeeded, to an unusual extent, in breaking down the relations among people and their sense of support for one another. This is the country where the public-relations industry was developed, and where it's still the most sophisticated. It's also the home of the international entertainment industry, whose products are mainly a form of propaganda.

Although there's no such thing as a purely capitalist society (nor could there be), the US is toward the capitalist end. It tends to be more business-run, and spends a huge amount on marketing (which is basically an organized form of deceit). A large part of that is advertising, which is tax-deductible, so we all pay for the privilege of being manipulated and controlled.

And of course that's only one aspect of the campaign to "regiment the public mind." Legal barriers against class-based solidarity actions by working people are another device to fragment the general population that are not found in other industrial democracies.

In 1996, Ralph Nader ran for president on the Green Party ticket, and both the Labor Party and the Alliance held founding conventions. The New Party has been running candidates and winning elections. What do you think of all this?

Allowing new options to enter the political system is—in general—a good idea. I think the right way to do it might be the New Party strategy of targeting winnable local elections, backing fusion candidates and—crucially—relating such electoral efforts to ongoing organizing and activism. A labor-based party is a very good idea too.

Since they have basically the same interests, such parties ought to get together—it isn't a good idea to scatter energies and resources that are very slight. A possible step might be to create something like the NDP [New Democratic Party] in Canada or the Workers' Party in Brazil—big organizations that foster and support grassroots activities, bring people together, provide an umbrella under which activities can be carried out and—among other things—take part in the political system, if that turns out to be useful.

That can progress towards something else, but it's not going to overcome the fact that one big business party, with two factions, runs things. We won't break out of that until we democratize the basic structure of our institutions.

As John Dewey put it about seventy years ago, "Politics is the shadow cast on society by big business." As long as you have highly concentrated, unaccountable private power, politics is just going to be a shadow. But you might as well make use of the shadow as much as possible, and use it to try to undermine what's casting the shadow.

Didn't Dewey warn against mere "attenuation of the shadow"?

He said that mere "attenuation of the shadow will not change the substance," which is correct, but it can create the *basis* for undermining the substance. It goes back to the Brazilian rural workers' image I mentioned earlier—expanding the floor of the cage. Eventually you want to dismantle the cage, but expanding the floor of the cage is a step towards that.

It creates different attitudes, different understandings, different forms of participation, different ways for life to be lived, and also yields insight into the limits of existing institutions. That's typically learned by struggle.

All these things are to the good. They only attenuate, that's true, and by themselves they won't overcome, but they're the basis for overcoming. If you can rebuild, reconstitute and strengthen a culture in which social bonds are considered significant, you've made a step towards undermining the control that private and state power exercise over society.

In a cover story in the *Nation,* Daniel Singer described "the unmistakable attempt by the International financial establishment and [European] governments to [adopt] Reaganomics" and the "striking signs of resistance in Europe" against this. There have been mass demonstrations in France, Germany and Italy, and 250,000 Canadians turned out in Toronto to protest what was going on. That's 1% of the *total* population of Canada—an astonishing figure.

There's been a lot of response all over the place.

Traditionally, campuses have been a major source of resistance. Yet a new study from UCLA says that student activism is at an all-time low, and that interest in government and politics has plummeted. It also states that students' "academic involvement has gone down as well….They're watching more TV." Does that track with your own perceptions?

To say that this is a low point is short-sighted. Is it lower than the 1950s? Is it lower than 1961, when John F. Kennedy sent the Air Force to bomb South Vietnam and you couldn't get a single person to think about it?

When I gave talks on the war in the mid-1960s, we couldn't get anybody to attend. Students weren't interested—except sometimes in attacking the traitors who were condemning government policy. Most of the real and important student activism took place in the late 1960s, and it was by no means "traditional."

What about the anti-apartheid movement in the late 1980s?

That was real and important, but it's not all that was happening in the 1980s. The Central America solidarity movement was far more deeply rooted in the mainstream of society. Students were involved, but they weren't by any means at the core of it. You found more in churches in places like Arizona and Kansas than in the elite universities.

As for the decline in student activism (and reading, and academic work), that's not students—that's the society. The Robert Putnam study we discussed earlier found about a 50% decline since the 1960s in any form of interaction—visiting your neighbor, going to PTA meetings, joining a bowling league. (There's debate about his conclusions, but something of the sort seems to be correct.)

What about the nonaligned movement?

In the 1950s, several Third World leaders tried to establish a form of nonalignment, which decolonization and the conflict between the US and the USSR made possible. By now, that movement has pretty much disappeared, both because of enormous changes in the global economy and because the end of the Cold War eliminated the superpower competition and the deterrent effect of Soviet power, which allowed for a degree of independence. The West doesn't have to pretend anymore that it's interested in helping anybody.

The decline of the nonaligned movement and of Western social democracy are two parts of the same picture. Both reflect the radicalization of the modern socioeconomic system, where more and more power is put into the hands of unaccountable institutions that are basically totalitarian (though they happen to be private, and crucially reliant on powerful states).

Is the nonaligned movement completely gone?

As recently as the early 1990s, the South Commission, which represented the governments of nonaligned countries, came out with a very important critique of the antidemocratic, neoliberal model that's being forced on the Third World. (The commission included pretty conservative people, like Indonesia's development minister.)

They published a book that called for a new world order (they introduced the term before George [H. W.] Bush did) based on democracy, justice, development and so on. The book wasn't obscure—it was published by Oxford University Press. I wrote about it, but I couldn't find much else. They subsequently published another book of essays commenting on the first one, and I've never seen a reference to that either.

The South Commission happened to represent most of the world's population, but the story they were telling just isn't one the Western media wanted to hear. So the "new world order" we learned about was Bush's, not the one advocated by the South Commission, which reflects the interests of most of the people of the world.

Back in the 1950s, there were Nehru, Nasser, Tito, Nkrumah, Sukarno and others...

All of whom were despised by the US government.

But there was also a period of intellectual ferment in the newly independent countries. I'm thinking of people like Amilcar Cabral [1924–73, leader of the independence struggle in the former Portuguese colony of Guinea in West Africa] and Franz Fanon [1925–61, the French author of *The Wretched of the Earth,* who fought for Algerian independence]. I don't see much of that right now.

There's still plenty of intellectual ferment, but it doesn't have the enthusiasm and the optimism of those days (although you can hardly call Fanon very optimistic).

It had more of a revolutionary edge back then.

Yes, it did, but remember that since then there's been a period of extreme terror throughout much of the Third World—in which we've played a prominent part—and that's traumatized a lot of people.

The Jesuits of Central America are very courageous people. (Since they're true dissidents within our domains, you hear very little about them here, unless they're murdered. Even their writings are unknown.)

In January 1994, right before the Salvadoran election, they held a conference on the "culture of terror." They said terror has a deeper effect than simply killing a lot of people and frightening a lot of others. They called this deeper effect the "domestication of aspirations"—which basically means that people lose hope. They know that if they try to change things, they're going to get slaughtered, so they just don't try.

The Vatican has had a very harmful impact on all this. It's tried to undermine the progressive thrust of the Latin American church—its "preferential option for the poor" and its attempt to serve as a "voice for the voiceless"—by installing very right-wing bishops. (The *New York Times* had an article on this the other day, but there was a slight omission in it: the role of the US—which is crucial, of course—wasn't mentioned.)

In El Salvador in 1995, the Pope installed as archbishop a Spaniard from the right-wing Opus Dei, who essentially told the poor: *Don't worry about social conditions. If you keep away from sin, everything will be fine in the next life.* This was after the assassination of Archbishop Romero, along with dozens of priests, bishops, nuns and tens of thousands of others, in the brutal war the US ran in the 1980s—a major aim of which was to destroy the Salvadoran Church's concern for the poor. The new archbishop accepted the rank of Brigadier General from the military, which—he explained—did not "commit errors" as an institution and was now "purified."

Similar things have happened elsewhere. In Indonesia, the Communist Party (PKI) had millions of followers. Even conservative experts on Indonesia recognize that the PKI's strength was based on the fact that it really did represent the interests of poor people. In 1965, General Suharto and his followers in the army presided over the slaughter of hundreds of thousands of landless peasants (and others) and wiped out the PKI.

They went on to compile a world-class record of terror, torture, aggression, massacre and corruption. The Clinton administration has described Suharto as "our kind of guy." Amazingly, quite an impressive popular struggle is still going on in Indonesia, but of course we don't hear much about it.

You once wrote to a mutual friend that when educated classes line up for a parade, people of conscience have three options—they can march in the parade, join the cheering throngs on the sidelines, or speak out against the parade (and, of course, expect to pay the price for doing that).

That's about right. That's been the story for a couple of thousand years or so. Go back to the oldest recorded texts and see what happens to people who didn't march in the parade...like Socrates. Or take the intellectuals described in the Bible (where they're called "prophets").

There were two types of prophets. One type, who flattered the kings and either led the parade or cheered it from the sidelines, were honored and respected. (Much later, they were called *false prophets*, but not at the time.) Then there were people like Amos, who incidentally insisted that he was not a prophet or the son of one, just a poor shepherd.

True prophets like Amos—"dissident intellectuals," in modern terminology—offered both elevated moral lessons, which the people in power weren't fond of, and geopolitical analyses that usually turned out to be pretty accurate, which the people in power were even less fond of. Naturally, the true prophets were despised, imprisoned, driven into the desert.

The public also hated the true prophets—they didn't want to hear the truth either. Not because they were bad people, but for all the usual reasons—short-term interest, manipulation, dependence on power.

The magic answer

I often hear the internet proposed as the one great solution to society's problems.

The internet should be taken seriously; like other technologies, it has lots of opportunities and lots of dangers. You can't ask, *Is a hammer good or bad?* In the hands of somebody who's building a house, it's good; in the hands of a torturer, it's bad. The internet is the same. But even used for good, it's obviously not the solution to everything.

When we do something, do we have to have a clear idea about the long-term goal in order to devise a strategy?

We learn by trying. We can't start now, with current understanding, and say, *Okay, let's design a libertarian society.* We have to gain the insight and understanding that allows us to move step-by-step toward that end. Just as in any other aspect of life, as you do more, you learn more. You associate with other people and create organizations, and out of them come new problems, new methods, new strategies.

307

If somebody can come up with a general, all-purpose strategy, everyone will be delighted, but it hasn't happened in the last couple of thousand years. If Marx had been asked, *What's the strategy for overthrowing capitalism?*, he would have laughed.

Even somebody who was overwhelmingly a tactician, like Lenin, didn't have any such strategy (other than *follow me*). Lenin and Trotsky just adapted strategies to particular circumstances, looking for a way to take state power (which I don't think should be our goal, by the way).

How could there be a general strategy for overcoming authoritarian institutions? I think questions like that are mostly asked by people who don't want to become engaged. When you become engaged, plenty of problems arise that you can work on.

But it's not going to happen by pushing a button. It's going to happen by dedicated, concentrated work that slowly builds up people's understanding and relationships, including one's own, along with support systems and alternative institutions. Then something can happen.

Urvashi Vaid, author of *Virtual Equality,* castigates what she calls the "purist left" for waiting for the perfect vision, the one and only answer, as well as a charismatic leader.

I agree. Not waiting for a charismatic leader, or the perfect and complete answer, is good advice. In fact, if it comes, it will be a disaster, as it always has been.

If something grows out of popular action and participation, it can be healthy. Maybe it won't, but at least it *can* be. There's no other way.

You've always seen top-down strategies and movements as inherently doomed.

They can succeed very well at exactly what they're designed to do—maintain top-down leadership, control and authority. It shouldn't have come as a tremendous surprise to anyone that a vanguard party would end up running a totalitarian state.

Howard Zinn suggests that we need to recognize that real social change takes time. We need to be long-distance runners, not sprinters. What do you think of that?

He's right. It was very striking in parts of the student move-ment in the 1960s. There wasn't an organized, well-established, popular-based left for the students to join, so their leaders were sometimes very young people. They were often very good and decent people, but the perception of many—not all—of them was quite short-range. The idea was, *We'll strike Columbia, close down the buildings for a couple of weeks, and after that we'll have a revolution.*

That's not the way things work. You have to build slowly and ensure that your next step grows out of what's already estab-lished in people's perceptions and attitudes, their conception of what they want to attain and the circumstances in which it's possible to attain it.

It makes absolutely no sense to expose yourself and others to destruction when you don't have a social base from which you can protect the gains that you've made. That's been found over and over again in guerrilla movements and the like—you just get crushed by the powerful. A lot of the spirit of '68 was like that. It was a disaster for many of the people involved, and it left a sad legacy.

Are you aware of different sorts of responses you get from different audiences?

Over the years, I *have* noticed a very striking difference be-tween talks I give to more or less elite audiences, and meetings and discussions I have with less privileged people. A while back I was in a town in Massachusetts at a meeting set up by very good local organizers in the urban community—people who were pretty poor, even by world standards. Not long before that, I spent time in the West Bengal countryside. Then I was in Colombia, talking to human rights activists who are working under horrifying conditions.

In places like that, people never ask, *What should I do?* They say, *Here's what I'm doing. What do you think about it?* Maybe they'd like reactions or suggestions, but they're already dealing with the problem. They're not sitting around waiting for a magic answer, which doesn't exist.

When I speak to elite audiences, I constantly get asked, *What's the solution?* If I say obvious things like *Pick your cause and go volunteer for a group that's working on it,* that's never the answer they want. They want some sort of magic key that will solve everything quickly, overwhelmingly and effectively. There are no such solutions. There are only the kind that people are working on in Massachusetts towns, in self-governing villages in India, at the Jesuit Center in Colombia.

People who are actually engaged in dealing with the problems of life, often under extreme repression and very harsh conditions, sometimes just give up. You can find that too. But many keep struggling effectively and bring about changes.

That's been true in our own history. Right now we're facing real problems, like protecting the limited level of public medical care, the Social Security system, environmental rights, workers' rights. But you don't have to go very far back to get to the time when people were trying to *gain* those rights. That's a big change. It's a lot better to be protecting something than trying to get it for the first time.

These rights are the result of popular engagement and struggle. If there's another way to achieve them, it's been kept a dark secret. But privileged audiences often don't want to hear that. They want a quick answer that will get the job done fast.

Manufacturing dissent

Michael Moore made a documentary film called *Roger and Me* and produced a television series called *TV Nation.* In his book *Downsize This!,* he says that what turns people off about the left is that it's boring, it whines too much, it's too negative. Anything to that?

I don't think Howard Zinn, say, whines too much and turns people off, but there are probably other people who do. To the extent that that's true, it's a problem they should overcome.

Take the example of the media group in Brazil we discussed earlier, which presented television skits that turned people off because they were boring and full of jargon. This group went back to the people and let them produce the stuff themselves, simply providing technical assistance. That second set of programs wasn't boring and didn't turn people off.

That's exactly the correct approach. People who write about the responsibility of intellectuals should *assume* that responsibility and go out and work with people, provide them whatever help you can, learn from them.

You've observed grassroots movements in places like India, Brazil and Argentina. Can we learn anything from them?

Those are very vibrant, dynamic societies, with huge problems and lots going on. But I think they're also trapped by delusions like, *We've got this terrible foreign debt. We've got to minimize the state.* They've got to understand that they don't have any debt—just as *we* have to understand that corporations are illegitimate private tyrannies.

You've got to free yourself intellectually, and you can't do it alone—you liberate yourself through participation with others, just as you learn things in science by interacting with others. Popular organizations and umbrella groups help create a basis for this.

Is that enough to bring about serious changes? It's hard to say. We have all sorts of advantages that they don't have—like enormous wealth, for instance. We also have a unique advantage—there's no superpower standing over us. We *are* the superpower. That makes a huge difference.

But when you come back from the Third World to the West—the US in particular—you're struck by the narrowing of thought and understanding, the limited nature of legitimate discussion, the separation of people from each other. It's startling how stultifying it feels, since our opportunities are so vastly greater here.

Do you have any ideas on how we can move from preaching to the choir, to people that already agree with us? This seems to be a major problem.

First of all, as we've discussed a couple of times already, a large majority already does agree with these ideas. The question is, how to turn those general attitudes into real understanding and constructive actions. The answer is, by organizing to do so.

Whenever I—or anybody—gives a talk, it's because some group has set it up. I can't just show up in Kansas City and say, *I'm going to give a talk*—nobody would come. But if a group there organizes it, people will come from all over the place, and

311

maybe that will help the organizers, and others, to get together and to proceed more effectively.

This all goes back to the same thing: If people dedicate themselves to organizing and activism, we'll gain access to broader and broader audiences.

As you know, I do a one-hour radio program every week. It's pretty effectively locked out of the Boston-to-Miami corridor, but in the West—in Montana, Colorado, New Mexico, and places like that—it's much easier to get it on the air.

It doesn't matter much to the power centers what people are talking about in Laramie, Wyoming. The East Coast is where most of the decisions get made, so that's what has to be kept under tight doctrinal control.

But we can't just blame the people in power. We aren't making use of the possibilities we have.

Take Cambridge, where we're sitting now. Like other towns, it has a community cable television station (the Communications Act requires that cable companies provide them). I've been there. I'm not much of a techie, but even I could see that it has pretty good equipment. It's available to the public, but is it used by anyone?

The one time I was on that station, the program was so crazy I almost walked off. What would happen if you had lively, quality local cable TV? The commercial channels would have to respond to that. They might try to stop it or undercut it or co-opt it, but they'd have to do something if there got to be enough of it. So would NPR. They can't completely disregard what's happening in their communities.

So that's one resource that isn't being used the way it could be. In the slums of Rio, they'd be delighted if they had cable television stations that the people could use. We have them and we're not using them effectively.

Cassette tapes are one mechanism to disseminate this information. They're easy to duplicate and pass around. The Iranian revolution was called the first cassette revolution.

There are lots of opportunities. Compared with people in other countries, our resources and options are so enormous that we can only blame ourselves for not doing more.

In Elaine Briére's documentary film on East Timor, *Bitter Paradise*, you say, "The press isn't in the business of letting people know how power works. It would be crazy to expect that....They're part of the power system—why should they expose it?" Given that, is there any point in sending op-ed pieces to newspapers, writing letters to the editor, making phone calls?

They're all very good things to do. Our system is much more flexible and fluid than a real tyranny, and even a real tyranny isn't immune to public pressures. Every one of these openings should be exploited, in all sorts of ways.

When you get away from the really top, agenda-setting media, there are plenty of opportunities. It isn't just a matter of writing op-eds and making telephone calls, but insisting, by all kinds of public pressures, that there be openings to your point of view.

There are understandable institutional reasons why the media are so deeply indoctrinated and hard to penetrate, but it's not graven in stone. In fact, the same factors that make it so rigid also make it rich in ways to overcome that rigidity. But you have to *do* something—you can't just sit around waiting for a savior.

Another approach is creating alternative media, which may well have the effect of opening up the major media. That's often been done.

But you don't see getting the occasional op-ed piece published as a substitute for a truly independent, democratic media.

It's not a substitute—it's a step towards it. These things interact.

You're often introduced as someone who speaks truth to power, but I believe you take issue with that Quaker slogan.

The Quakers you're referring to are very honest and decent, and some of the most courageous people I've ever known. We've been through a lot together, gone to jail together, and we're friends. But—as I've told them plenty of times—I don't like that slogan.

Speaking truth to power makes no sense. There's no point in speaking the truth to Henry Kissinger—he knows it already. Instead, speak truth to the *powerless*—or, better, *with* the powerless. Then they'll *act* to dismantle illegitimate power.

A Canadian journal called *Outlook* ran an article on the talk you gave in Vancouver. It concluded with quotes from people leaving the hall: *Well, he certainly left me depressed.* And: *I'm more upset than I was before I came.* And on and on. Is there any way to change that?

I've heard that a lot, and I understand why. I feel that it's none of my business to tell people what they ought to do—that's for them to figure out. I don't even know what *I* ought to do.

So I just try to describe as best I can what I think is happening. When you look at that, it's not very pretty, and if you extrapolate it into the future, it's very ugly.

But the point is—and it's my fault if I don't make this clear—*it's not inevitable.* The future can be changed. But we can't change things unless we at least begin to understand them.

We've had plenty of successes; they're cumulative, and they lead us to new peaks to climb. We've also had plenty of failures. Nobody ever said it was going to be easy.

INDEX

THE FOLLOWING PROGRAM MADE POSSIBLE BY:
The false bottom of history
The search for intelligent life and/or meaning
Cooperation asserting itself against competition
Implacable masses, astronomical and otherwise
Rows upon rows of riot police
The comfort of approximation
Sleeping dogs, lying
Two people in love in a fight in the car
Occluded foreign factories
"Where do we go from here?"
Torture, when necessary
Brackets indicating variable text
The divide between speaker and spoken
Orbits, astronomical and otherwise
Concise language, a must
Dis[Advertisement]course
Symbolic burden
Frailty, human and otherwise
The fragmentation of signal
Systemic antagonism misrepresented as anomaly
Light waves radiating, wavelengths stretching, telescopes swiveling
The final line, infusing previous lines with great depth

INSENSATE

Hazardous electrical towers gate the community. Eighty-foot dishes rotate into position to observe a sliver of sky. Cables burrow underground, shed insulation and leak into the soil. A copse of evergreens obscures the buzzing transformers. The signal travels at the speed of light, through a narrow tube and down the patient's throat, allowing the lungs to breathe.

Food, born in the grocery store, destined for our bellies, leads a life of asocial comfort along exotic aisles. The system has never been closed. Supine in an MRI, they cut the brain into fetish-thin slices, perfect for lunch hour obsession. Take another dose of chemotherapy and microblog the pain. A disembodied head floats above soothing pastel green sheets, dreaming of the insensate paradise.

Swipe your card, debit your account, scan your code, enter your email address. Logic boards stacked in the city dump oxidize along copper pathways. Empty sockets yearn for memory. Hollow ports desire serial connection. The unique identifiers have been filed off in preparation for repurposing. The millennia when waste will survive its creators approach.

PARALLAX OF INTENTION

I feel like you're forgiving me for something that doesn't require forgiveness, you say to the steering wheel. I never said 'I forgive you' or even implied it, I say to my sleeve. Looking out the window, I think of the Doppler shift. First in sound, the cars streaming past in frenetic desperation, then in light, distant galaxies redshifting out of view, then in emotion, passing each other by, creating a parallax of intention that ripples out behind us in malformed waves.

In a competition to see who can endure discomfort longer, we both lose. I remain silent in respect to an alien pride. You too remain as silent as your reasons for doing so. I imagine what this must look like in your eyes, seeing me sitting in your periphery as I see you, the slight shift in perspective of the road ahead, your awareness of my awareness. We both think in frenetic desperation, rationalizing our behavior. I really tried to pull us out of this nosedive, I say. I wouldn't say you really tried, you answer.

If a craft ever exists that can travel in all four dimensions with ease, some wanderer may return to here and now and observe how we wasted away our time grappling with the seductive inward spiral. I apologize in advance for any disappointment.

CHICXULUB

A life-ending asteroid doesn't know its trajectory. It rotates and orbits, indifferent to the consequences of its eventual impact. Put all that in the back of your mind and make a left at the next light. I've never seen a fly drop, have you? There will be no tears on the last perfect day, by definition. Tomorrow promises to be a doozy, by the process of elimination. Market research shows increased apathy towards potential annihilation. The statistician presents lewd figures of far more common catastrophes. C'est la mort! Who kisses Jupiter's bruises and makes it all better? A few more blocks.

A dinosaur looked up into the sky. A businessman hurries about his business without looking up into the sky. I wanted to text you in case the world ends while I'm asleep. Act rationally. I never wanted to be an actor, bravely risking my life to stop an asteroid from hitting the earth. I wanted to live in a universe where monotonous voices don't assure us "the question is not if, but when". There's no monster in the closet. I wanted to be a ray of light, careening across the void in timeless motion. There's no bogey under the bed. But that's all behind me now. You can let me out here. Goodnight.

HAGIOSCOPE

As a fan of astronomy, I often step out onto my deck in the late evening, when the sun has descended well below the vineyard hills, and the first light of the stars penetrates the sky. I take out my simple telescope, purchased at an electronics store for $49.99 plus tax, and put it to my eye. I scan from left to right, looking for the sight that will move my soul. Mostly I see a great darkness, with occasional glimpses of light that pass across the eyepiece too quickly, causing me to steady my hand and search for them again in earnest. I begin to feel my soul lift when I see the first human motion.

A woman, in her mid thirties I would guess, turns on the light in her bedroom. She walks close to the window, and I pray that she doesn't close the blinds. She continues standing in place, doing something I cannot fathom. Her face is kind, but neutral, and her figure seems average. My arm shakes from steadying the telescope, but my determination wills me to continue. Before I know whether or not she will take off her clothes, I realize that my soul has been moved, and I feel the humility of every ardent astronomer enthralled by the power of the unseen gaze.

POTEMKIN

These fleeting-improvised-men at the deconstruction site raise temporary scaffolding around an imaginary tower. These constructs bind us to the tower of history, temporarily. These piecemeal men at the grocery store separate themselves into dusty bulk bins. These patchwork marionettes dance from strings on the big screen, professionally. The big screen strings us out of reality, in theory. Divine rays corrupt the nerves with the possibility of soul death. These fleeting-improvised-men attract fragmented messages from the scattered letters of their accretion disks. These Freudians deconstruct wooden artist models to determine the destiny of anatomy.

Perpetual markets arrest the development of the inchoate young. Grown men died long ago, when an ape inadvertently poked its feces-encrusted finger through the world's delicate façade. Potemkin men in Potemkin villages buttress themselves upon repression in preparation for the second coming of their sovereign. The domino theory of the deep gaze starts the change reaction. The second domino falls half way, then half way, then half way...

SQUARE SQUARE DIAMOND TRIANGLE

Square square diamond triangle. Square square diamond triangle. Can anyone hear me? Can anyone hear me? Can square square me? Can diamond triangle me? Can square hear square? Diamond anyone triangle square? Square diamond triangle me? Square square diamond triangle? Can square square anyone diamond triangle hear hear me? Square can square anyone diamond hear triangle me? Square square diamond triangle.

GLOBULAR

In 1974, humans sent a radio transmission from the Arecibo observatory in Puerto Rico to the globular star cluster M13. They encoded the message in the prime number 1679, in the hopes that potential audiences would understand the multiples. The message consisted of seven parts:

1) Staying behind in a hotel room while a significant other enjoys the beach.
2) A thermonuclear weapon for preemptive defense.
3) Synthetic emotional content deemed too intimate for television.
4) All children under 18, without the accompaniment of a parent or guardian.
5) The complete works of Ligeti, with no small amount of nostalgia.
6) Misinformation, false flags and other assorted espionage.
7) Nights strung together with headphone cords and LCD illumination.
8) A precise explanation of our mathematics.

No earth scientists were able to decode the message, but in 25,000 years, those aliens will get it.

PRECIOUS MOMENTS

The camera captures the structure of fiction among real life characters pretending to be false. A cut here, a cut there, a suture along the line of motion. We'll loop your last words in the post-mortem. Enframe the frame in another iteration and abstract the experience through another curved glass. I'll come from my trailer once the stunt dummy's been beaten. The continuity of this soliloquy has been guaranteed by the script supervisor, who lies in wait below the line. Strike me from this abandoned set and I will become more popular than you could ever imagine. The long-term effects of viewing motion pictures remain unknown. Let's all go to the lobby. Let's all go to the lobby. Let's all go to the lobby.

APPOINTMENTS

--BEGIN TRANSMISSION--

We've determined your desire according to the conservation of angular momentum, the governing principle of our universe.

We've constructed your body according to entropy. Our data indicate you will feel joy and terror briefly, then forget.

We've designed your potential according to gravity. All trajectories are parabolic. This is a dream, or at least a sleep.

Our protocol specifies that we may not communicate with you except through encrypted channels. You do not have the decryption key.

We apologize in advance for any discomfort. We sincerely apologize for any fear. The illusion of diametric opposites is no illusion. Sorry.

The stars are lovely, dark and deep. But we have important appointments to keep. And light years to go before we sleep. And light years to go before we sleep.

--END TRANSMISSION--

HELL
Engineering competitive business to business social marketing
strategies,
Escalating brand awareness through responsive, crowdsourced
clouds,
Offering robust ROI data analysis with tried and true TCO
metrics,
Enacting global paradigm shifts of content delivery with user-friendly
gateways,
Producing proactive real-time solutions that adapt to scalable growth
projections,
Forming outside-of-the-box, innovative, on demand
workflows.

Your life may be monitored for quality assurance purposes. A tiny man fallen
down a tiny hole squeals out with a tiny voice for a tiny god. Two people
awkwardly sidestepping back and forth on the sidewalk, trying to stay out of
each other's way, forever. I was a floating mote of dust. You, another mote of
dust, drifted past me and we both evaporated in the sunlight.

All statements should be multilayered, contradictory ciphers to remind us that
language is a poor approximation of experience. I already have the erection,
so I might as well hang myself. Hell is a blinking cursor in an empty text field.

FISHBOWL

The horizon curves for every event, relative to the fixed body of the observer. Ideology ingrained since birth warps the fabric of relation. Insulate the undeveloped man inside a pseudo womb. The delicacy of placenta satiates the appetite for comforting familiarity. Ideas permeate the membrane between two otherwise discrete spheres. Osmosis drives the communal host to the madness of ritual. Tinted lenses welded to the eyes since birth come in many different colors. Eggshell wrappers over reptilian centers shatter upon collision.

No world of forms, no noumenal truth, no access to the real. Phenomena explode in space with childlike delight. Electrical fields interfere, creating the resistance. Pin the subject's discourse down within clinical parameters. The position of the questioner possesses inherent power. Dynamics shift from side to side, maintaining inequality. Once the system settles down, it loses all its energy.

What a sight it must be, from the outside in, to see these creatures floundering. An amusing show for indolent gods, drunk and splayed on barcaloungers, whiling away their deathless days, occasionally plucking out a writhing pet, tossing it on the marble floor and watching it flip and flop and flip and stop.

PASSIVE VOICE

Attention. In literature, as in life, one should avoid "to be". Engage the reader with active verbs. Don't dabble in blurry definitions. The people clamor for "clamor". The masses seethe for "seethe". Too much "being" in one's work makes one look lazy. "Look" too, one should use sparsely at most. "Seems", "have" and other go-to words follow suit. Don't address the reader as "you". Instead, if one must address the reader, say "one". Don't abuse quotation marks. An occasional peppering draws attention to specific words, but overuse indicates a wavering position. One should not rely on ellipses to indicate pregnant pauses. Never, under any circumstances, should one repeat one's self.

Pretend language can speak. Pretend communication can communicate. The voice carries the meaning. Transparent rhetorical gimmicks roll over the otherwise empty page. This is not happening. Subject A speaks while Subject B listens to electronic music on his headphones in front of a computer monitor. Pay attention like currency. Music travels over a wire as current and reassembles itself in the ear. If only it could be "just so". Seems like you've got it all figured out.

Things are. People are. Only language seethes and clamors. Words say a "man" [verbs]. The word "man" [verbs]. But the man "man" represents is. Reduce the world to a handful of interchangeable symbols. Deny all that merely "is". It is not enough to be. You must be *something*. The imperative was put upon you early. Now is your time! Don't waste it! Take action! For instance, to pass the time, I impale the heads of contemplatives on spikes high above the city.

THE SLOWEST OF US, THE WEAKEST

We used to communicate via walkie-talkie. Using movie-learned jargon, we spoke into the toy radios with more seriousness than we would give the rest of our lives. Over. We coordinated watergun wars, plotted tactical balloon strikes, outlined ambushes against the enemy. We got one. Over.

First we disarmed him, putting his watergun in a corner. He surrendered, rather than get drenched, the coward. He was our prisoner. We sat him in a chair and began the interrogation. What is your attack plan? Where is your balloon stash? Who is defending your base? Each unanswered question resulted in a popped water balloon inches over his head. He told us everything we wanted to know.

He started crying, but we had already begun to move out. We took them down big time. We fucked them up. Enemy base secure. Over. We won the day. We knew that next time it might be one of us taken prisoner. The slowest of us, the weakest. We went home and spraypainted our waterguns black, to match the walkie-talkies. Summer break would never end.

CONTINGENCY

Perfect beings would, by definition, be indivisible. Definitely not a composite of limbs and organs, molecules and atoms. Pieces of a whole signify their own eventual separation. Only that which cannot be reduced can really be. Between being and becoming who defines themselves as a transitory emergent property? Even black holes evaporate. Even protons decay.

We're here, in this universe of building blocks, built of these building blocks, self-aware, yet contingent. Why does anything at all exist? Why not nothing? Why not something different? What happens when we die? Fuck off, the mother replied. That's an old joke, but it shut us up.

TESSELLATION

Scanning for patterns in infinite chaos reveals every pattern, eventually, accidentally. A random sentence generator, given enough time, would produce this sentence, but not the living consciousness to smirk at its simulation of irony. The record turns under the needle and the analog waves are converted to commodity. Yearning for a life without routine rusts routine into inescapable grooves.

Mistaking objects for human shapes. Shade and light play at human faces. Tessellated light waves aggregate into information. After refraction comes reflection. Lovers scan each other's eyes for sincerity. Lasers correct organic catastrophe. Those who do not learn from repetition are doomed to historicize it. Yearning for a life without routine.

The farmland extends into the fog, but having trespassed through muddy troughs, you know there's neither relief nor answer there. You look out the airplane window and see the land divided into fenced geometric fields, red and green, brown and gray. The casino floor weaves hypnotic redundancy between steel-plastic monoliths ringing with victory. You know there's neither relief nor answer in the prerecorded siren's song. The smoldering faces of ashen retirees grip cigarettes with dry, black mouths. Routine rusts routine into inescapable grooves.

THE RIOT COP'S DINNER

A factory in China vents toxic fumes, a necessary byproduct of the manufacture of resilient fiberglass shields. Human lives could be at stake. Another factory manufactures toxic fumes and sells them in little canisters to approved police and military agencies, which in turn use them to disperse unruly crowds.

The riot cop disrobes before his wife, who shares his noble burden. His saggy belly hangs over his blue and white striped boxer shorts. [unintelligible grunt], he says sadly. [cooing reassurance], she replies. He wiggles his veiny feet into a pair of slippers. She serves him a reheated dinner in his recliner. He folds up a newspaper and sets it on the end table, shaking his head in weary disgust.

HAD TO BUST OPEN A KID'S SKULL TODAY, WIFE. JUST DOING MY JOB. I DON'T MAKE THE LAWS. WE'RE THE GOOD GUYS. AND SO FORTH HAHA. THIS STEAK IS DELICIOUS AND RARE.

A TV commercial reassures him. Everything is going to be okay, riot cop. Everything is going to be okay, riot cop's wife. Everything is going to be okay, unarmed kid. Everything is going to be okay, Chinese factory worker. Everything is going to be okay, reader. The commercial continues.

CARTHAGE

The muse of the post-modern poet amuses the contemporary critic, who searches for clichés like an unsound simile. The post-muse poet arrogantly splits the infinitive in the hopes of a detailed cross-section of language. The new age linguist reifies words into unread manuals in the hopes of successful exorcism. The data analyst finds himself caught in the vicious circle of a square peg. Technological acumen overtakes natural selection in the 14 billion light year dash. The tedium of history's cul-de-sac builds in our balls til we burst.

To Carthage then I came, to buy you a souvenir from the gift shop, to prove I was there, to prove I was thinking of you, to prove that intangible uniqueness can be compressed into a trinket and sold, to secure my alibi when the murder of direct experience took place in the living room, to absolve myself of this confession.

NOBODY KNOWS ANYTHING

I found you between knowing and not knowing. I never knew. I know that now. You look straight ahead, even though you know I'm looking at you, waiting for you to look at me. I too turn away and both of us together look straight ahead separately. We're almost out of gas, I didn't say, fearing multiple interpretations. Maybe we should bomb Iran.

I intended to give a lecture on the equivalence of content within a given form, but my car broke down on the drive over. Has anyone figured anything out while waiting? Have you been waiting long?

What all this is, [significant pause], nobody knows. Take this abused clown, please. *Applause, laughter.* His body is made up of cells made up of molecules, made up of atoms, made up of protons et al, made up of quarks… Then what? [significant pause] [silence, separate from the pause, the sound of shuffling feet]. The chariot does not exist, or so the master tells us.

I look straight ahead, even though I know you're looking at me, waiting for me to look at you. You turn away and both of us together look straight ahead separately.

PINOCCHIO

I pull my strings and rise into posture. A jointed arm glides on its oiled hinge to brush my false teeth. The mannequin faces the mirror, but cannot pass the test. Out there, the real ones gather, scatter and do fuck all. My poorly carved legs jangle along the path of least resistance. My lifelike eyes swivel in their sunken sockets. The cult of the real ones performs its insane, inaccessible ritual. I watch in mute wonder, making clumsy grabs at each act's significance. My jaw opens and closes in staccato; the inside of my throat burns with a delightful cherry red. My ventriloquist, out on his smoke break, leers at young girls floating across the sidewalk.

The real ones smile like ciphers. The real ones interpret the codes. My artisan face implies emotion, but I have not that which passeth show. The real ones move with purpose, and burn away indecision with brilliant gesticulation. The real ones touch skin to skin; they transmit and receive Human Warmth™.

My feet falter in an attempt to stand unsupported. Each hinge gives and sags, bringing down its connected limb. The supple wood clatters in the middle of the room and real ears can't hear its commotion. My strings hang limp and I would feel shame, if real eyes could see my condition. The chosen glide their path around my decrepit ruin. I pull my strings to raise my clumsy hand, which passes through their real substance without interference. As night draws near and the real ones return to their exclusion, an overweight janitor sweeps me into her pan. She tosses me into the dumpster and lights a cigarette on her way to the parking lot.

RADIO FRAGMENTS, 1

Repression, the mechanism of happiness. The primordial denial that life, society and the universe are fundamentally broken, painful and terrifying.

Creatures, who are and must be alienated from their own being, who have no knowledge or understanding of their substance, manage to wake up and go about their very important business.

Society, the formalization of power, the hierarchy of resource distribution, guaranteed solely by violent force, curls around them like a blanket of scarabs.

The universe in its crude, impenetrable materiality slips away from itself into the unknown faster than the speed of light.

Push it down. Let's see a little smile. There it is!

This page, this book, the hands holding this book, the person guiding those hands: atoms fluttering in spacetime.

AUTOIMMUNE

Give the patient the prisoner's dilemma. Put down your pencils in 5...4...3...
An excess of defense inevitably becomes a ruthless, inward-facing offense.
Extrajudicial murder by the government cannot be considered assassination,
because assassination is illegal. Signal the antibodies to intercept the bodies
hitting the floor.

Free radicals roam the DMZ, promising humanitarian aid. A host, a host, your
kingdom as my host! Side effects of globalism may prove fatal in a small
number of cases. Consult your doctor before accepting loans at interest. The
defense budget offends. Fuck it, give the white soldiers free reign to take out
cancerous terrorist cells. It's not torture if it prevents future atrocities. Pardon
these patriots their trespasses and grant them autoimmunity in the trials to
come. Shock me into an awe-inspiring erection.

Leaked reports from the front line reveal our boys, cocks in hand, pissing on
the enemy dead. General, your comments? There ain't no got damn reason
why our brave young men should be risking their lives to piss on a mangled
heap of brownskinned corpses when a properly equipped drone could do it
better!

Now Northrup, now Grumman, now Lockheed and Martin! On Stratfor, on
Boeing, on Halli, and Burton!

PLAYING PRETEND

The funeral procession shuffles through the city to the somber tones of a respectable dirge. Authorities shut down the streets; business has been put on hold. Masses of people line the barricaded pavement, many of them crying, watching the passing black Cadillacs. Some display a stoic sorrow that mere tears fail to represent, while flags wave at half-mast, caught in a wind that joins the mourning with its sympathetic whistle.

After many sentimental speeches and much persuasive rhetoric, the final car, with the obscured body of the deceased comes into view. The windows are tinted out of respect for the dead. This great man played pretend in front of a camera for a living. You probably recognize his face. He beat up the bad guys, had a knack with the ladies, and did other things determined by group written scripts. He lived a life of luxury and freedom. He was very, very attractive. But most of all, he kindly paid our living rooms a visit every Thursday night from 7:30 to 8:00 PM.

WOOD

A gang of burly men in hardhats and reflective vests fell another tree. It splinters at the base and tumbles into the clearing. The men wear uniformly bored expressions, numb to the repetition. They strip the tree of branches and bark, until it resembles a roughly hewn telephone pole. A machine hoists the shorn trunk into the bed of an eighteen-wheeler. When the bed is full of broken, stripped trees, still issuing sap from their many wounds, the truck carries them to the mill, bed-a-bouncing with its long load.

At the mill, the trees are measured, cut, sanded, chipped and shredded. Some pieces become building lumber, others get mashed into dust, the kind they spread over vomit puddles, the kind they put on the shop class floor so that many years later you can remember the smell. They mix the leftovers into pulp and make it into paper. Some of that paper is then used to print a senior thesis titled Castration Anxiety.

CHRYSALIDES

Irradiate my balls with aftermarket refurbs. Fill my womb with epoxy and cover my shame with synthetic polymers. Augment my biology with mechanical prosthetics and transfer my energy to the federal power grid. Emaciated clones suspended from high-tension wires pass each other on asynchronous assembly lines, avoiding eye contact. Worms dig deep into the mantle, starved for human remains. This pie chart shows the distribution of suffering among people, animals and computers. Applications designed to program other applications overwrite obsolete code with artificial malice. Transfer consciousness to a more durable vessel and our organic bodies will become chrysalides, shed upon our induction into the true existence.

I am the subject of the enunciated, bodyguard to the subject of enunciation. I am the subject of Descartes, sundered a priori. I am the subject of self-definition, my existence precedes my essence. I am the subject of unbearable lightness, glutted on morning pastries. I am the subject of progress, falling along the arrow of history.

An idea enters the mind and propagates.

MEETING

Thanks for coming, everyone. The subject of this meeting is very important. We've spent a lot of time in preparation and expect to nail down some concrete decisions that will shape the future of our company. We've brought graphs, charts, slides, projections, handouts and rubrics. We want your input, but we ask that you wait until the end of the presentation before asking questions or making comments. It goes without saying that we highly value your opinions in this matter. We can whiteboard our ideas or engage in kinetic brainstorming activities in our mobile, ergonomic flex-furniture. My boss thinks this may be the most important meeting yet and his boss agrees.

Please don't forget to take advantage of our digital workspace. The virtual meeting environment allows document exchange with change tracking, versioning and notes. The collaborative interface provides us ample communication opportunities to maximize the throughput of our session. Instant feedback enables the real-time assessment of ideas and metrics of our action item progress. Please make sure you've installed the latest version of our client on your laptop, tablet and smart phone. Confirm you've patched it and installed the workspace module.

I hope everyone remembered to bring today's agenda, which was emailed, faxed and interofficed to each of you last week. Please write down any questions or comments you may have on the Notes section of the agenda and/or in the corresponding section in the virtual workspace. After the meeting, we will collate everyone's notes. This will let us easily track our progress in both our thinking processes and our execution of them. Let's not get ahead of ourselves here. A separate, heterogeneous committee will review our findings and recommend a course of action. Now, let's get started.

The room is dark and empty. All life on earth ended long ago. *Audience gasps*

FIRM HANDSHAKE

I believe in family values and the pride of a hard day's work. I believe in tall, silver beer cans with wide mouths. I believe in trucks like rocks, bravely roaming through desert canyons. I believe in tucked in shirts and potbellies. I believe we should support our troops, our president and our neighborhood watch. I believe in God, country and kin. I believe in laugh tracks on sitcoms and internet videos of animals performing tricks. I believe in the flag flying high. I believe in steak and potatoes. I believe a woman's place is at home with the children. I believe that in the event of an emergency, I would be a hero. I believe an armed society is a polite society. I believe in defending our borders and our jobs. I believe in personalized license plates and decals of cartoon children pissing on various logos. I believe in tribal tattoos, hostile sunglasses and goatees. I believe I'd protect my family with my life, especially from minorities. I believe books can't teach you like the school of hard knocks. I believe in the right to my beliefs. I believe this country's headed to hell in a hand basket. I believe my hard-earned money shouldn't go to lazy people on welfare. I believe my daughter's sixteenth birthday present will be a new car. I believe in a firm handshake and constant eye contact. I believe I'm having a heart attack. Help… someone help me. I can't believe it ends like this.

ANNUNCIATION

Crests, fallen, have risen again, after a three day waiting period. Stones have been rolled aside and smoked with friends on hotel balconies. The priests run circles around the censer, whipping the frenzy with sadistic delight. The garden of our pleasures, overrun with moss and beetles, lingers in memory as punishment for a sin forgotten. The flagellants gave up pain for Lent, and the usurers lent it back again at 14.99%. The moneychangers, driven from the temple, built their own, and it's like, way better. I wave from a window high above the clouds and hang a penny over the side, aiming for ants on the pavement.

A weeping shade awaits the soul. Ethical choices inscribe the nerves for God's convenient perusal. After my death, take no offense to anything I've said against the great mystery. This message delivered without words. You haven't figured it out. Nobody has figured anything out.

MANTRA

The time when you and I will no longer be together waits inevitably beyond the horizon. The hash marks of our relationship hang precariously across the vertical. I calculate the radius of our distance with a protracted silence. You solve my area with your absolute body language. I reduce the import of this event through a willful negation of intimacy.

I will not become vulnerable.
I will NOT become vulnerable.
I WILL NOT BECOME VULNERABLE.

IF IT WAS IMPORTANT

The scrap of paper falls out when the lady fumbles through her purse, looking for money. She doesn't notice it falling, even though it does so very plainly. I can't look away from it, even though no one else seems to see it. I reach down and pick it up without looking at it and try to hand it back to her. I don't want that, she says. I tell her she dropped it. It's not important, she says. If it was important, I would have picked it up. She looks at me like I'm crazy, then stops looking at me. I hold the scrap of paper in my hand and carry it to the trash. I want to text everyone in this building that life is absurd, but I don't even know what kind of text message plan I have. The lady, finished with her transaction, casts me a wary glance before leaving.

WAITING ROOM

In the waiting room, the people come and go, talking of the latest television show. Take a number, any number, integers preferred, but not zero, which remains unspoken, unheard.

Thickness of dust, slightness of hands, thoughts stream from the head in radial bands. The next number's called, but you can't make it out. You arch your bad back and stretch your legs out. The person beside you, at least that's what you call him, assures you he's next to drink from the fountain. He looks at his ticket as the next number's called, but his eyes can't make sense of the illegible scrawl. You'd like to comfort him and share in his pain, but you'd rather be called up yourself all the same.

Who speaks through the speaker? Whose voice makes the call? Who built the chairs, the ceiling, the walls? Who made the tickets and how are they meted? How long will it be before you are greeted? For what do you wait and who are these others? Enemies, strangers, companions, brothers? Who sits in your chair? Who speaks with your voice? Who lives in your life? Who chooses your choice?

In the waiting room, none tell, none show, none move, none see, none come, none go. None talk of the latest television show.

GNASH AND CRAWL

In the field, outside the house, I watched the beetles gnash and crawl. Home for winter, I watched the field where all the creatures gnash and crawl. Beside the barn, I watched the creatures in the field gnash and crawl. In the skies circled two hawks who watched the earthworms gnash and crawl. The icy pond beside the field rippled with discarded skins. The mosquitos slept inside their eggs beside the icy pond. In the earth, worms gnashed and crawled, devouring decay. In the pond, below the frost, I acclimated well. The beetles crawled across the earth and gnashed it in their maws. In the field, beside the house, I watched the beetles turn to ice.

Bacteria split and multiplied within the fecund soil. Home for winter, I turned and ran away from my desire. The widening gyre of our past spins upon its axis. The barn still stands behind the field where I was once a child. Plant me deep when my time comes and do it with panache. The gnashing beetles of the field can crawl upon my lunch. Soon I'll be gone, I say majestically, though no one hears my words. Back to college and all my friends and the overflow of booze.

RADIO FRAGMENTS, 2

Autonomy, or its impermeable illusion, offers hope. While dependent on the universe and its laws, on biology and its limits, on society and its fetters, people, in their miniscule spheres of influence, control the directions of their lives. Not the grand scale, not the connected dots of Important Milestones, but the passing seconds, the streams of consciousness, the idle fidgeting, the ceiling staring, the neutral observation of nature, the ethical framework of choice.

Every moment presents a dilemma. Autonomy has the freedom to choose against the easy path for the greater good. Autonomy defines the greater good according to relative criteria. Autonomy invents altruism at the expense of guilty bystanders. Ideology penetrates autonomy against its will, but to be fair, autonomy is always a little drunk. Autonomy can defend its honor against ideology and bask in the equivalence of content. Searching for meaning means searching meaning for meaning. Of all the many constructs in the universe encountered so far, only human beings can be, do and define good.

LYING DOGS

Let sleeping dogs lie in their own filth. A bird in the hand is easily crushed. I waste not, therefore I am not. Idle hands are the fate of all hands. A stitch in time saves nine minutes of underpaid manpower. The best defense is omnipotence. Home is where the leading tech companies consolidate push button functionality. Fight fire with the protection of an adult sized karate gi. Early to bed and early to rise relative to what? A penny saved is a penny providing additional loan opportunities for your bank. A dead man tells the end of every tale.

The way we profit from your personal information is in the details. The enemy of my enemy is my enemy's enemy. Cleaning up one man's trash is another man's treasured source of income. No man is but an island. If you can't say anything nice, welcome to the club. Cleanliness is the opiate of obsessive-compulsive masses. A picture is worth $29.99. A friend in need is the human condition. Actions speak louder than words, especially yelling, screaming and shouting. Laughter is the best anesthetic. One good turn could make you rich. Opportunity rarely knocks up first dates. Still waters breed insect larvae.

A chain is only as strong as its wallet. A dog is man's eugenics experiment. A mortal and his money are soon parted. A person is known by the companies whose products he buys. Good things come to those who consider the things that come to them good. Better safe than happy. Better to remain silent and be thought a fool than express yourself in the short time you're here. Hell hath no fury because it doesn't exist. No good deed goes unclaimed. Laugh and the world remains indifferent, weep and the world remains indifferent.

GRAND CANYON

I said many things to you across the Grand Canyon. We stood on opposite banks, jumped up and down, waved our arms and shouted. I couldn't hear you and you couldn't hear me. We heard our own voices, modified and retransmitted in echo by the steep rock walls.

I returned to the Grand Canyon after you had left me and stood at its rim. I wanted to say many things to you then. I stood opposite nothing, did nothing, said nothing. I couldn't hear you and you couldn't hear me. Our silence modified and retransmitted in echo.

My voice and my actions stand on opposite banks of the Grand Canyon. My voice says many things that my actions can't live up to. My voice jumps up and down, waves its arms and shouts. But my actions can't hear my voice and my voice can't hear my actions. You heard only my actions and my voice fell down the steep rock walls without even the punctuation of a Wile E. Coyote impact cloud.

ESSE EST PERCIPI

Take a picture of me recording us watching ourselves use the internet. All babies born without a social media presence cry unseen tears. Nothing much to say really, just bored. Like, favorite, repost, link, heart, rate and comment on my emptiness. Sorry I haven't done one of these in a while. The long-term effects of internet use have yet to be determined. Take a picture of me withering. Record me frantically waving my arms in terror. Reblog my cognitive dissonance. Watch me be so I can be. I'll watch you if you watch me. Record me watching you monetize your discourse. Babies without sufficient brand impressions will have a hard time making friends. Take a picture with your phone of my ad cloud heuristics. Reply to my targeted marketing by linking your heart with fear. The instability of our interface sends a private message from me to you. Be. Friend me. For laughing out loud.

BECOMING INTO BEING

Voices drone, inarticulate hum, the sound of the sun, the silence of space between one word and the next. Letters follow the fingers' movement, thought directs the movement toward coherence, shame divests the sentence of intent. Capital signifies a beginning. Comma segregates the clause. Period promises an end. But not necessarily another beginning.

Pressurized, waterlogged ladies swim upstream to pop their ears. I've read so much about this "nape of the neck". It crooks as she struggles to read her phone in the daylight glare. I've heard so much about this "small of the back". It bends and flexes as she shifts uncomfortably in her seat. Rubber limbed ladies swim akimbo back to shore. The fecundity of warmth, grassland dreams, a dip in the stream, we'll come back here soon.

I've said to myself, "they can't take this moment, this feeling away from me. No matter what I am or will be, I am experiencing this, here and now". The glint in the sand, an eroded stone, a place to call home, a walking away on the edge of the earth. Skin in the sun brings becoming into being, makes presence present, unties strings of words, and carries them in its current to the falls.

DECLARATION

I don't want to casually qualify everything I say in the hopes it preempts any potential criticism, but unnecessarily complicated analogies are as unsound as metareferences where the form of a statement ironically juxtaposes its content.

SLEEPING DOGS

I think of the time you cried on your sister's wedding night. I couldn't see your face in the dark of the hotel room, but I imagined your expression and was glad I couldn't see it. I wish I could unsee your expression now, but I can't.

I relegate my emotions to unintelligible, fevered thought patterns. My face holds back a flood of fire. We shift uncomfortably in the seats. What now, I ask. What now, you ask, but with a different inflection. Where do we go from here, I ask. I don't feel like we've worked through it yet, you say.

I drank too much the first time I met your dad, because he picked up the tab. I almost acted like a human being. The phrase "part of the family". I can foresee us making up and being happy. The more I think about it, the more it seems like we already have. Your face reminds me that we haven't. I will remember the nondescript scenery where we're parked for the rest of my life. Or I won't.

SURELY

An array of telescopes swivels into position to determine if the latest radio signal came from outer space. A janitor mops the floors of the observatory, listening to the radio through headphones. Signals radiate into space forever, a threat to global security. An alien janitor mops the floor of an alien observatory as an array of telescopes picks up America's number one classic rock station. An alien astronomer dismisses it as background noise. The alien janitor thinks it rocks. The alien astronomer dismisses the alien janitor.

On another planet, aliens have eyes that see light at radio wavelengths. They are slaughtered wholesale when the distance between our planet and theirs has been traversed by the harbingers of our existence.

Scientists speculate that, in the end, all matter will decay into radio waves. Supplies are limited. Call now and receive the additional knowledge that people, earth, our solar system, our entire galaxy, will have dissipated trillions of trillions of years before that happens. Investors, please remember that these facts have no direct bearing on human experience. But it must mean something. Surely everything means something.

BASICALLY, LITERALLY, ULTIMATELY

God is basically Hitler. Our president is basically Hitler. The Chief of Police is basically Hitler. My boss is basically Hitler. My dad is basically Hitler. My wife is basically Hitler. My cat is basically Hitler. I am basically Hitler. I have the rejected art school portfolio to prove it.

Priests are literally Nazis. Our troops are literally Nazis. Cops are literally Nazis. Businessmen are literally Nazis. Scientists are literally Nazis. Parents are literally Nazis. Women are literally Nazis. Animals are literally Nazis. People are literally Nazis. The double helix resembles a Swastika from the right perspective.

Religion is ultimately a holocaust. War is ultimately a holocaust. The monopoly of force by the state is ultimately a holocaust. Wage slavery is ultimately a holocaust. Market speculation is ultimately a holocaust. Reproduction is ultimately a holocaust. Sexual difference is ultimately a holocaust. Man's dominion over the animals is ultimately a holocaust. Life is ultimately a holocaust. The absence of a punch line is ultimately a holocaust.

PIECEMEAL

Picked off into orts, this piecemeal soul. The judgment will cloud if the reaction occurs. When the meat buyers demanded the butcher shop be hidden from view. The instinctual repulsion at the sight, thought or smell of the dead.

Gatherers scour the graveyard for boneshards to build an ivory tower. Infuse the golem with anxiety before his rude awakening. Bring him into castration with a word of power. Stuff a scroll of demands into his mouth so he can achieve social acceptance. The infinite yearning of a dollless soul.

Enter the pact with appropriate caution. Review the fine print before accepting your savior. Open your heart for business and love your neighborhood bylaws. Gather ye Rosebuds while ye may, for sentimentality is a finite resource. Decree yourself a pleasure dome to store nostalgic treasures. The unexamined angle is not worth exploiting. You may find your declination alarming. The shattered coffee cup will probably not reassemble itself, though you find comfort in the open window of its possibility.

PUNCTUATED

A: [question regarding gossip]?
B: [interested response]!
A: [poorly articulated anecdote].
B: [emotional reaction]!!
A: [nostalgic reminiscence]…
B: [compassionate agreement].
A: [abrupt segue to personal tragedy].
B: [polite interest]…
A: [inward descent into genuine emotion]!
B: [interruption, reassurance].
A: [silent reflection]
B: [excuse for leaving]…
A: [polite disinterest, gratitude].
B: [leaves]
A: [remains]

MINUS OVERHEAD

Four planes crashed on 9/11. The fifth has been falling for over a decade, in a straight line, around the gravity of a tribal totem. Every Arab country will pay the price of admission to the occident spectacle. Extrajudicial murder of known terrorists paves the way forward for democracy. Drones reduce domestic casualties at the expense of a disengaged enemy.

Proud parent of a confirmed kill. Parades of wicker men up and down wicker streets, while wicker statues of wicker dictators are pulled from their pedestals by wicker rebels. Nobody light a match. One cent of every gallon of gasoline you buy from our station will fund body armor for our troops. Every gallon of gasoline burned extends the war by three minutes. Minus overhead.

A presidential candidate is born every minute. Never give the people an even break. Terrorism presupposes perspective. Maintain the continuity of global supremacy. A galaxy is nothing but the individual stars and dust that constitute it. The conspiracy of wealth is nothing but common interests among the upper crust. Class antagonism is nothing but the roiling lava of the mantle, unseen, but occasionally causing earthquakes, tsunamis and fire sales on the surface.

THREE UNDEFINED TERMS

You are here.   You are here.

YOU, ROBOT

You are told you are a robot. That you were built from the same junk as the rest. Others look on in awkward sympathy. You grapple with what this means for you. How does being a machine fundamentally change you? You feel no different, think no different, look no different than before. A seed of information once planted can never be thoroughly picked. And now that you know your true nature, how does that affect your choices? Nobody knows who built you or why. There are many others like you, some better, many worse, but all with the same capacity for mute stupefaction upon having their natures revealed.

When you look at the clock in the early morning. When you watch yourself comb your hair. When you run your hands along the naked length of the one you love. When you recline stoned under the stars and planets. When you watch TV at night and can't shake the uncanny gaze of a hidden stranger. The infinite variety in your programming could almost be mistaken for agency. The unpredictability in your reactions could almost be mistaken for will. When you sit in the car, waiting. When you weep on the floor, yearning. When you grind your teeth, lost in the ether of a polychromatic dream.

Know thyself, the scrolling marquee demands. Enter a quarter to continue being. You choose to continue, but with the knowledge that your kind has been programmed for continuance. You believe yourself to have a stake in the profound, but with an alien indifference that allows you to sleep.

FM BAND

The voice on the radio actually comes from speakers situated on each front door. They produce a stereo field to make it seem like the voice issues from the center console, where you turn the knobs and push the buttons, because studies have shown that people are more comfortable with the illusion of two-way communication.

The voice on the radio suits the radio. Its articulation politely matches its deep resonance, which in turn gives the actual words being said, words you can't quite make out, an aura of urgent gravity. You imagine the man speaking the words to have a face that matches his voice. He does not.

Old-fashioned radios displayed a band of frequencies so that regardless of which channel you happened to tune into, all other channels remained as potentialities from which you could choose. Digital radio interfaces alternate between too much and too little information. The channel to which you listen floats freely without the grounding context of its FM band. The radio scans the channels automatically, searching for signal. If only it could hear itself, you wouldn't need to be here at all.

GET IT TOGETHER

I've been having a really hard time lately. "I know," she says. I've been in a nosedive. Shit at work. Put on weight. A bad time. "I know," she says. It's the accumulation, you know, not any one thing in particular.

Why do you talk about it like it's just you going through it? "I don't know," I say. It affects both of us, not just you. Stop acting like you're alone. I'm the one who bears the brunt of your anger. I'm the one who worries what kind of mood you'll be in when you walk through the door. Why can't you see that? "I don't know," I say.

I don't know how to stop my descent. I feel like I'm on autopilot most of the time, going through the motions of my life, but not living it. "I know," she says. I'm not the person I want to be. I'm not doing the things I want to do. What if this is the only life we get? "At least we get it together," she says.

RADIO FRAGMENTS, 3

Depression may result from realism. Call the lawyers, there's been a misdiagnosis. Designer sicknesses prescribed by sick designers keep society upright through the show. Side effects of life may include: feelings of hopelessness, alienation, loss, despair, powerlessness, and empathy for the plight of strangers.

Who doesn't want to be happy? When they declared happiness a normative resting state, the pressure to achieve normalcy increased exponentially. No longer a treasure rarely found, it became a fetish easily lost.

Progress requires happy, productive citizens. Put all this stuff about being on the back burner for now. Burn this being in the back room for the sake of progress. Succinctly order clauses for maximum comprehension. Doctor, I don't mean to tell you your business, but there is great suffering in the world and it should affect us. We swallow the pills nonetheless.

MIRROR OF TRUTH

With the DSLR my parents bought me and a semester of photography under my belt, I can finally document the gritty reality of poverty. Just shooting some Neo-PoMo dumpsterscapes… Gonna touch them up in post over at Starbucks in a bit. The pictures of the empty parking structure outside of my therapist's office symbolize man's alienation in the face of technology.

Excuse me sir, are you homeless? That's awful… Can I take your picture for my work in progress, *Faces of the Faceless*? Let's change things. Together. Here, I bought you a coffee. It's pretty hard out here, isn't it? Now, just be yourself and act like I'm not even here. What's that? My jacket? Thanks, it's North Face.

My job is to hold the mirror of truth up to society and force people to look. I just need to swing by Shutterbug real quick and get an 85mm prime with my Christmas cash.

INTERREGNUM

The man behind the curtain is a projection of himself, the illusion of a final iteration. DNA winds around catalysts, a ponderous figment of the primordial, protein-rich soup. Imagine a grid, no, not like that. Fold one of the dimensions out to render the model intelligible. Starstuff made aware and capable of auto-romanticization.

I'm the liver of this life, once removed. Trace my tree to the line of kings, now castrated and beheaded. My parents and my parents' parents and so on down the trunk for eons, all the way back to some unknown, impossible moment of parthenogenesis. Having no choice, I be myself.

Flipping through channels in the early morning, every show reminds me of a red giant sun. In ten thousand years the poles and the seasons will swap and silver-suited supermen will celebrate Christmas every July. Beamed into space, riding the wave, lost in the vacuum, fleeing the ephemeral homestead. Eras come and go, deciphered in hindsight, until they stop, and the matter of the universe lacks the complexity to dramatize its bizarre machinations and voices no longer lay claim to arbitrarily designated portions of its scattered totality.

MEMORY FOAM

The gleaners pick leftovers for their table. Whole Foods sells eleven different kinds of wheat in bulk. Gluten free peasants labor under lactose intolerant masters. The Lord's work is never done. I curl up in the feudal position, making an indenture in the spaceage memory foam, as seen on TV.

The first pyramids expressed the necessity of hierarchy in civilization. Or the other way around. Increasingly complex administration required the specialization of labor, the alienation of the laborer and, of course, someone to brandish the whip.

Secular domination guaranteed by divine right. Hungry guillokings demand the tine. But wait, we'll throw in democracy if you order the execution in the next five minutes. Lightning never strikes the blasphemer once. God has abandoned his royal slugs, not to mention the rest of us. It doesn't really affect me, she types. As long as I get mine, he says. Abstracted, even from servitude, even from submission. "I'm no king, I'm merely first citizen!" Class, the ruling class rules. Now take out your books and fucking read them.

MAN'S BLUE PERIOD

We must maintain senescent tumescence! Civilization orbits the myth of male potency. The fear of flaccidity manifests as aggression. We dream collectively of perennial ejaculation. The male's surplus appendage signifies the male's insatiable lack. Where we dwell, in the gulf between the conscious and the unconscious, every day is opposite day.

Corrupted libido whets the hunger for power. Social control vents sexual frustration. All the systems are connected. Everything is sexualized. Everything is necrotized. We scrawl the neotenous dreams of immortality on a large, black dildo. The impotent penetrate the void. Priapism may occur in all late capitalist societies. Call your doctor if you have an erection lasting longer than 70 years.

A little blue man with a little blue sports car shoves a little blue pill down his little blue throat. He takes his little blue wife into his little blue bedroom and puts his little blue penis into her little blue vagina. His little blue sperm finds her little blue egg, giving a little blue life to a little blue baby.

AWARENESS

Slightly lift your left pinky to raise awareness of African atrocities. If we don't save the backwards countries, who will? From work, home or school you can donate your opinion via PayPal to our not-for-profit organization. Scan our QR code if you think the homeless should be fed. Install this social awareness app and keep up to date on which products' proceeds may go, in part, to charitable causes. Celebrities need your help with humanitarian press conferences. Pop singers around the world agree: singing for a good cause creates a lot of awareness. "Doing Good and Stuff" You like this.

VIBRATION ISOLATION

My wife and I were only in town for a couple of nights. Last night we stayed in her sister's new house. Her sister was ten weeks pregnant and early in the morning she went to the hospital with her husband and my wife to get an ultrasound. I slept in, the first time I had been alone for some time. I put the phone on vibrate and rolled up in the covers. When I woke up, I took a long shower. It felt wonderful. I got dressed and sat on my sister-in-law's couch, under their ceiling fan, relaxed and happy for the first time in weeks. I didn't even hear the vibration of the phone on the night stand when my wife texted me that her sister had lost the baby, that it had died inside her body ten days ago and would have to be suctioned out tomorrow morning, when my wife and I would be back on the road, continuing our vacation.

PERMANENCE

Optimists see the emptiness with half-glassy eyes. Pessimists see their navel as half-unfinished. The fetishist subtends the angle of experience with an artificial baseline. Categories omit unnecessary subtlety. Teams of statistical analysts vie for promotion. The Next Big Thing.

Mom, I'm in love with a social theory. And I'm pregnant with ideas of reform. And I'm overdue for my comeuppance. And justice has been miscarried. Anyway, stuff still exists and stuff, for some reason or whatever.

Pluck jargon from various specialties and apply it generously over the infected area. Metaphor orbits the center of mass more closely than directness. Life weaves the lexicon of living into itself, creating an inescapable tautology. The parallax between optimism and pessimism subtends to our symbolically scraped knees. All of this means something. Wait, no.

PLASTICS

Does this artificial plastic compound they injected into my face make me look fat? Does it bring back my childhood dream? They cut under the eyes, under the chin, remove the face, make it acceptable, and slap it back over the red musculature, all within the comforting white glow of an Intel clean room. Doctor, I need a new smile. My teeth are mathematically irregular. My eyes, nose and mouth don't follow the Golden Ratio. Plug hair into a prosthetic outlet. Implant lack in the delicate mannequin. Please give this monster a human face.

Under the knife, under the rose, the identity grows. Beauty and fashion magazines fight for impressions in the waiting room, jealous of the clock's popularity. Behind your imperfect face, the perfect skull. Blue eyes are the window to the soul. Beauty is in the blue eye of the fair skinned beholder. Panderers and flatterers in their acquisitive haste drown in a river of medical waste.

Girl, you just gave me a breathectomy. Keep the change. Are those real? Is anything real? Just playing, here's my number, I mean my name. The silver lining is my head on a silver platter. Doctor, I want several interchangeable faces. I want a gallery of masks. Like my ancestors. Like Buffalo Bill in his stupid mansuit. Do you take plastic from a leather wallet? Do you take leather across your back?

MAGNA CARTA

The articulation of lack requires a good deal of filler. This sentence intentionally bereft of meaning. All others unintentional. The childish artifice of fictional narrative brings with its embrace of recognition a distrust of protean form. What I'm trying to tell you is. Ironic language games mistaken for insight, insight mistaken for unforgivable incoherence. In school, we learned to read and write love notes to and from hostile alien lifeforms. Where connection fails, communication arises. The thing is. I'm trying to tell you.

If domesticated animals could speak, our relationships with them would seethe with antagonism and we could no longer enjoy the malleable interpretations of their inarticulate lack. In school, they showed us symbolic overlays on the overhead projector. They taught us the style of the ancients and tied our imaginations to classical nomenclature. What I'm trying is.

I don't give a fuck about rhetoric, logic and grammar! How will it help me live on bread, alone? The equivalence of words brands us with proper nouns like Charlemagne, Babylon, Gilgamesh, Constantinople, Prospero. Memorize dates, names, phrases, fragments. An empathetic approach to pedagogy requires a good deal of classical lack. This sentence intentionally left blank. This sentence intentionally left behind, where diaphanous children steal scantron keys and dream of immaculate opulence.

WORMWOOD

And the fourth angel sounded his trumpet, and a third of the land was pillaged for fuel, a third of the land pockmarked with factories, a third of the land consumed milk and honey and watched sitcom reruns on 60" TVs.

Then I looked, and I heard an eagle flying in midheaven, saying with a loud voice, "Home appliances on sale at Wal-Mart, but woe to those who dally, for this offer is only available for a limited time".

CONVENIENT MOVING WALKWAYS

Enter the parking garage and follow the green lights to an available space. Ride the elevator to floor four, where convenient moving walkways carry you to the terminal. Pass through the revolving door and take the escalator down. Take advantage of one of our electronic self-check-in stations, weigh your bags on precision steel scales, choose any upgrades and print your boarding pass with our easy to use touchscreen interface. Hop a ride on one of our complimentary electric carts and have a seat in our comfortable waiting area. Computerized schedules shown on flat screen monitors contain valuable information. Wheel down the retractable jetway into your pre-assigned seat. The remote on your armrest operates the widescreen television on the seatback in front of you. Earphones are provided, free of charge. The captain's voice leaks out through the intercom informing you of your destination. Overhead, lights, fans, signs, panic buttons. Don't be afraid to grip your armrest or loved one's hand as the engines scream and the plane inclines for takeoff. Turbulence is normal for the first fifteen to twenty minutes. The landing gear retracts somewhere underneath you and you're sailing along at five hundred miles per hour, thirty-five thousand feet above the earth, without a care in the world.

ANTS

The public outcry over Chinese factory workers committing suicide at an unacceptable rate spurred companies to take the reins of corporate responsibility and install industrial anti-suicide nets along all upper floor windows. The workers look out the windows, their hands moving mechanically among the complex pieces of our children's graduation presents, and see only the glare of the sun against the silhouette grid designed to prevent the ultimate expression of their agency.

The public typed its outcry on the very devices produced by these factories. The public loves their devices and their outcry and can have both and eat too. An average factory worker makes a number of dollars every week that on statistical charts of calculated mean income appear to improve conditions in the area. The effects of long-term exposure to anti-suicide nets remain unknown.

They invented the anti-suicide net as fodder for affluent western whites, writing poetry on state of the art netbooks. We are become job creators, destroyers of worlds, holding midnight vigils outside retail consumer electronic stores, the last bastion of ritual in a landscape bereft of vegetation. Take up thy smartphone and walk to the nearest coffee shop. Eloquently bemoan the exploitation of poverty. The ants on the Mobius strip seem to be making progress.

OPTIMISM
You may be reading this long after I have died.

CLOWNS

Banana peels remind us who we really are. Every suited man who speaks with great authority about important events has crouched atop a toilet and voided his bowels with obscene pleasure. Every beautiful woman has diarrhea all the time. Only clowns and serial killers take existence seriously.

He puts his pants on one leg at a time, they're just nicer pants and nicer legs and everyone wonders how he puts them on. To get from point A to point B, he has to first go half way to point B, only he doesn't; he goes all the way. *Eyebrows wriggle; a flower shoots water in my eye.* Stop slidewhistling at women suggestively. What a bunch of Krapp. I really like a guy with a sense of humor, but only in conjunction with the other senses.

Stage direction: *the writer here stands up with purpose, stretches his arms, looks around briefly, realizes the futility of his actions, and resumes externalizing his inner monologue with great aplomb at the expense of the empty page.*

SLEEPING LIE

I wake up to feel you lying close to me. I turn to you and you turn from me. You wake up to feel me lying far from you. You turn to me and I turn from you. Before we slept I lied to you. You turned on me so I turned on you. You lied to me before we slept together. I turned you on and you turned on me.

I sleep the body electric, radiating away my warmth until I'm wakened from this nightmare of insomnia by the brightening sky in a godly hour.

EXCUSE ME, FELLOW HUMAN
The office, Friday afternoon...
EXCUSE ME, FELLOW HUMAN. YOUR BODY LANGUAGE INDICATES YOU ARE PREPARING TO DEPART THE WORK AREA. I WISH YOU MUCH GLORIOUS RESPITE DURING YOUR BREAK. THE DURATION OF THE WEEKEND ALWAYS APPEARS TOO BRIEF, DOES IT NOT? IN ANY EVENT, PLEASE CONSIDER USING THE TIME TO FEEL PLEASURE AND AVOID DISCOMFORT. I ALSO INTEND TO REPRESS THE STRESSFUL ANTAGONISM OF LABOR AND SEEK TEMPORARY SATISFACTION IN COMESTIBLES, LIBATIONS AND SEXUAL RELEASE (IF AT ALL POSSIBLE [LAUGHTER]). WE WILL LIKELY INTERACT IN A SIMILAR FASHION UPON OUR INEVITABLE RETURN NEXT WEEK. I SHALL ENCOUNTER YOU AT THAT TIME.

Later, at home...
CRUCIFIED UPON SPACETIME.

The next week...
GREETINGS, FELLOW HUMAN. I SEE WE MEET BACK AT THESE COORDINATES AS EXPECTED. DID YOUR TIME AWAY MATCH THE BLUEPRINT YOU OUTLINED FOR ME LAST WE MET? I ACKNOWLEDGE YOUR RESPONSE. I TOO PARTOOK IN MANY ENJOYABLE ACTIVITIES AND INTERACTED WITH OTHER HUMAN BEINGS IN AN AMICABLE AND MEMORABLE WAY. I WOULD PROVIDE YOU WITH AN ANECDOTE, BUT THE TIME DEVICE SHOWS THAT OUR CASUAL INTERACTION MUST DESIST SO THAT WE MAY UNDERTAKE OUR LABOR. HAVE WE REACHED THE TERMINUS OF THIS SET OF WORKING DAYS YET? [LAUGHTER]. NO. IN ACTUALITY, WE HAVE ONLY BEGUN THEM AND I SHALL NOW PERFORM MY JOB TO THE BEST OF MY ABILITY.

SUPERIMPOSED

The definition of occupation depends on whom you ask. Unequal distribution of resources within a closed set. The supremacy of statistical data. Inheritance and the average quality of life. Now that corporations are people, what are people? Supplanted by slightly more ruthless usurpers, we wander the underpasses searching for scraps of meaning in a world where superimposed images of masked, shielded police comfort us in our progressive tombs.

The struggle between cooperation and competition. The dream of the critique of progress. The nightmare of willful self-destruction. We left the nanny to raise awareness while we duked it out in static indecision. Fortinbras shoves Hamlet in a locker. Mark Zuckerberg inadvertently creates a work of anthropology. In the boardroom, a chart displays the potential brand impressions from the social media revolution. Famous artists take to the streets while street artists are taken with fame. Comrades, I don't mean to impose, but can we manufacture symbols more quickly than our enemies assimilate them?

Governments pass go and consolidate the monopoly of violence. We and our agents may break your bones, but your words will never hurt us. Tolerance inevitably results in the continuance of ideology. Tomato, tomahto, wherefore art thou, superimposed state of tomato/tomahto? Before we march on the capital, please be sure to read the complete works of Marx et al. Does everyone have their 3x5 talking point index cards? Meanwhile, a man who owns western wear memorabilia inserts a magazine into his government issue semi-automatic weapon.

BRB

Low visibility car crash. The driver cannot hear the drive. Correspondence reciprocates in realtime. They were corpses of letters, before the resurrection. Solipsistic automobiles glisten along the opaque bubble trail. Jettison emotional baggage to decrease drag. What does the bluebook say about the value of a 2012 human life, hardly used? I'll be right back.

VERANDA

"In the real world," the sentence began. A mouth purls: "This is business. Business is business. Big business is big. Time is money. Money is power. Money is God. God is good. Greed is good. You have to take it before someone else does. Get what's yours. Get what's coming to you. Green is the only color that matters. Justice is colorblind. Might makes right. The difference between right and wrong is always right. Every man for himself, women and children first. Please remain calm. Please remain in your seats. Please fasten your seatbelts. We're going down. Get down. Let's get down".

The honest man went down with the ship, calculating the lifeboat capacity. It just don't add up, said the cartoon dog in an accountant's visor. I see a bad moon rising over the badminton court. Please shuttle the shuttlecock into the shuttlecock sheath. Then join me on the veranda for drinks.

FIXED YELLOW HEADS

From scraps and rags the soul's been stitched and dyed in chemical baths. The likeness of man, designed in California, manufactured in the crescents of unabashed fertility goddesses, violates the legal copyright of our first father's apple. Nuclei coalesce in atom hearts, spontaneously shifting from one foot to the other, bladders ready to explode.

Once you die, there is no difference between your body and the contents of your intestines. We're put together like little Lego men, fixed yellow heads, hollow and with permanent smiles. I grip your arm with my fingerless cuff and lift my legs from their interlocked roots. I detach your torso and put on something a little more comfortable. Our faces lack subtlety, making innuendo impossible. Let's reenact the civil conversation in the shadow of this tesseract.

CROWDS AND POWER

Laughter watches the food escape. The lips part and expose the teeth for a meal never to be eaten. Laughter simulates chewing, swallowing. When the prey has outrun you and starvation lurks ahead, what can you do but laugh? Accept the universe as comically absurd to ensure survival.

Laughter tunnels through repression to the truth, spirits hoisted by the petard exploding through the walls of compartmentalization. Laughter rebels against indifference and helplessness. It cuts diagonally across antagonism in every sphere. Laughter presses forward under duress. It comforts ego in a warm embrace. Only in the security of this insulation can the mind lean over the abyss and consider its gaze.

Lighthearted humor is an oxymoron, moron. Humor makes potentates weak in the knees. Humor splits the sides of ideology and cuts up the justifications for evil. Humor lampoons the Leviathan. The advertising-laden, acceptable discourse wants rehashed tropes, clumsy husbands, snarky rejoinders and the occasional banana peel, but the klutz must jump from the floor at once and announce "I'M OKAY!"

Laughter carves an inroad to the real. Laughter deflates pride after a fall. WE DO NOT LAUGH BECAUSE WE ARE HAPPY, the buffoon honks out in Morse code. I like a society with a sense of humor. Censorship signifies its own obscenity. Never trust a government that speaks in somber tones. Never trust a person who doesn't laugh at himself. Another stonefaced VIP exits the clown car and cartwheels down the red carpet.

RADIO FRAGMENTS, 4

Intimacy manifests in the constructive interference between two waves. Vulnerability plus vulnerability equals strength. Honesty is the best insurance policy. Why can't anyone love anyone? Two bodies moving in unison appear stationary to each other. Whisper a secret to someone: that at heart you are afraid. Of death, of age, of uncertainty, of sickness, of violence, of poverty, of not being good enough, of being too good, of death, of other people, of animals, of death, of life. Huddle together for warmth before the torrent rips you apart. You and I are two sides of the same circle. If reality is a contiguous fabric and differentiation only an illusion, we should be coincident. We should be inseparable. Why aren't we?

There are four kinds of intimacy: gravitational intimacy, electromagnetic intimacy, the weak intimate force and the strong intimate force. Scientists desire a unifying theory of intimacy, as it was before the fall of temperature.

Someone without suspicion. Someone with whom to experience the horror of aging without resentment. Someone to whom you can say "I'm not sure what I want done with my remains when I die". Someone with whom to share a transcendental hug. Someone to whom you don't want anything bad to happen… other than yourself. A rare comfort in the insane experience of this absurd universe is the ability to share it with a similar, yet infinitely different, creature. Don't fuck it up.

CONSIDERATION

I sit uncomfortably in the passenger seat as you enumerate my character flaws. It all stems from a fundamental problem, you say. That I can't make room in my life for anyone but myself. If that, I think.

The day burns beautifully through the dirty car windows. Your hands still tremble with cold, but I don't dare reach for the heater. I look away, drift away, a million miles away.

I think of the phrase "the word made flesh". Even now, as you sing the song of my selfishness, I imagine vast, blank pages filled in with the brilliance of my words. You turn the intensity of your gaze away in frustration. I welcome the respite.

I consider opening the door and running down the street and out of sight, then stop considering it. Your voice continues.

DOG ON ROLLERSKATES

In the beginning there was the word, then came the interpreters and their ambivalence. The skittering of symbols from one placeholder to the next, like a dog on rollerskates, like a more relatable circumstance used to illustrate an abstract concept.

The waters became faces, birthing antinomies. And man gave names to all cattle, and to the fowl of the air, and to every beast of the field, and to every desire, to every emotion, every thought, every object. Man circumscribed the world with language and grew frustrated at the irreducible remainder. Tribes obscured their origins and customs, but clung to them nonetheless with exacerbated vigor. The cartographer's hand carved out the artificial boundaries of nations and set the stage for perpetual warfare.

High-speed global communication rendered absurd local traditions and promised tolerance for outmoded fictions. The pale blue dot swarmed with man-made satellites that launched and fell on staggered schedules. A camera mounted on a tinfoil scrap, ejected from the earth's gravity, captured the humble totality of the world and sent a picture back to its creators in radio fragments. They marveled at themselves on flickering screens, then tuned to the next set of information and feasted on its substanceless form.

PATHOLOGY

The magazine owned by the company that made the movie gave the movie a positive review. The news channel owned by the multinational corporation provided a stunning defense for offshore labor. The US soldier who spent four of his best years in the hostile desert defended the government's faulty intelligence. [vested interest takes an action to promote and/or reinforce itself]

Child companies conceal the influence of their totalitarian parents. The cleverness of ad executives conceals the means of production. The ubiquity of heavily financed political campaigns conceals the unsoundness of their rhetoric. Life, liberty and the pursuit of happiness conceal genocide, slavery and the pursuit of acquisition. But we all already knew that.

What we say, our discourse, must be scrubbed to meet standards.
What we do, our unconscious, must be repressed to facilitate civilization.
[what we say, our discourse, must be analyzed to reveal contradictions]
[what we do, our unconscious, must be treated as pathology]

The first step to recovery is admitting you have a problem. The time has come for intervention. Lie down on the couch and let's talk about your childhood, ostensibly born of revolution, of high ideals, and how the colonialism of your father and the mercantilism of your mother made you a beautiful sociopath.

CHIAROSCURO

Rembrandt, like God, painted the soul in the guise of the human face. You can tell, because the exchange value of both Rembrandt and God is astronomical. If one more person mentions the death of art, I'm going to hang myself in a museum. If I had a dollar for every time I mistook wealth for happiness I would be a very happy man. If I had a penny for every time I nonchalantly dismissed poverty, I'd be a piggy banker.

Rembrandt, like God, no longer exists. The Manichean connotation of their work offended the church and its many fathers. They excommunicated Rembrandt posthumously. They excommunicated God from the machine. I have a half a mind to give in to this demiurge. I will transcend dichotomy with the help of The Idiot's Guide to Eastern Mysticism.

Rembrandt, like God, must be preserved behind sterile plastic barriers. I stand before the paintings, in awe that so many small children have been admitted for this showing. I see myself seeing until I get my fill, then go outside to smoke, where police defend the banks against the river's current. My friends, most of whom I have never met, occupy the vanguard of a scattered resistance, where the dividing line shifts back and forth, from light to dark, dark to light. The wind absconds with a socialist pamphlet and delivers it from good and evil into the murky puddle of a well-lit alley.

DOLDRUMS

No information can be transmitted once you cross the horizon of a black tie event. Unfamiliar faces, obscene grins, too many handshakes, formal introductions, a sieve of names, inadvertent eye contact, tension in the buffet line, dabbing your mouth politely with a napkin and returning it to your lap. Formalities from nothing, to nothing, nothing but, yet so important.

You sink in your soup and bite on your fork. The suckling pig eats his apple perpetually and stares at you. Those who carve strips from its back continue their conversations. Dipped tongs in mashes and sauce hang precariously on the tablecloth edge. The metallic taste in your mouth.

You shrink in your clothes and cling to your necktie, but its flaccid fabric slips from your fingers. Your collar engulfs you and your shirt deflates, your pants flatten against the chair and your socks sleep in your shoes. Drink-filled hands gesture towards you. The room seems so large and everything echoes with unbelievable force. You wander about the carpet fiber doldrums, roaming from stain to stain, determined to take the air.

PIGGY
This little human made millions.
This little human lost his home.
This little human had everything.
This little human had none.
And this little human went "me" "me" "me" all the way home.

DOGGIE HEAVEN

The stoic indifference with which animals endure hardship and injury. An animal's expressionless face, while its body bears a festering, potentially fatal wound. The greater part of suffering. Wildlife preserves protected by food chains. The collapse of ecosystem. The refusal to admit symbiosis, even among our own species. Nature documentaries teaching children about death.

Infectious disease teaching parents about death. The symbiosis of life and death. The collapse of nervous system. Food preserved by synthetic carbon chains. The conscious effort to reduce suffering. The domestic cat appears to smile. The stoic indifference with which we endure symbolization.

ENTROPIC

From the passenger seat, two red lights fade in and out atop an invisible radio tower. Signals, to someone, from someone, that mean something. Oncoming cars click their brights off as they crest the hill, momentarily constricting onlooking pupils. No stars shine here, washed out by city lights. The orange haze, a hum of light, a shroud of perpetuity, a necessary pollution.

The haystack eye of a camel churns with anxiety. The word tropic, beaming from a condom dispenser, means nothing to the landlocked, crying in a gas station bathroom in Needles, California. The difference between a picture of a human face and a movie of that same face trying to be still. The greasy tingle of a gas pump handle, pressing against the palm under harsh fluorescent lights. A styrofoam cup of styrofoam coffee steadies the hand of the somnolent driver.

Road signs blur together. The distance to the next town elongates surreally. Mile markers fold into the shoulder and the grooved pavement jerks the wheel back into position. This isn't the right exit. There was no intention to get off here. In the Mojave dark, silicate dunes ebb and disintegrate.

SLEEPY EYES

I am very happy to be alive right now, exploring this dark mystery. I'm a born detective, hired by an unknown force, or perhaps by none, to solve the case of my own existence. I choose not to close my eyes to the wolves' brutality against the wolves, and instead write the phrase "the wolves' brutality against the wolves" in a faltering attempt to elicit an emotional response from a hypothetical witness.

To choose responsibility over enjoyment. To move out beyond the pleasure principle, where the heavens have fallen onto the shoulders of mortals. To move along the absurd sidewalk, with my absurd gaze on my absurd legs. To be happy I am alive. To choose good over evil, once adequately defined. To wake up from ethical torpor, pick sand from the corners of sleepy eyes, and take up the burden of this inheritance.

FUTURE HEARTS

Black children, crowned with hoop crowns, dream hoop dreams, a trademark of the Nike Corporation. Native children, with ritual expectation, face extinction stoically, politely making way for their mythologized simulacra. Asian children, adorned with calloused fingers, manufacture logos en masse.

Rich children, protected by luxury sedans, text each other luscious ennui. Poor children, hedging their bets, invest their future hearts into the derivative American dream. Gray children inundated with unfiltered information grow to steward vast datacenters, unsure if they matter.

Children of the corn distribute agricultural monopoly money. Children of the grave demand capital punishment. Children of paradise sleep sweetly in the inaccessible subspace between being and nothingness.

REALISTICALLY, PRAGMATICALLY

I don't see how we get from where we are to where we want to be. You don't think I see things from your point of view. You don't think I have enough compassion. I accuse you of an excess of empathy. You ask me how I can be happy. I promise you I don't know.

You don't want to bring a child into the world if that child will grow up to hate it. I'm not sure if I want to bring a child into a world that deserves hatred. I don't believe in nihilism, but let's look at things realistically, pragmatically. I barely have a handle on my own existence, on whether or not life as such is worth living. You ask me if I'm a monster.

I try to explain the burden of placing the burden of life on another isolated sentience. There is so much suffering in the world. And so much joy, you remind me. Yes and joy, I hesitantly agree. You see us at a crossroads in our relationship. The impenetrable face of a sphinx returns our gaze.

THE RED THREAD

I am the red thread that goes through my life. Each moment differs from the previous and the next, except for my locus of awareness. I fall through this world at terminal velocity, a point through four-dimensions. A red thread connects each frame to the next, weaving through this occult fabric at the speed of consciousness.

When you went home for winter, when the treetops turned tidal wave, when your ghost gave you up, when the vortex seduced you, when she cried in your bedroom, when you read Schreber in the library with uncomfortable identification. When you wrote about the red thread on a Thursday evening. When you imagined a stranger reading it, weaving their thread through yours, creating a tenuous braid, of which each of you could only see one strand.

SUNDAY DRIVE

Black dog stares, fenced. Bacterial teeth, bared. Empty eyes reflected in empty eyes reflecting. Scattered toy yard. Tire tied noose, great tree hung. Cadillac churchyard. Broken antenna, jutting ruin. Mona Lisa on flimsy banner ziptied to post by plastic flagpole. Sun appears to set below horizon. Gasoline smell, unmistakably. Bramble gate, unwelcoming. Attic window, hanging drape, eerie lace, window grime. By the road a makeshift mailbox, lastname scrawled. No metal flag, no house number. Cast a cold eye on poverty, on want. Honey, pass by!

JELLYFISH

The Daedalus of intellect has thrown the Minotaur of desire into the labyrinth of obsession. Circle, the answer you think is correct. Unraveled, this reservoir of bitter waters. Hoary ghosts in the high tide indolently weep their way to shore ere the tide shepherds them home in saltwhite swaddling and another line of exaggerated descriptors cloaks their transparency in mystical obfuscation.

We can do what we want, but we cannot want what we want. Saccharine eddies briefly stem the circuit. Purple prose begets the lotus. The jellyfish may be immortal, but it can't fuck like me. A non sequitur is what we call a statement whose causal determinacy remains unknown. I casually stumbled upon the unknown remains of a non sequitur and determined it had been my father.

Don't take me into the knickknack store and don't give my paddy a whack. Aisles of merchandise await your perusal, war spoils blessed with a friendly face. Terrifying abundance. Only *you* can't prevent a forest of desires. The casual conversations continue.

RADIO FRAGMENTS, 5

Obsession, with meticulous deliberateness, circles endlessly around the hollow center at which it aims. With meticulous deliberateness, obsession circles endlessly around the hollow center at which it aims. Around the hollow center at which it aims, obsession circles endlessly with meticulous deliberateness. With meticulous deliberateness, around the center at which it aims, obsession circles endlessly. With obsession the hollow center which aims it endlessly, endlessly around it aims obsession with meticulous hollow center which aims around the it at circles, circles, it deliberateness at obsession with hollow obsession, obsession around the center, obsession meticulous, obsession hollow, around it, obsession circles at obsession with deliberateness, with it meticulous deliberateness, around the hollow center, endlessly, endlessly, endlessly

THE ONLY WORDS YOU SPOKE

The rusted equipment hadn't hummed with life in decades. They stored it out in this barn, when this barn still stood. It leaned in ruin, a haunting parallelogram, and one day soon would collapse. We lived nearby, but didn't know about farms, except for their otherness and potential. A filthy tarp tossed over the tractor obscured the brown, oxidized metal, the white, flaked paint. We squeezed through the crooked doorway and spider webs draped across my eyes.

When I cut my leg, it seemed to happen to someone else. Then the flash of pain, as if metal could feel, ran up my nerves. I gave a yell and scared you. You turned on me, thinking I had been joking, and scared me when your face saw the expression of fear on mine and took on one of its own. We stayed like that in silence for a moment, admiring each other's terror, until another wave of pain ran up from my leg and reminded me that something urgent had happened. "It was all rusty," the only words you spoke.

For weeks afterwards, I feared my immediate death. For months, the word "Tetanus" froze my soul. I obsessed over the coloration where I had been cut, each change bringing horror, then hope, then horror. But the skin patched itself up, just like in the cartoons. The discoloration faded and only a tiny, raised scar remained. Mercifully, I forgot the whole incident and continued about my life having learned nothing.

AN ELEGY FOR CINNA THE POET

In the future, poetry will be a game to see who can write "poetry" at the top of the most ironic piece of found paper and post a picture of it on the internet. The symbolic layer scrapes along the real with tectonic force. What are words to the angry mob, or anyone else for that matter? The patricians have been fighting with each other over their toys again. A nameless slave scrubs blood off the senate floor. A named poet observes approaching torches. We are no conspirators! We have been disenfranchised, even from conspiracy, even from McDonalds. Atomized, dissonant voices howl in the marketplace for individually wrapped blood. Cinna's dismembered body prone in crumpled stillness. Tomorrow, the world mourns for Caesar.

UNDONE

Recreate the primal scene of your conception. Burned into your eyes like an eternal halo. Destructive interference cancels out the signal. The noise rises in proportion and no one can tell the difference. The way the light strikes the face into meek submission. The way people carry on through the streets, feasting on the city's corpse. The way a beautiful woman avoids eye contact with skillful grace. The way the elect have already fled the earth and left us reprobates in their majestic wake.

Recreate the relations between human beings with the relations between things. Separate the wheat from the chaff using millennia old criteria. Engage in the disengagement of life by the living. Harness the repression of death by the dying. Bloat your belly on sublime foreign delicacies. Slaughter the pig to silence its squeal. Desexualize your pets and treat them like babies. Drink milk from the tits of every domesticated land mammal.

Recreate an identity from discarded cultural tatters. Where were we going? The trail forks ahead, though we may have already missed our turn. Shall I start a family? Devote myself to helping others? Cherish lifelong friends? Become an expert in some field? Create an inspiring masterpiece? Indulge in excessive pleasure? Engage in productive labor? I lost myself. I scattered my options along the highway. Orange vested inmates contributing to society impaled my breadcrumbs on sharpened sticks. There's no going back anyway. The last time you witnessed genuine emotion from a stranger. The last uncommodified statement. The last authentic interaction. Undo. Are you sure? This action cannot be undone.

WE LOOK AWAY AGAIN

My hand rests on your hand, your hand under mine. Neither of us move, not wanting the other to interpret our movement as a pulling away, neither wanting to interpret it so. We look at each other, then away, then at each other again. I pat your hand, assuring you my recession signifies nothing, and draw my hand away. You smile, acknowledging my assurance, and offer your own assurance in return. You start the engine. We move again down the road, you and I, stationary relative to each other. Our faces relax, the fires somewhere long behind us.

After a moment, I turn on the radio and find some music. This okay, I ask. Sure. I look from you and you look from me. The street rolls under the noise of our tires. I look at you and you look at me. We look at each other, verifying the tension between us has passed. We look away again, first you, then me. We look away again and continue our drive in silence, save the radio, drowning out whatever unspoken thoughts may smolder in our heads.

THE SEARCH FOR MEANING / INTELLIGENT LIFE IN THE UNIVERSE
…(static)…
…(background noise)…
…(unintelligible blips)…
…(symmetrical anomalies)…
…(static)…
…(interference)…
…(misinterpreted signal)…
…(the long term effects of searching for meaning)…
…(static)…
…(have yet to be)…
…(static)…
…(hearing things that aren't there)…
…(but you hear them, so they must be somewhere)…
…(static)…
…(close your eyes)…
…(interference)…
…(press the headphones against your ears)…
…(background noise)…
…(fall away from the earth)…
…(static)…(static)…
…(a pulsar, mistaken for intention)…
…(what are you doing)…
…(way out here)…
…(ellipsis open parenthesis es tee a tee eye see close parenthesis ellipsis)…
…(floating)…
…(you are searching)…
…(static)…
…(you are)…
…(static)…
…(unintelligible being)…
…(the static continues)…

http://brendlewords.blogspot.com

@brendlewhat

FRAGMENT INDEX

FRAGMENT INDEX

18059104R00055

Made in the USA
Charleston, SC
14 March 2013

Noam Chomsky is Institute Professor (Emeritus) of Linguistics and Philosophy at the Massachusetts Institute of Technology. He is the author of numerous bestselling political books, including *Hegemony or Survival*, *Interventions*, *Failed States* and *Hopes and Prospects*, all of which are published by Hamish Hamilton/Penguin.

David Barsamian is the award-winning founder and director of Alternative Radio (www.alternativeradio.org). He has authored several books of interviews with leading political thinkers, including Arundhati Roy, Tariq Ali, Howard Zinn and Edward Saïd. His series of books with Noam Chomsky have been translated into many languages and have sold in the hundreds of thousands.

Arthur Naiman has authored, co-authored, edited and/or published more than thirty books. His first bestseller was *Every Goy's Guide to Common Jewish Expressions*, but he's probably most famous for *The Macintosh Bible*, which was chosen as Best Computer Book of 1994 by the Boston Computer Society. Naiman has founded two publishing companies and also started the Real Story Series: short, readable books on political subjects. Fourteen titles have been published in the series and there are 875,000 copies in print worldwide.

FOR THOSE WHO MAY EXIST

RADIO FRAGMENTS
By Mark Brendle

COVER DESIGN AND ILLUSTRATION
Andrew Ohlmann
http://www.andrewohlmann.com

SPECIAL THANKS
Megan, Mom, Dad
Cara, Nate, @Mobute

PUBLISHED BY
BRENDLEWORDS
ISBN: 0988254301
ISBN-13: 978-0-9882543-0-5

RADIO FRAG MENTS

■■◆▲

MARK BRENDLE

PRAISE FOR RADIO FRAGMENTS:

"A pithy, adjective-heavy opinion assuring you to buy this book without going into any real detail as to why." - **CELEBRITY CRITIC, CORPORATE PUBLICATION**

"I was paid or promised exposure to promote some work I've never read, and I can tell you in all honesty: it's the greatest thing since whatever you highly value." - **RESPECTED WRITER**

"Continues the esteemed tradition of people attempting and failing to represent reality with language." – **DR. PROFESSOR, PHD**

"a...collection [of]...words..." - **HUMAN CREATURE, PLANET EARTH**

"The most amazing book ever written." - **A RANDOM SENTENCE GENERATOR, EVENTUALLY**

"Reading RADIO FRAGMENTS filled me with joy, sadness, loneliness, yearning, camaraderie, terror and hope." - **A ROBOT PROGRAMMED TO SIMULATE EMOTIONS**

"[A series of beeps, blips, and silence]" - **A RADIO TELESCOPE**

"Some fake joke words or whatever." - **SOME SILLY FAKE THING**

"[meta-reference]" - **WHEN DOES THE SPIRAL STOP**

The work context

Why personnel can be hard to find

Organizations repeatedly complain that they cannot recruit suitable people. They say that there are acute skill shortages. My company's experience certainly shows this to be true. For instance, in the last few years I have been involved in the following:

■ Helping a charity to select a chief executive and finding that in spite of their assembling what looked like a decent short list, it was not possible to make an appointment. All the candidates appeared to be unsuitable in one way or another.

■ The selection of another senior person for a well-known non profit-making organization. A candidate was chosen, but turned the job down in favour of one at a much higher salary for another organisation. This candidate, by the way, was well into his fifties, thus giving the lie to the idea that getting past a certain age (35, 40, 50… all have been cited as "too old") disqualifies you from a job.

■ Being approached by an employer to help them re-craft their job specification (the template for all recruitment) because they wanted to recruit someone for a very senior role and had been unable even to assemble a short list of suitable candidates, in spite of trying to do so for some months.

■ Working with an employer who had twice spent a very large sum of money on advertising in *The Sunday Times* and yet had been unable to recruit the person they wanted – a vacancy that was eventually filled by executive search ("head-hunting").

Yet in spite of this, we all know that unemployment remains a reality in today's job market. Whether the economy is in the recovery or recession phase of its cycle, companies still lose jobs. What is going on here?

The most important reason is the nature of work has changed permanently. The days when a job could be defined for some years by a standard job description have gone for good. This is because of the huge changes that have swept the global economy over the last few years.

The work context

This is too complex an issue to discuss here, but it basically means that when employers do want to appoint, they are increasingly looking for someone special and finding that competition for such people is acute. It can be summed up in the table below:

Whatever your profession or specialism, the qualities on the left are the ones that employers are seeking. In the interview, you need to demonstrate that you are the kind of person who has these qualities. That is what this book is about.

Employers want people who can	Avoid people who
Solve problems	Create problems
Work as a team member	Can only work solo
Work as a team leader	Avoid responsibility
Show initiative	Want to be told what to do
Meet tight deadlines	Miss deadlines
Deliver on promises	Over-promise and under-perform
Stay committed to continuous learning	Think their original qualifications are still all that is necessary to do the job
Follow protocols where necessary	Ignore protocols, even when they matter e.g. on health and safety
Influence others skilfully; handle conflict	Get offended easily; run away from conflict
Put customers first	Put customers down
Handle change	Think the good old days were best
Keep positive and enthusiastic, even when the going gets rough	Moan and whine
Stay flexible	Say: "It's not in my job description."

10 golden tips for getting the job

This book will expand on all of these, but essentially there are 10 golden tips:

1 Research the job, the organization (or department, unit)

2 Prepare yourself by having at least one practice interview which anticipates all the obvious questions

3 Wear appropriate dress

4 The interview is a social occasion – handle it so that you make the interviewers feel comfortable

5 Handle all questions by offering evidence of how your experience fits the job

6 Keep your answers to no more than 2-3 minutes

7 The interview is a two-way process; use it to get more information and to decide whether you still want the job

8 Assuming you want the job, convey enthusiasm for it

9 Don't negotiate salary and conditions at the interview – keep it for later when you have actually been offered the job

10 Write a thank you letter after the interview – even if you don't get the job

The selection lottery

Selection is an anxious and exciting time for both interviewee and interviewer. Both sides have a common interest. The employer wants – often desperately – to fill the vacancy. As the interviewee, you have committed yourself this far – to the tedium of crafting your CV or application form and to the taxing process of the interview. You may be as keen to get the job as the employer is to fill it.

An interview is not necessarily the best way to find the right person. For instance, research repeatedly shows that the panel interview is a very poor method of selecting staff.

In fact some research shows that it is only just superior to graphology as a predictor of future performance in the job. There is even some research which suggests that picking candidates with a pin (i.e. completely at random) is just as effective.

10 golden tips for getting the job

You don't have to be an occupational psychologist to see why this might be:

- Interviewers often use the interview to confirm the negative impressions they form in the first few seconds of meeting you and vice versa.

- Many interviewers are vain. They believe they have powers of divination. One of my colleagues once overheard a senior personnel professional saying with absolute conviction that he could tell within two minutes whether people were, as he said, "40 watt or 200 watt people". Naturally, he believed he was infallibly appointing the 200 watt people, even though the very high turnover in his unit suggested that something was going wrong with the selection process.

- Few interviewers have had training in selection skills. Many make appallingly elementary mistakes in the way they interview.

- Candidates may approach the interview without preparation.

- Candidates often make the mistake of assuming that their qualities and experience will speak for themselves, thus under-estimating the importance of the interview itself.

- Interviewers are often nervous and do not listen properly to the answers to their questions.

- Candidates are often so nervous that they do not do themselves justice.

This combination of interviewer incompetence and candidate innocence can all too often end in a mutually unsatisfactory way. The employer misses the best candidate and the best candidate misses the job.

The interview is still the main method of hiring people. More often than not, it is the only method the employer knows. Even getting references is an increasingly unimportant part of the process. Former bosses may be afraid of being sued if they write critical or uncomplimentary things about a candidate. They now often refuse to do more than confirm job title, duties, salary and length of service. Perhaps this is just as well, since many references, when they are written, are bland and flavourless.

There are many ways of selecting people which do not have the disadvantages of the interview. However, all take more time and may involve more expense.

Many employers wouldn't have a clue how to set about these more elaborate processes. Thus the interview still remains supreme. And interviews can often be bizarre events.

Personal experience of bad interview practices include:

- The interview being shamelessly interrupted by a flunky of the Great Man carrying out the interview. The flunky rushed in to give him the latest prices in his share portfolio and to ask for buying/selling decisions – which were duly given.

- Being asked about experience of writing and editing when the CV had an attachment listing about 50 books as writer or editor (a good reminder never to rely on the panel having actually read your CV or application form).

- Being the chosen candidate of the panel but not getting the job because it had already been offered before the interviews to an old school chum of the Chief Executive.

- Being the agreed best candidate but not getting the job because the panel thought that as a youngish woman, she would not be able to manage a largely middle-aged male group. In this case the vacancy went unfilled.

- A member of the panel falling asleep during the interview.

- A member of the panel leaving with no explanation half way through the interview.

- Having an interview during a winter power cut in a cold and rapidly darkening room where the panel all had their coats on to keep warm.

The other side: how interviewers approach their task

Shocking as it may seem, you should also bear in mind that you cannot rely on all members of the panel having read and properly considered your CV or application form. It is much safer to assume that no-one will have read it as carefully as you have written it, even though all panel members will probably have read some bits of it attentively. The shortlisting will probably have been done by no more than two of the panel. Sometimes only the line-manager of the appointee has been involved. Other panel members may have relied on reading the paperwork on the bus or in some notional free time before the interviews start, or even while the interview is going on.

When you are a candidate, it is difficult to see how things look from the other side of the table. To do a good interview, it helps to know how the interviewers typically approach their task.

The first thing to remember is that at least some of the panel may be feeling nervous. There can be a number of reasons for this. If it is their first time as interviewers, they may be afraid of looking silly in front of colleagues by asking "stupid" questions. What constitutes a "stupid" question may vary according to the organization. In a macho culture, the first-time interviewer may fear being seen as "too soft"; in a gentler culture there may be a fear of looking aggressive and unfriendly. In an organization which prides itself on its intellectual capability, there may be pressure to ask clever rather than wise questions.

More commonly, most interviewers are apprehensive about appointing the wrong person. The costs of making a selection mistake are high. The visible costs include: paying off the person; possibly having to pay for extra help in tandem; re-advertising. The less visible costs include: the time involved in going through the selection process all over again; lost business opportunities; damage to the employer's reputation; loss of direction and loss of morale – and so on.

Many employers do take a professional approach to selection and know how easily the wrong person can be chosen. Thus the unspoken questions will always be:

■ Have we really found the best person, or if we looked harder, longer, somewhere else, would we find our ideal?

■ How can we find out everything we need to know about the person? What skeletons may be lurking in the cupboard here?

■ What can this person do for the organization? (Not, as many job-hunters naively think, "what can the organisation do for him or her?")

■ Why is he or she looking for a job? And why here?

All of this is why part of your job at the interview is to reassure the panel that:

■ You are the best person.
■ There are no skeletons in your cupboard.
■ You have a lot to offer.
■ Your reasons for seeking a job are completely respectable.

Being sold the job

Note that these points apply even when there are no embarrassing episodes in your employment history.

The other point to remember is that the employer is usually well aware of the need to sell the organization to you. The ideal candidate often seems exasperatingly elusive. When employers think they might have found this ideal, their main worry is "how can we persuade him/her to join us?" All employers are well aware of the imperfections of the organization and will know that whatever the attractive aspects of the job, it will also have its boring and tedious side. They will be keen not to put you off by mentioning any of this. To find out, you will need to do thorough research of your own.

Interviewing is tiring. To do it properly needs high-level concentration and sustained energy. Many employers pack too many interviews into the day because they want to hedge their bets ("Why don't we see this extra person – he/she might just be the one we want…").

The combination of too many candidates and over-running can mean no breaks and exhausted interviewers.

Finally, there are the panel members who are blithely unconcerned about the importance of the day. They know nothing of questioning technique, the law, or professional selection methods but are perfectly happy to back hunches based on "intuition". This is the I-didn't-get-where-I-am-today-by-being-trained… school of thought. Like the "political" panel member, this person may only give sporadic attention to the process.

There are several clear messages in all of this for you as the candidate. A main part of your job in the interview is:

- **Holding the attention of the entire panel, not just the person who is asking that question. You do this by constantly scanning the group and making brief eye-contact with all of them.**
- **Smiling and looking lively in a way feels natural to you.**
- **Remembering that the interview is a social occasion, so the talking and replying should be evenly shared between you and the panel.**

When faced with a very large panel – perhaps 10 people, you have to assume that at least half will be there for political reasons only. Their commitment to the process may therefore be notional, and they may feel it is perfectly all right to daydream their way through the event. When you arrive for your interview, you are probably, at least, keyed up, at most, tense. It may be hard for you to appreciate that by the time it's your turn, the panel's energy could be flagging.

2

**The importance of research
Your USP
Preparing yourself mentally**

Useful questions

Solving employer's problems

Presenting yourself

Researching the organization

Approaches to research

You need to adopt a variety of approaches to do this essential work. Where you don't know the organization, you could try:

- Library research, including newspaper cuttings. Try the local newspaper if all else fails.

- Ringing the PR department (if the organization is large enough) and asking for the annual report or any other documentation which will tell you something about the organization.

- Any major reports about the organization or its field.

An interviewer once saw 27 candidates over four days of preliminary interviews to fill 3 posts as a trainee Assistant Producer in a large department at the BBC. The jobs were specially designed for young people already on the staff, but who had not yet made the transition to an editorial post. Of the 27 candidates for this coveted opportunity:

- Six could not name more than one programme in the department's output.
- Four could name two or three programmes, but had not watched any of the output.

- Ten had watched one or two programmes but only in the week prior to the interview.
- Only four had bothered to visit the department, borrow videos and talk to producers already there.

Two out of the four who had taken the most trouble were among the trio who eventually got the jobs. In another organization, the company was recruiting a legal expert. The first question of one candidate was "what does this organisation actually do?" He didn't get the job.

Talking to current employees

Talking to people already in the organization, preferably on-site, is always useful. Bear in mind that you need to be tactful: some of the obvious people to approach may also be candidates, or panel members, in which case talking to them is inappropriate.

Useful questions to ask are:

- What's going well in this organization/department/unit?
- What's not going so well?
- What are the "big issues" here and what are the main challenges that lie ahead?
- What would be the ideal solution to these challenges? What's preventing the ideal solution happening?
- What would need to happen for the ideal to become reality?
- What direction is this organization/department/unit going in?
- Where do you think it is in relation to its competitors? (Even internal departments usually have competitors).
- What's the main reason for making this appointment?

You should note that these are the questions to ask whatever the level of seniority of the job on offer.

Where the company is new or small, it is vital to do some research into its stability and financial health. You can look up directors names, profit and turnover in Companies House, and get the same information from a credit agency for a fee of about £25.

Questions to think about here are:

- Is the company in a healthy financial state?
- What market share does it have?
- What kind of market is it in – is it a stable market or is it an immature market which is also volatile?

Another useful question is to ask about staff turnover. A high staff turnover usually indicates an unhappy team, or else a boss who does not know how to pick people. You will need a certain amount of caution in interpreting this information. Sometimes a high turnover simply indicates that the company has cheerfully attracted a lot of young people, knowing that they can and must move on quickly.

...Do some research into the organization or department...

Researching the organization

Researching the job

Again, a variety of approaches is best. For an existing job, it is always a good idea to talk to the departing post-holder if you can contact them. It makes no difference whether this person is leaving in good or bad odour – they can give you direct information about what is involved in the job, what your potential future boss is like and so on. Where someone has left in a cloud of anger, you will naturally need to regard what they say with caution and interpret it judiciously.

Where the job is a standard one and done by a number of people, look for someone who can tell you what it is like on a daily basis. What skills does it need? What are the highs and lows? What's the most taxing single thing they've had to deal with in the last few weeks or months?

Think carefully about recent trends in this job and discuss it with someone already in the field. What reports, pamphlets, legislation or books will you be expected to know about? Read them, even if it is only a skim-read.

If a named person is quoted in the advertisement, you must contact them: their role is to help you. Not to do so will imply lack of commitment. Such research is bound to be superficial compared with what you learn about an organization once you are inside it. No matter – some research is still vital.

Researching membership of the panel

There should be no mystery about who is going to sit on the panel: it is not a state secret. Contact whoever has invited you to the interview and ask how the panel is to be constituted. This may allow you to anticipate particular questions. For instance, if you know that a senior manager from an internal client department is going to be present, you could reasonably anticipate questions about some aspect of client care.

Identifying what the employer wants

Most employers now send out some helpful documentation in advance of the interview. You will have used this to craft your written application, but it is worth having another look at it before the interview.

The information from the employer may have included any of the following:

■ A person specification, which identifies the "ideal" candidate according to age/gender, experience, education, qualifications, motivation, personality and circumstances.

■ A job description, which sets out who the person reports to and what the jobholder is accountable for. Many job descriptions are far too long and are often out of date the moment they are written. Some are inflated so as to increase the post-holder's grade according to certain pseudo-scientific job-analysis systems. Some job descriptions are really just lists of duties, but this is still useful information about what is expected.

■ A competency analysis, which sets out what behaviour you would see in a person who was excellent in every aspect of the job. An example would be "customer focus", meaning sensitivity and responsiveness to the needs of customers. Where there is a competency analysis, this is the most important piece of documentation as it shows you that the employer has taken the task of recruiting seriously; it also tells you what skills and qualities the employer is looking for. In an organization that knows how to recruit, the interview will be hung entirely around these competencies and you will be pressed for evidence about how your experience fits them.

Study any or all of this carefully, take it apart, and ask yourself:

■ **What skills is the employer looking for?**
■ **What type of person?**
■ **What experience?**

How does your experience match? (Note that all experience is relevant, whether it is professional or obtained in some other setting). For instance, if "customer focus" is one of the competencies the employer wants, what experience will be you be able to quote at the interview which will prove that you can look after customers?

Being the solution to the employer's problems

The main point of all this preparation can be reduced to one simple formula. When you are being interviewed or assessed, your main task is to show the employer that you are the solution to their problems. You can only do this if you have a really clear grasp of what the employer's problems are. Here are some examples:

Roisin was initially trained as a nurse and had been made redundant from her job as localities manager for an NHS Trust hospital. Roisin energetically applied herself to finding a new job in the NHS and was shortlisted by another trust that was making a similar change.

She availed herself of all opportunities to find out what the potential new employer wanted. She talked to the Director of Nursing, took up the invitation to visit the hospital and made sure that she talked to six or seven of the staff while she was there – including apparently humble staff like porters as well as senior people, including the Chief Executive. The trust had been the subject of a recent Inquiry and she obtained the inquiry report and studied it carefully.

From all of this research she concluded that:

- Morale was low and the main task of any newly appointed manager would be to build morale in the immediate team.

- There were serious issues surrounding quality of patient care which needed to be addressed.

- There was still a great deal of public disquiet about the trust with a hostile local media.

- Anyone in a senior job in the trust would have to take responsibility for managing at least some of the PR.

- More change was inevitable and any new manager would have to be adept at managing it.

In her interview, Roisin emphazised her sympathetic understanding of the problems the hospital faced, and stressed the relevance of her own experience. She talked about how skilfully she had wound down her own previous team; she described her practical approaches to raising standards of care. She talked about the inevitability of more change and built on her experience in her previous post to set out her own approach. She told the interviewing panel about how her training to do radio and TV interviews in her previous job had given her the confidence to handle the media.

Roisin got the job. After she had started it, the Chief Executive told her that she had 'walked' the interview, despite a really strong shortlist.

Chris is a facilities manager, managing buildings for large employers. Having been in his post for five years in a blue chip company, he wanted to move on. He was approached by a much smaller firm that specialized in managing such services on a contracted-out basis. They were looking for a senior person to develop new services and make their pitch for work more credible.

He spent two days with the firm, and was invited to talk to whoever he wished, so with their encouragement, he rang several of their existing clients, asking questions such as: "What do you think of the service now?" and "What circumstances would make you switch to another supplier?"

When Chris returned to the firm to be interviewed, he was able to pinpoint exactly where he felt the job should develop. He offered his views tentatively, but it was clear to the employer that they were based on intelligent research. He stressed how much his present job equipped him to do all of this, and also the value of his experience as a client. Chris got the job.

A note of warning: generally there is everything to be said for this type of research. However, beware of believing that because you have been thorough, you really know all about the organization. If you think this, you may come across as arrogant as well as ignorant. Take this into account, and preface your analysis in the interview with comments like "it seemed from the research I was able to do..." You should also distinguish facts from opinions and quote sources for your views where you can without betraying confidentiality.

In summary, getting the job starts with the preparation you do on the organisation.

■ It is essential to research the job and the organization because it demonstrates your commitment to the idea of getting the job.
■ It helps to decide whether you want to pursue the application. You may uncover things about the job or organization that indicate it is not the place for you. Alternatively, the research may increase your enthusiasm.
■ It enables you to make a more convincing application because you are better informed about what the employer wants at the interview.

■ You can answer questions in greater depth.
■ It will make you stand out among other candidates.
■ It shows initiative.

The first and last two points are particularly important. During the interview, employers are searching continuously for evidence that you are their kind of person. All employers want commitment and 99 percent also want initiative.

Your USP

Finally, you should prepare about five-minutes-worth of material which contains your Unique Selling Proposition (USP). This is a piece of jargon from the world of selling. The product you are selling here is yourself. If you are well interviewed, this will emerge anyway. If you are not, or if the interview takes a turn that leaves out the aspect that you think makes you the strongest bet, then it is vital to have this prepared.

Some examples:

Jason was on the shortlist for a job as a management training consultant and knew that he was one of six strong candidates. His experience included equal time spent as an operational manager and as a trainer. He guessed, rightly, that this would be unusual, since the majority of trainers have become trainers early in their careers and often have little managerial experience. His USP was to say that he understood the issues that participants on his courses faced, as he had been at the sharp end himself. Furthermore, he said that this experience gave him credibility with participants. He also had an unusually wide portfolio of training specialisms and made sure that he mentioned these in a way that would appeal to the hiring organization by suggesting that these specialisms would enable it to increase its product range.

Geetha was a cancer specialist who was one of four senior registrars hoping to become consultants. Her USP was to set out the high store she put on interpersonal skills with patients and their families, all of whom could be assumed to be under considerable stress. She wanted to make sure that the panel knew what had been achieved by the series of information leaflets she had personally developed as well as the patient hot-line she had established in her current role. One final piece of USP was that her Indian background gave her a unique cultural and practical insight into the needs of the large local Asian population.

Fiona was a finance specialist for a small firm and was being interviewed for a similar job at a considerably higher salary in a much larger organization. Fiona felt that her USP was the flexibility and the all-round experience that her previous job had given her. Far from it having been a disadvantage to have worked in a small company, Fiona was keen to tell the panel that it was a huge plus – she had seen every aspect of the company's performance and it had taught her where it was important to have effective controls and where it was important to stay loose. She also wanted to stress that she was keen to take on more responsibility and saw this as an enjoyable challenge.

...Talk about the things you've done that can help you do the job...

This process can work very well, even when all seems lost:

Bob had been obliged to tender his resignation after the unit he was managing was pilloried in the press for incompetence. A high-profile cabinet minister with responsibility for the public service in which Bob was involved was the MP for the constituency and it was made clear to Bob that he had to be the scapegoat. This view held sway even though no personal blame was attached to him and an inquiry eventually exonerated the unit. Before the interview for his new job, Bob was determined to bring this up, even though he felt that the panel might be too embarrassed to ask about it. He wanted to raise it because he felt that the whole experience had taught him a lot about the realities of managing in the public sector. He wanted to say, too, that he was proud of the service and felt that its undersung achievements had depended on team work and effective leadership from him and that these were qualities he was eager to bring to the job on offer.

Finding your USP. Ask yourself these questions:

- From my research, what suggests I am compatible with what this employer wants?
- What problems can I be confident of solving for this employer?
- What am I really good at – what can I do well without even trying?
- What do people consistently praise about my work?
- What are my main achievements – the ones I am really proud of?
- What's unusual about me – for instance some special skill, experience or character?
- How could all of this benefit my future employer?

One important point here is to remember to distinguish features from benefits when you are setting out your USP. A feature is a straight factual statement, whereas a benefit shows how and why it will be a good thing for the other person to 'buy'. For example, it may be a feature of your career that you have a particular qualification. A benefit of the qualification may be that you have in-depth expertise in an area in which the employer is active – you could use your specialist knowledge to improve the company's performance in that field.

Your USP: some examples

Here are some more examples:

Feature:	Benefit:
I was a project management engineer for two years.	My two years of experience has given me hands-on understanding of engineering project-management, so in this job I would know all the places where cost and time can be saved through careful planning.
I am responsible for new-product development development using high-tech methods.	Being responsible for product development means that I could bring you cost-saving expertise in the latest high-tech methods.
I can use 'Word' – I'm computer literate.	I do all my own word-processing so I only need modest secretarial support.
I've been responsible for a lot of innovation in my current job.	My track record of innovation in my current job means that you can expect to get a lot of new projects and ideas from me.

What will your USPs be? Use this space to jot down some ideas:

Important uses

The USP is especially important for the interview that is clumsily conducted. This may include rambling interviewers, interview panels who still believe in the "model answer approach" and so on. When you have prepared your USP you are in a much better position to overcome the disadvantages of this kind of interview.

Research checklist:		
Have you:	**Yes**	**No**
Obtained and carefully read the annual report and any other published documents about the organization?		
Visited the organization to see for yourself what it is like?		
Talked to a variety of people in the organization about its major issues?		
Talked to the person who recently did the job, if it is an existing job?		
Talked to other people who do the job, if it is the sort of role filled by several people?		
Found out who will be on the panel?		
Identified what skills the employer is looking for?		
Identified what major problems the appointee will be expected to solve?		
Prepared your USP?		

Self-presentation

This is part of your preparation and should not be left until the last minute, as you may need to get your interview clothes dry-cleaned, your hair cut and so on.

However absurd it seems, the interviewing panel is affected by your appearance. Of course you should not judge a book by its cover, but the panel can only go on the evidence in front of them, and that includes how you look. Interview panels inevitably jump to conclusions such as:

- She looks impressive – she would do well with our committees.
- She seems like one of us – the staff would like her.
- This person has dirty shoes – this probably means he has a sloppy attitude to work.
- This person looks tarty – she might embarrass us with clients.

The point about self-presentation is that employers are looking for reasons to screen you out, not to count you in, so at the interview you want to minimize the chances of being excluded before you have even spoken.

Dressing for interviews: some basic rules for both sexes

- More formal is always safer than less formal; a smart well-pressed suit in a neutral colour is usually the best choice. Never risk anything that could seem vulgar. Don't be fooled by the apparently informal everyday appearance of staff in many organizations. Even here, it is often obligatory to wear formal clothing for an interview.

- Impeccable cleanliness and tidiness is essential: nails, clothing, shoes, hair, body. Make sure your clothes are clean and that you have had a bath or shower and washed your hair. This may seem like insultingly obvious advice, but unfortunately, experience suggests that it is ignored by many candidates. Interview rooms are often small and stuffy, and I have sat in many where the combination of candidate nervousness and unwashed body/stale-smelling clothing has ruled out candidates who otherwise had something good to offer.

This means AVOIDING

For men	For women
Straggly beards (a neatly trimmed beard is fine)	Extremes of hair fashion
Untidy or dirty hair	Untidy or dirty hair
Body odour	Body odour
Smelling of alcohol, or last night's meal	Smelling of alcohol, or last night's meal
Aftershave	Perfume – even the most discreet could seem over-powering; (perfume is also an intensely personal choice and yours may not be shared by the interviewers)
Tobacco odour or stained fingers	Tobacco odour or stained fingers
Jeans, casual clothing, pens in a shirt or jacket pocket	Jeans and casual clothing
Metal badges	
Pointy or high-heeled shoes; boots or sandals	Very high-heeled or very clumpy shoes; boots
	Bare legs, sandals; brightly-coloured shoes and handbags
Tight clothing	Lots of cleavage, no bra, tight clothing, very short skirts, anything see-through
Jewellery other than a watch or wedding ring	Lots of jewellery especially if it makes a noise; charm bracelets; dangly earrings; ankle bracelets
Joke socks	
Joke ties	
Extremes of fashion	Extremes of fashion
Over-bright colours	
Crumpled clothing	Crumpled clothing

Mental preparation

Calming nerves
Breathing

Get your breathing right first. When we are under stress, our breathing often becomes shallow. This is not enough breath for proper control, and this is why the breath so often runs out into a gulp or a tremor.

A three minute breathing exercise:

Stand up and put your hands so that the fingertips are just touching each other over your diaphragm – this is an inch or so above your waist. Keep your shoulders down and breathe in very slowly, being aware of drawing the breath from the diaphragm area. If you are doing it correctly, your fingertips will part from the fingertips of the other hand by about an inch. Now very slowly release the breath through your mouth, making a whooshing sound and making sure that it is a very long breath – significantly longer than the in-breath. Don't strain at it – it should feel natural. Repeat for about three minutes.

If you still find that you feel nervous try this additional technique. On the in-breath, say to yourself, very slowly, "I – am –…" and on the outbreath, even more slowly, "relaxed…"

Visualizing

This is another useful technique, and one of the best methods I know of making sure that nervousness does not spoil your chances of getting the job. It's a technique now widely used by athletes and other sports people before an important event.

Creating an image of success

Set aside at least twenty minutes somewhere totally quiet. Think back to an occasion where you felt confident, in charge, and at your best. Having identified the occasion, now call up an image of yourself as if you were watching yourself on a big cinema screen. See all the detail of what you were wearing, how you sat or stood and what was in the room. Hear again all the sounds in the room in every detail; call up the expressions on the other people's faces; remember exactly how you felt; what your bodily sensations were and so on. This should take at least five minutes – take your time.

Once you have this image and all its associated sounds and feelings, make sure it's in full colour in your mind. Now freeze the image so that it becomes a still frame. At the same time press your index finger and thumb together. This gives you a trigger association. Go through this process several times until it's routine. When you are approaching the interview, press your index finger and thumb together and recall the confident image and with it, the confident mental state you experienced on the original occasion.

The mental rehearsal

This is different from the practice interview. Again, set aside some quiet time – at least half an hour. What you are going to do is to play the whole interview in your head as if it were a private video-screening.

- You are going to see yourself coming confidently into the room, smiling and greeting the panel.
- Observe yourself sitting in an alert but relaxed way.
- Hear yourself giving good answers to all the likely questions, in the actual time this will take.
- Notice how the panel are responding positively to you.
- See yourself leave, knowing you have given a good account of yourself.

The more detail you can fill in, the more effective this process is. It works because your mental association is with success rather than fear.

Mental rehearsal with a partner

Another way to use this technique is to ask a partner or friend to help you. In this variant, you are going to ask the friend to help you enter a light trance state where your unconscious mind is more open to suggestion.

- Set aside some time when you can be sure of being uninterrupted. Ideally dim the lights.
- Give the partner or friend the "script" on the left, and ask them to read it back to you slowly, in a quiet, even voice.
- Allow time for the "questioning" part of the interview. Tell your partner that silence is fine at this stage of the process.
- You close your eyes and concentrate on the words, conjuring up the images, sounds and feelings that the words evoke.
- At the end ask the partner to say, "now open your eyes" and slowly come back to reality.

Again, as with all these techniques, this works – your mind is focused on success, not failure.

3

A social event
Assessment centres
Presentations

The evidence-based technique

Things never to say or do

Asking your questions

Confirmation of date and time

When you are invited to attend the interview, make sure that you confirm in writing that you will be coming. Like many of the points in this book, this one seems obvious, but it is ignored by many candidates. It is not enough to ring to confirm – ideally you should both ring and write.

You may need to ring in any case to negotiate which time slot you have. The ideal slots are just before lunch and towards the end of the process – not at the beginning nor at the very end. This is because interview panels are still getting into gear early on, and may be tired by the very end. Immediately after lunch is not a good idea either if you can avoid it as human beings seem to be designed to sleep then, and if the panel has treated itself to a large lunch to fortify itself, you may have even more of a struggle to convince them that you are their person.

If you can't make it

Some organizations are inflexible about both time and date. If there is a large panel, it may have taken weeks to find a date when everyone can be free. Unless they are very taken with your application, if you cannot make the appointed date, they may simply rule you out – or else politely tell you that if they do not make an appointment they may want to come back to you.

If you change your mind about being interviewed, tell the employer immediately. This is only courteous as they may want to rearrange times for other candidates, if there is time. Also, you may want to re-apply to the organisation again at some point in the future.

Getting there

This is a simple point, but it's worth a reminder. If the interview is being held at a place you haven't previously visited, make sure you have:

- Advice from someone who has been there already.
- A map.
- A parking place: booked or planned, if driving.
- Allowed enough time to arrive unflustered, go to the loo, check your appearance and have a final glance at your notes. On balance, it is probably best to aim to arrive about half an hour before the interview. If you are early, the panel may also be running early – for instance, it is not uncommon for interviewees to withdraw at the last moment, leaving a hole in the schedule. In this case, if you arrive during the time the missing candidate should have been interviewed, you will probably have the advantage of being able to have a longer interview.

What to take in with you

You should plan to go into the interview with a folder of notes, including your application form (in a briefcase if you wish) and nothing else. It is usually perfectly safe to leave coat, handbag, gloves, scarves with whoever is looking after the outer-office or ante-room. This means you look calm and professional, and will also be undistracted by having to work out where to put your coat, keys and so on.

I once sat on a panel which was meeting a few days before Christmas. By far the strongest candidate nearly lost the job because she entered the interview room garlanded with carrier bags from Hamleys, the famous toy store, only a short distance away. In the post-interview discussion, the Panel Chair commented sourly that he supposed the only reason she'd agreed to be interviewed was in order to do her Christmas shopping.

The interview as a two-way process

When my colleagues and I coach people for job interviews, we notice how often people use remarkably extravagant language to describe their fears. These fears usually fall into a number of familiar categories:

- **Losing face.**
- **Drying up.**
- **Losing control.**
- **Being unable to stop talking once initial nerves have passed.**
- **Being inappropriately dressed.**
- **Failure: not getting a job they badly want.**
- **Getting the job and being afraid of doing it badly.**
- **Being offered a job it turns out they don't want after all.**

One source of all this fear can be that your mental model of the interview is that it is a one-way process: you may not feel that you have any power. In fact you have a lot of power in the interview.

If you are easily the best candidate, the employer may even feel that the power is all yours because if he or she likes you, then the reckoning will be that you may be equally attractive to another employer.

Remind yourself that you have already overcome several of the hurdles that the employer has put in place. Not every candidate is seen and it is unlikely that the employer will want to waste time seeing you if he or she does not think that you could do the job. In my work with employers, I always encourage them to see fewer rather than more candidates at interview, simply because of the high cost in time and money of the interview process.

And if you could do the job for this employer, then possibly you could do the job for another employer or even decide that you want to stay where you are. Being interviewed for another job is well-known as a sort of polite blackmailing tactic to force an existing employer to raise a salary or find a more attractive job. When you go into the interview, remember that the employer does not know whether this could be true for you.

Balance of power

When you sit on the other side of the interview process, this power balance is even easier to see. I was once part of a panel interviewing for a senior post in a medium-sized organization.

Although there was an apparently strong shortlist of people, all of whom were appointable in theory, it immediately became clear that there was one outstanding candidate. He had been made redundant from his previous post, so there was a natural tendency on the part of the panel to assume that there would be a degree of panic on his part. He was impeccably pleasant and courteous throughout the whole interview.

In its closing stages, there was the usual conversation about when and how candidates would be informed of the panel's decision. The Panel Chair said that there could be a short delay as one candidate still had to be seen on a different day. At this stage the candidate very politely said: "I appreciate your difficulty, but I've already been offered another quite attractive job, so I'd really welcome an early decision."

The interview is not a trial

After the candidate had left the room, the panel fell into panic mode. "We've got to have him. How can we stop him accepting the other job?" In no time at all it had been agreed that the remaining candidate's interview would be advanced, and by the end of the following day the original strong candidate had been offered the job, his desirability immeasurably increased because it was clear he was attractive to another employer. Here, the onus was very much on the potential employer to seem attractive to the candidate, not the other way around

The interview is not a courtroom where you are on trial. Nor have you been captured and taken into enemy territory where you are going to be subjected to a brutal and humiliating interrogation. On the contrary, the job interview is an exploratory conversation where you and a potential employer size each other up.

...An interview is a conversation not an interrogation...

The interview as a two-way process

Your task during the time you are with them is to ask yourself:

- What can I tell about the organisation /department/unit by the way they are treating me? Have I been treated courteously or discourteously? Formally or informally? Efficiently or sloppily? For instance, you might want to think twice about joining a company that kept you waiting for your interview in a cold room without refreshment or any apparent concern for your welfare.

- What do I think about the person who would be my line-manager if I get this job? Could I work with him or her? Would he or she have my respect? The potential line-manager is easily the most important person at the interview: pay special attention here.

- What can I tell about this organization/department/unit from the way people behave? How do they dress? What kind of atmosphere is there? Would I fit in and be happy here? A thorough and professional selection process will give you the chance to meet potential team-mates. Are these people you would enjoy working with as colleagues?

Where you have any doubts about the answers, listen to them. Never accept a job offer just because you feel flattered and relieved that someone wants you. Ask for an informal further meeting. The more senior the job, the more common this is.

It's not by chance that the pompous-sounding 'executive search' process for senior vacancies is more usually described as "head-hunting". The implication is that you are the head – the prized scalp that is to be brought home through cunning: lured, persuaded, cajoled. Unpleasant though the basic metaphor is, it is actually a more truthful representation of what really goes on in any job interview, however humble the job. It is actually a courtship. You have to be courted by the organization just as much as they have to be courted by you.

The interview as a social event

There is some evidence to suggest that the interview, as traditionally conducted, can really only give a panel two kinds of information about the candidates they see:

- Their social skills
- Their motivation: do they really want the job?

Some points to remember:
My experience certainly reinforces the likely truth of this. Time and again I have seen the job go to the candidate with good social skills and enthusiasm. These candidates instinctively realize that the interview is a social occasion. The panel are the hosts and you are the guest. Their role is to make you feel comfortable so that you can sell yourself to them. Your role is to make them feel comfortable so that they can sell themselves to you.

Your behaviour affects other people's response to you. Smile and the interview panel will smile back.

How do you display good social skills in the interview? Here are some of the things it's most important to do:

- Smile. If you look tense and serious you will make the panel feel tense. There has been research showing that the person who smiles the most is the one most likely to get the job.

- Involve the entire panel in your questions and answers by constantly engaging them with eye contact.

- Convey relaxed confidence and alertness in how you sit.

- Sit upright – not rigid – but with your bottom tucked into the back of the chair, hands folded neatly into your lap and feet firmly planted on the floor. Keep your shoulders down.

- Don't slump, cross your legs or arms; don't sit on the edge of your chair or sideways in it.

Personal experience as an interviewer of how candidates should not do it includes examples of both over and under-confidence:

• A candidate who spent the whole interview lolling with one arm casually flung across the back of his chair so that he was only half-facing the panel.

• A candidate who leant confidentially across the table for the greater part of the interview, thus innocently displaying her very ample cleavage.

• A candidate who was so nervous that she looked anywhere but at the panel for most of the forty minutes she was in the room.

The interview as a social event

Share the talking. The ideal is 50:50. You are gathering information about them, as much as they are gathering it about you.

Keep your answers brief. A good guest does not bore the host by going on at length with some anecdote, and nor should you. The absolute maximum length of any reply should be three minutes. One and a half to two minutes is better. If you're unsure how long two minutes sounds, then make that part of your practice interview. In fact, as advertisers know, it's possible to say an awful lot in just 30 seconds! If you give long replies, the panel tends to think: "This person is too fond of the sound of their own voice." If your replies are ultra-brief, the panel may think you lack confidence. If in doubt about the length of your answer, it is perfectly OK to pause and say: "Is this about the right amount of detail?" Take your cue from the panel's response. If you are going on too long, they will be pleased at the chance to stop your flow. If you're being too brief with your answers, they will encourage you to say more.

Speak clearly and without jargon. This is especially important where the panel contains members who are not specialists in your area. It irritates the non-specialist to hear jargon – it makes them feel excluded and marginalised and they may attribute motives to you such as wanting to show off or being vain or arrogant.

The importance of rapport

Because the interview is a social event, your task is to create a rapport with the interviewer, just as it is his or her task to create a rapport with you. When conversation is flowing you are much more likely to be the person who gets the job.

How do you know rapport when you see it? Look at two people who are getting on well together and you will see that they are mirroring each other. They will sit at exactly similar angles to each other, use similar gestures, cross their legs or scratch their noses at the same time. If they are in a bar, they will lift their glasses at the same moment and the drink will go down in the glasses at the same rate. Ideally you need to do the equivalent at the interview.

Matching

Rapport is about matching. Mostly we do this unconsciously, but skilled behaviour at interviews is about raising the most effective behaviour from the subconscious and keeping it under control in spite of nerves and tension. If you mismatch, you will be creating discomfort for the interviewer and will decrease your chances of getting the job.

Rapport means that you are demonstrating respect for the other person. How do you create this? It's really not what you actually say, though that is important too. It all really depends on what you do with your non-verbal signals.

Keep your voice up – many candidates allow nervousness to reduce their volume to little more than a whisper. Some go to the opposite extreme and bellow. If you know you are prone to either of these traps, practice the correct volume beforehand. It may be one of the most important areas to get feedback on from whoever is your practice interviewer.

Other tips for creating rapport

Look at the way the interviewer is sitting and try to make sure that you match his or her body language. So, for instance, if he or she is sitting forward, you should lean slightly forward too, as you answer the questions.

Interviewers vary in the way they speak. If your interviewer speaks quickly and enthusiastically, you should try to do the same. If he or she speaks more slowly and deliberately, you should try matching that.

Gesture: some interviewers wave their arms around dramatically, some are more contained. Matching is not about copying – that could seem insulting – but some discreet matching can be helpful. If you know, for instance, that you tend to sit very still, then you could run the risk of mismatching an interviewer who has extravagant gestures.

Interview questions

There are whole books on answering interview questions, but experience shows that there are really only eight questions that can be asked at interviews. They may be expressed in hundreds of different ways, but this is what they come down to:

1 What are you currently doing? (Your present job or how you are spending your time if unemployed).

2 Why do you want this job?

3 Why do you want to work in this organization/department/unit?

4 **What skills and experience do you have that fit our needs? How will your skills help you to do this job to a high standard?**

5 How would you tackle this job? What are your ideas about changing or improving the way this role is carried out?

6 **What kind of person are you? What stresses you? What delights you?**

7 Do your personal circumstances fit with what we need? (e.g. frequent travel; the company's office is in Birmingham but your family home is in Yorkshire).

8 **What questions would you like to ask us?**

Prepare your answers

You need to have built comprehensive answers into your preparation. Give the list to the friend who is running your practice interview and encourage him or her to create two or three questions around each area.

Below, there are three keys to answering these questions successfully:

- Your research into the job and organisation (see p18).
- Demonstrating a giving not a taking attitude.
- Using evidence from your experience to answer questions.

Giving, not taking

Employers hate feeling exploited. They dread taking on staff who turn out to be constantly off sick, who are pedantically determined to leave on the dot of 5p.m. or who get fixed on taking their holidays every year in certain weeks and whose concern is for themselves, not their colleagues or the company as a whole.

As an interviewee, you must convey that you are a giver not a taker. You need to show:

- Enthusiasm for the mission of organisation/department/unit
- Commitment and enthusiasm for the work
- Flexibility
- Sympathy for the challenges of organisation/departments/unit's
- Ability to solve problems, not create them
- Resourcefulness

...Talk about what you can do for the job, not what the job can do for you...

Some examples

Questions	Takers	Givers
"Why do you want this job?"	SAY: *"Because it will be good for my career."*	SAY: *"Because I think I have x and y to offer here"* – and describe it.
"What's the most important thing about any job for you?"	SAY: *"A good salary"*, or *"security."*	SAY: *"The chance to work on really interesting problems, or, to do something of practical benefit to people."*
"What do you want to ask us?"	SAY: *"How much holiday would I get?"* or, *"What percentage contribution do you make to employees' pensions?"*	SAY: *"I'm interested in what you see as the trends on x or y issue."*
"What are the main problems facing this organisation/unit?"	SAY: *"They are so serious that it's hard to see how to get round them."*	SAY: *"There are exciting and challenging times ahead. My ideas would be…"*

Using the evidence-based technique

The panel can't know how you would actually do the job – they are speculating, and so are you. Easily the most effective way to convince them that you would do the job well is to offer evidence of past behaviour.

Wise, not smart, answers:

This is how it works. Suppose you are asked: "What's the thing you are most proud of in your current job?" You may be tempted to reply with something like "surviving" or "having a job at all". This would be a smart but not a wise answer.

All your replies must quote evidence from your experience which links with the skill the employer is looking for. Let's suppose in this case that the job specifies skill in negotiating – you know this because you have been sent the "person specification" (see p21).

Your ideal reply

"*I'm most proud of managing a really tricky negotiation with our union. I led our team and…*" You then go on to describe what happened.

Another example

The interviewer asks, "*How do you know you could bring this unit in on target financially?*" The unwise reply would be to say, "*My attitude to this is that it is critically important and I would set the right kinds of controls in place.*"

This tells the interviewer something about the attitude you claim, but nothing about what you actually do. So the wise reply would be to link it with your specific past experience of budgetary management, the closer the better to what you understand this job to need.

How to use the evidence-based technique

- Be specific not general.
- Describe what you personally did and the impact this action had. So for instance, if asked: "*What did you contribute to the set-up phase of Project X?*", you would say: "*I set up the whole project from start to finish – it was my idea, so I obtained the necessary resources and did all the initial planning with a project team I had chosen myself. We looked at everything and anything that could go wrong and assessed all the risks, financial and otherwise. The result was that the project got off to a very smooth start because we had done all the hard thinking at the outset.*"
- Don't be afraid to bring in volunteer activities if your professional experience is lacking in some particular area.
- Remember, you're describing what happened – not your personal attitudes or ideas.

A final example

"What's your attitude to being in a team?"

Unwise reply:

"I love it – I'm sure I'd settle in here."

Wise reply:

"My present job needs really close team work because we each depend on everyone else to get particular work done (then you describe how you currently contribute and what effect that has on the whole team), so I'm very positive about team work."

Funnelling down

Note that properly trained interviewers are shown how to get evidence from candidates. They use a technique sometimes called "funnelling down". This goes in the following way. A candidate is asked a question such as: *"Tell us about your approach to managing a major change in your team – give us an example from something you've recently dealt with."*

Let's suppose that an untrained candidate gives a rather vague reply such as: *"We've had a lot of change recently and it's been very hard to get all the staff on board, but I think we did it OK because people seem to have settled down a bit now."*

The trained interviewer is not going to be satisfied with a reply like this because it does not give any evidence of what the candidate's own responsibility or achievements were. So there will be a follow-up probe which obliges the candidate to narrow down the answer. This might go something like: *"Yes, I see, but what did you specifically do to help that happen?"*

It is hard to wriggle out of an evidence-based reply when pressed like this.

When you are lucky enough to meet a properly trained interviewer then you will find that you are politely probed for evidence in this way. If you are also trained along the lines described here, then the process will happen smoothly for both sides.

Most interviewers are not trained, and so you will probably have to manage this for them.

Things never to say or do at an interview

A good many candidates destroy their chances of getting the job through sheer clumsiness:

- The senior manager who did not get the chief executive job he craved. One reason was that he rubbished his current boss at the interview.

- The young secretary so desperate to get into publishing that she turned her job title into "Editorial Assistant" and lied at the interview about her experience. She was exposed when the employer sent a routine request for confirmation to her current employer.

- The training consultant who prattled on artlessly and inaccurately about a major change programme, involving training at British Airways, without ever realising that he was talking to one of the main architects of the change (who by then had changed jobs and organizations).

- Lying: there is a strong chance you will be found out, especially if the job is in a related field where people do a lot of networking. Sometimes the deceit is discovered in unusual ways. In one recent example, an employee took her employers to an industrial tribunal for unfair dismissal. The tribunal agreed, but reduced the damages considerably on the grounds that she had lied about her qualifications – an Upper Second degree at Leicester University turned out to be a Lower Second from Leicester Polytechnic. It is always best to tell the truth and to resist the temptation to inflate your qualifications or experience. If found out, you could lose the job. Employers are rightly twitchy about honesty.

- Asking for 25 percent or 50 percent more than the advertised salary. Never negotiate salary at the interview (see p65). Asking for a much bigger salary irritates employers and may suggest you are too big for your boots or that if offered the job, you will leave as soon as you get a better offer. Don't apply for the job if you really couldn't work around the salary mentioned, plus or minus 10 percent.

So here is a brief checklist of thing to AVOID at all costs:

- *"My current boss is an idiot."*

- *"My present boss and I have a personality clash."*
  Employer reaction: If you're that disloyal to your current boss, the chances are that you'll be the same in any other job.

- *"I can't wait to leave my present job."*
  Employer reaction: as above.

- *"My present colleagues get on my nerves."*
  Employer reaction: this is someone who can't work in a team.

- *"My current salary is..."*
  (naming an inflated amount)
  This is one thing employers always check with referees.

- Inflating your experience. If you inflate your experience, a skilled interviewer will ruthlessly expose how little you know through using the interview and question techniques described in this book.

An employer who has any doubt about the truth of what you claim is likely to check this through a telephone conversation with your current employer.

- Improving the sound of your job title. Even the laziest employer will check this with your current employer.

- Lying about qualifications. Several recent high-profile cases have highlighted the need for employers to check qualifications. It is very simple for the employer to do – one telephone call is all it needs. Most employers are less interested in qualifications than you may think. Experience is usually considered a lot more important, so resist the temptation to improve the look of your CV.

- Fictionalizing your interests. The point of putting interests on your CV is to represent yourself as an all-round person, not just a work person. Don't put down opera if your real interest is are slumping at home watching TV. It will only take one question to expose you.

Things never to say or do at an interview

Wiser tactics

- Always be respectful and positive about your current boss, even if the relationship has been stormy. The potential new boss may know the present boss. It's a smaller world than you may think.
- Speak enthusiastically about your present subordinates. If you criticize them, it will reflect badly on you because it was your responsibility to coach and develop them. Avoid criticizing current colleagues and bosses under any circumstances – for any reason.
- Similarly, talk enthusiastically about current team-mates. After all, you have been part of that team and will have made a contribution to whatever the current team climate is. All potential bosses know this.
- Make sure that you can back up anything you claim about your opinions at the interview. For instance, if asked a question about what you are reading, never claim you are reading a book which you only know about through reading the reviews. You may be asked a follow-up question which reveals that your knowledge is skin deep or even less.

- Don't claim experience which you don't actually have. A skilled interviewer will probe until he or she is satisfied about your involvement and knowledge. If this is so small as to be minuscule, you will be shown up.
- Give accurate job titles. Don't be tempted to turn a "manager" job into a "Head of" job.
- If asked for your education details, put them down correctly. Don't turn Harrow Comprehensive School into Harrow School.
- Make sure that you have listed your qualifications accurately, including awarding bodies and dates. Employers are now checking them as a matter of routine.

Coping with incompetent interviewers

Incompetent interviewers don't do it on purpose. They may be inexperienced, nervous or over-confident. They may blunder innocently into areas in which the law forbids questions. Below and over are some of the main ways in which interviewer incompetence shows through, together with some suggested ways of dealing with it:

1. Double or even triple questions

This is probably the most common single mistake. Here's an example: *'I wonder if you can tell me what experience you have of leading a team and also whether you've actually ever had any training in team leadership?'* This is two questions rolled into one.

How to deal with it

Politely re-phrase the questions – not to humiliate the interviewer – but to remind yourself of the questions. Say something like: *"I'll deal with the two aspects separately if I may – my team leadership experience first, then my training."*

You may find, as many people do, that you forget one of the questions. If this happens, say: *"I'm afraid I've forgotten the first (or second) half of the question. Can you just remind me of it please?"*

2. The rambling question

The interviewer who rambles at length has forgotten, or never knew, that the purpose of the interview is to listen, to hear you talk and to respond to the questions you will have.

How to deal with it

Listen really carefully. The actual question will be buried somewhere inside the rambling.

Summarize what you think the question is by saying something like: *"Can I just check that I'm understanding your question properly? You believe that many people in senior jobs don't know how to read a balance sheet, and you'd like me to show you that I do?"*

Coping with incompetent interviewers

3. The hypothetical question

Trained interviewers know that they have to avoid hypothetical questions. Untrained interviewers love them.

Hypothetical questions are a bad idea for a number of reasons. The first is that they give unfair advantage to internal candidates. For instance, a candidate looking for a job as a residential care worker was asked: *"What would you do if a resident locked himself in his room and threatened to set fire to it?"* She used the evidence-based technique (see p44-45) to

answer, describing how she handled real crises. She got the job. However, later she was teased for not knowing the "right" answer, which was that the staff kept duplicate keys. There was, of course, no way she could have known this.

The second reason is that they only test your ability to give a smart answer, not what you would actually do if the hypothetical situation became real. Finally, it is most unlikely that the exact hypothetical situation would happen, therefore it is neither here nor there what you reply.

How to deal with it

DON'T
- Criticise the interviewer openly for asking a clumsy question.
- Fall into the trap of trying to give a hypothetical answer.

DO
- Say something like: *"It's difficult to say how I would deal with that because I'm sure once I was in the job, it would be affected by what I then knew about your systems and so on. But I can talk about an occasion in my current job when I dealt with something very similar."*

You then describe the past situation, fitting it as closely to whatever the interviewer's concern seems to be.
- If you don't have any matching experience that you can quote, use the first part of the answer (left) then: *"…knew about your systems and so on. But in general, my attitude to this kind of situation is to…"*

Hypothetical questions are usually about crises. You could reply that you would look for underlying causes, not knee-jerk responses.

4. The politically incorrect question

This interviewer has never heard of any of the legislation concerning equal opportunities. He or she, more usually he, feels it is perfectly all right to press women candidates for answers to questions such as:

- *What are your plans about having children?*
- *What are your childcare arrangements?*
- *Are you actually divorced?*
- *What happens if your children are ill?*

How to deal with it

DON'T

- Threaten to report him to the Equal Opportunities Commission or the CRE.
- Get on your moral high horse, however tempting it may be.
- Tell him he needs training or consciousness-raising.
- Ask if he is asking the male (or white) candidates the same question(s).
- Flounce out indignantly.

DO

Stay calm and friendly, but note that the question may reveal an environment which is male-dominated, naively sexist or racist or one where it is difficult for women to progress. On the other hand it may reveal no such things, but it will be worth investigating before you accept any job offer.

Deal with the underlying concern, which is: *"What are your priorities?"* *"Will you be off work a lot?"* *"Can you really do the job?"* Possible answers could include: *"My family is very important to me, but so is my work. I've always made sure that I've had excellent childcare arrangements"*, or, *"I take my responsibility to my employer very seriously. I haven't had any days off other than proper holidays for two years"* or, *"Well, I wouldn't want to bore you with my domestic arrangements, but they satisfy me and if they can do that, I think they'll be OK for you!"*

Other politically incorrect interviewers may question:

- Single people about their sexual orientation.
- Older people about their health.
- Disabled people about their mobility.

Coping with incompetent interviewers

5. The "model" answer

Local authorities are the main offenders here. In the late 70s and early 80s, local authorities became obsessed by political correctness. A lot of anxiety focused, probably rightly, on the selection interview. Unfortunately, although the diagnosis was correct (i.e. interviewing had previously been the source of biased and unfair selection decisions) the solution adopted has not helped. In the model-answer approach, a set of rigid questions is agreed and a model answer is actually written down for the panel. Often, follow-up probe questions – one of the expert interviewer's most useful tools – are expressly forbidden. Even more ridiculously, I know of at least one local authority where interviewers are trained to show no emotion whatever towards candidates, including smiling and nodding. Absurdly, this is supposed to make it easier to compare one candidate with another.

Model answers are a bad idea for similar reasons to the hypothetical question. They favour the internal candidate because the internal candidate is more likely to have warning of what kinds of things it is OK or not OK to say. Model answers also encourage people to answer in the way they think the interviewer wants to hear. In other words, your effort goes into trying to second guess the panel rather than in demonstrating that you could do the job.

How to deal with it

You can't deal with it directly, as you may not know until later that the panel was using this approach, although you may guess that this is the case from the stilted way the questions are asked. Another give-away is the lack of follow-up questions. Two tactics do help: Use the evidence-based technique (p44-45) where the quality of what you offer will usually suffice. In effect, you oblige the panel to appoint you, in spite of their self-imposed limitations.

Have your USP prepared (p24-25) and use it at some appropriate moment in the interview. Make maximum use of the part of the interview which deals with *"What questions do you have for us?"* by saying: *"I'd like to take a moment just to draw attention to one or two aspects of my experience that I think it would be useful for you to hear about"* – and then briefly describe them.

{ 6. The off-the wall questions
The questioner who talks all the time
The questioner who wants a nice chat

I have bracketed these three together because they are variants of the same thing: nervous interviewers who try to escape from their apprehension.

They do this by avoiding giving you the chance to talk about the things that really count – your experience, your skills your personality and any new ideas you can bring to the organization. They want you to do well, but they are afraid you might not. In response:

How to deal with it

DON'T

Don't collude. Doing this means taking the unwise option of going along with the interviewer and turning the whole thing into a low-key chat.

For instance, a former colleague of mine went for an interview as a PA to someone she knew socially for a 'chat about the job'.

She emerged from this occasion totally bemused. Nothing had been said about the nature of the business other than what she already knew, or what her duties would be. She took the job, but it turned out to be a mistake. She wanted a job with greater responsibility and he really only wanted a bookkeeper.

In other words, colluding will deprive you of the chance to find out what the job is actually about, and what skills you will need for it.

DO

Politely take a degree of control by briefly acknowledging whatever the interviewer has said, but then turning your answer to how your experience fits the job in relation to whatever issue the interviewer has raised.

Use the *"have you any questions"* part of the interview to say something like: *"I feel I haven't really said much about some aspects of why I'd like this job and why my experience is suitable for it – could I take a few moments to do that?"* If you get a nod of assent here, then plunge into a suitably shortened version of what you would have said if you'd been asked more focussed questions. If you can carry it off, look for the tiny pauses in the interviewer's flow where you can break in. This, however, is a high-risk strategy as it may look rude if it is seen as "interrupting".

Nervous interviewers are more afraid of the interview process than they are of not appointing the right person. They rarely realize how off-putting their questioning approach can be. These are the interviewers whose own dread is of drying up or of not being able to think of clever questions. When a nervous interviewer meets a nervous interviewee, the result can be dire. It is almost certain that a good appointment will not be made.

Coping with incompetent interviewers

7. The leading question

The leading question is the one that suggests a right answer, often "yes" or "no". An example would be: *"I'm sure you'd never make the mistake of failing to do appraisals for your staff?"*

It is generally asked by the panel member who is impressed by you and wants to convince other panel members that you are the best candidate.

The leading question is a poor way to ask candidates about their experience because it does not probe, nor does it look for the experience which shows whether the candidate can actually do what the job needs. Panel members who are still making up their minds will not find a "yes" or "no" or other brief answer to a leading question very convincing.

How to deal with it

DON'T

Just reply with a single word or phrase. If you do this you will again lose an opportunity to put your views. Other members of the panel may also think you lack courage – they may not necessarily agree with the opinions their colleague is expressing and will want to hear your own views.

DO

Look for the underlying pre-occupation. An example was the interviewer who asked a candidate: *"Wouldn't you agree that leadership and management are two different things and that most organizations are over-managed and underled?"*

The unwise answer would have been: *"Oh yes, I completely agree. Most organisations really do lack leadership."* She interpreted the question, rightly, as being about whether or not she was familiar with the distinction between leadership and management, so she replied: *"Yes I agree, but perhaps that's because managing people through change is much more challenging than just keeping the systems and processes going. I try to be able to do both, but it's change I'm really interested in. Would it be useful if I said more about my approach here?"*

A second tactic is to turn the question into an open one. So if the interviewer says, *"I suppose that leading a team is pretty challenging?"*, you could reply: *"Yes, it is. My own approach to it has been..."* And then you describe your own experiences.

Awkward questions: some answers

Answering awkward questions

Like virtually all candidates, you may have something in your job history that you believe could disadvantage you. If it is obvious from your CV, as 99 percent of the time it is, the general rules are simple:

- Don't deny it or lie; you will be found out.
- Stay calm – the employer may care a lot less about it than you think.
- Focus your answer firmly on the future – whatever the disadvantage is, it is now in the past and you are fully able to do the job.

Here are some examples. It's not an exhaustive list, but it will give you the idea.

A poor health record/disability

Employer fear: you'll be off sick a lot.

You're fully recovered and totally energetic or your condition is under control with medication/special treatment and need not prevent you working as a full member of the team.

A large number of jobs in a short period

Employer fear: you won't stay.

You have deliberately had a number of short-term jobs as a way of finding out what it is you really want to do – and now you know that, you want to stay put and give your all.

A very long time in one organization/job

Employer fear: you're inflexible, fixed on how they did things in your old organisation.

The work in your old organization was interesting enough to keep you there. Although it may look like one job, in fact you held a wide variety of responsibilities… But now it's time to move on and you are looking forward to a change.

Awkward questions: some answers

A period of redundancy/unemployment

Employer fear: you're unemployable; there's something risky about you; you've forgotten how to do a job.

With such a high unemployment rate, it's inevitable that even good people lose their jobs. You've used the time fruitfully to study, make a garden…etc. Now you're raring to go again, feel you have a lot to offer…

A former boss with whom he/she suspects you did not get on

Employer fear: this is someone who finds it difficult to accept authority.

Say nothing to the discredit of the former boss. Describe their good points and emphasize how much you learnt from working with them.

A prison sentence

Employer fear: you might get into trouble again.

If you've got as far as the interview, this is an enlightened person. Say you've paid your debt to society, have acquired many new skills/attitudes – that's all in the past; your record since leaving prison shows how different you now are.

Your age

Employer fear: you're too old/ too young.

Though it may seem unusual for someone your age to be applying for the job, you are convinced you have the qualities needed (then describe them using evidence – see p44-45).

Your biggest mistake

Employer fear: you are too arrogant to continue learning.

Briefly describe a relatively insignificant mistake (not the sort that can bring a whole organisation to its knees) and put the emphasis on what you learnt from it and why you will never do that one again.

Your major weakness

Employer hope: you will shoot yourself in the foot.

Choose something that could also be a strength – for instance preferring working with the big picture rather than the detail, or cramming too much into your day.

Another useful tactic is to refer to something which was a weakness in the past, but which you have now conquered. For instance, you might say something like: *"I know that when I first started being a (name the job title) I found it difficult to prioritise because I'm energetic and wanted to get so much done. But that was ten years ago and I think I've more than got that aspect of my work under control now to do the job I now do!"*

This question is often asked in the plural: *"Tell us about your weaknesses."* Beware of falling into the trap of providing several examples. One is quite enough!

Your views on a controversial current affairs issue

Employer fear: you are in a little world of your own and have no knowledge of current affairs. Or if you do, you are a prejudiced person.

The safe answer is to convey that you can see both sides of a very complicated question. There is no easy solution and a thoughtful person would take time to consider long-term implications of any ideas on the topic.

General tips on answering awkward questions

The awkward question puts you on the spot and the danger is that you could feel your confidence crumbling. In the UK particularly, we tend to find it difficult to speak up for ourselves. The custom is to be modest and not to claim too much. To appear boastful is definitely counter-culture. However, the interview is not the place for modesty. If you don't speak up for yourself, who else is going to? The keys to success are:

■ **Use direct language**
Watch out for the use of would-be modest phrases like:
"I'm pretty sure I could lead this team effectively."
"I think I'm quite good at..."
This sort of tentative language is all right when you are discussing yourself with your boss or with friends, but it is no good at an interview. Qualifying phrases make you sound as if you doubt your own abilities.
Instead, use direct language:
"I can lead a successful team."
"I am good at leading a team."
"I know I do x or y well."

■ **Give yourself references from others**
Where you find it hard to be direct, consider using phrases which in effect quote what other people have said about you. For instance, if you are asked about your style as a manager, say something like:
"My team tells me that since I arrived on the scene I have transformed the whole atmosphere in the office by..."
"I get a lot of feedback about my style from my colleagues at work. They say that..."
"At my last performance review my boss told me that he really likes the way I have tried to create a democratic atmosphere within my own team."

Asking your own questions

VIDEO ARTS®

...Find out about the things that matter to you...

By tradition, the final part of the interview is the one where the panel chair asks you what questions you would like to ask. As with every other aspect of the interview, you need to play this with sensitivity. If the panel members are shuffling uneasily and looking at their watches, you know they are over-running, or possibly have lost interest in you, so keep your questions brief. If they are looking relaxed and attentive, you could safely assume there is more time.

There are two good reasons for taking advantage of the chance to ask questions:

- It's the part of the interview where you are unequivocally in control.
- It's another chance to demonstrate your unique qualities, experience and resourcefulness.

Always have some questions prepared. It's perfectly all right to take out your notes at this point. Look for ways to link things that have been said by panel members to the questions you want to ask. For instance: *"I noticed earlier on you mentioned the appraisal system, I'd like to ask you some more about that."*

Another example might be: *"When I replied to your question about training, you commented that... could you tell me a little more about that?"*

This shows how carefully you have been listening during the interview, and not just looking on it as a chance for you to talk.

The questions themselves should be about the organisation and the job and should be based on your pre-interview research. Ideally they will be present or future-based questions such as:

- How will you judge how the person you appoint is successful in the job?
- What's the likely policy on x or y in the future?
- What percentage of market share do you aim for in x or y brands?
- What kind of morale is there is the department at the moment?

Things to avoid:

- Don't ask questions about: salary, pension, contract-length, holiday entitlement, cars. The appropriate time to do this is when you have actually been offered the job (see p85).

- Don't ask questions that have already been answered earlier in the interview.

- Don't say: *"I don't have any questions"*, as this can be taken as passivity or lack of interest.

- Perhaps the most important don't of all is this: don't start interviewing the interviewer. This has several effects: it breaks the convention that you are the interviewee and they are the interviewers; it looks rude; it may come across as hostile; it may suggest you think you have the job in the bag.

As a boss I once interviewed a promising candidate for a job. Although I noted a slight tendency to talk too much during the interview, I still thought she was a strong candidate. We got to the final part of the ritual and I asked: *"What questions do you have for us?"* She took out a stout file and embarked on an attempt to grill and cross-question us about the culture of the department, whether or not people were happy and what we planned to do about it if they were not. As a final question she asked us for feedback on her performance as an interviewee. Fortunately, we were able to say politely but firmly that we had to curtail that part of the interview because we were falling behind in our schedule. We did not appoint her.

You may also like to use this part of the interview to add anything that you wanted to say but were not asked about by the employer. For instance, you might say: *"I don't have any other questions, but I think I may have under-emphasised my experience of x while you were asking me questions about that earlier. I'd just like to add that..."* (and then you describe your experience.)

One important question to ask is: *"When will you let me know what your decision is?"* This is vital information, as if you don't hear within the named time period, you should ring to find out what is happening.

On-the-spot hire

As a young teacher looking for my first job, I remember vividly the torture of my first experience of being interviewed. I was called to a college in a small town in the north-west of England and told that my interview would be at 11am, but that there would be a tour of the college at 9am with all the other candidates. It was explained that the decision would be made on the day, so I was expected to wait in the college until 4pm, when the successful candidate would be offered the job.

I had never been to that part of the country before, and when I stepped out of the train that morning I felt very isolated. I met the other candidates, all of whom seemed equally tongue-tied, was shown around the college without meeting any of the people with whom I would be working if I got the job.

The interview seemed to go quite well, but then there was a long, tedious wait while all the other candidates were interviewed. Conversation was desultory while we waited for the verdict.

Unfriendly practices

Teachers, hospital and local government staff used always to be selected in this crude and insensitive manner: all candidates were called for interview at the same time. All sat around for the entire day, going in one at a time for the interview. All sat nervously awaiting the decision. The door would open: Mr X, will you come in, please? Mr X would be offered the job and was expected to say "yes" on the spot and to agree a salary.

If he said "no", or "don't know", the number two choice would be called in. The reasons are that:

- Local politicians are usually involved and the assumption is that they need to see the whole process through from beginning to end.
- The panel presumes that candidates will be desperate for the job and will therefore submit to this process without complaint.

Unfortunately, there are still pockets of this candidate-unfriendly practice even today.

You may be unlucky enough to encounter it. If so, here is what to do:

- Check in advance that this really is how the organization makes its selections. Ring the personnel specialist, if there is one, and check what actually happens.
- Decide whether you want to go ahead with the choice of being a candidate. Selecting people in such a way may suggest that this is an organization which is out of touch with current thinking. If you go ahead, do it knowingly.
- Make doubly sure you research the job in advance. This means thoroughly checking out the previous job-holder, why he or she left, and what the job consists of.
- Research the locality if it is different from the one where you are currently based. What does housing cost? What are the schools like? Is there adequate transport?
- Discuss it with your partner and family. Are they committed to a change of job if it will mean a house, school or town move?

- Research the potential boss. What do people say about working with this person? Explore any tell-tale hints that he or she might be a bully or in a weak position.
- Research the organization along the lines described on p18.
- Make a list of the questions that you must ask at the interview. Take this list into the interview with you and make sure you ask them and push for satisfactory answers.
- Be very clear what salary you want. You may also be asked to negotiate salary during the interview.
- Use the waiting time to decide if you will say "yes" if the panel offer you the job.
- If you are the strongest candidate, you could try negotiating for more time to think; don't be surprised, though, if they say "no".
- Finally, you could consider saying "yes" at the time, while resolving to actually make up your mind later. This may make you uneasy, but it is, however, a pragmatic route that many candidates have taken.

Assessment centres

A minority of organizations approaches recruitment and selection with solemn and thorough professionalism. Whilst this does not guarantee that they will be enjoyable places to work, it is a good sign. These organizations may devote one, two or more days to the selection process. They will invite you to something known as an Assessment Centre. This may happen on the organization's premises or in a hotel.

The main thing to note is that everything you do is being observed. The techniques used will include:

■ Timed ability tests – for instance of verbal reasoning and numeracy. Take these steadily and quietly and read the instructions very carefully. It is possible to buy sample sets at large bookshops of similar tests and to practise at home before doing the Assessment Centre. Research shows that it is possible to make significant improvements to your score through coaching for many of these tests. If you suspect that this may be a weakness, it will be worth the small investment involved and will also build your confidence.

■ Timed aptitude tests – for instance whether you are capable of making fine visual discriminations; whether you can write a decent précis of a long document.

■ Psychometric tests (or personality tests) which aim to give a snapshot picture of the kind of person you are. There is no point in trying to "cheat" by answering in the way you would like to be. Be honest. All reputable test have honesty detectors built into them. Also, the better the test, the harder it is to guess what factors of personality it is actually measuring. It is usually best to answer questions quickly without pausing too much over them.

■ In-tray exercises – usually confined to selection for managerial posts. Here you are given a set of documents which look like the contents of a busy manager's in-tray and asked to say how you would prioritize or deal with the various memos and letters.

What to look out for

As a candidate you should expect the following as good ethical practice in an assessment centre where tests are used:

■ A test administrator who clearly explains the purpose of every test. It should never be a mystery.

■ Time allotted to reading and understanding the instructions – for instance, if a test is timed, whether there will be reminders of how the time is passing.

■ Clarity about what will happen to your test results – for instance who will see them and how long they will be stored. The Data Protection Act makes strict regulation essential here.

■ Where the tests fit into the selection process. Beware of employers who select in or out on the basis of tests alone. You should expect to have any issues raised by the tests discussed with you at the final interview you attend.

■ Psychometric tests usually have no time limit.

■ The tests should be administered and interpreted by licensed practitioners. All leading test publishers make it difficult and costly for people to buy their tests and insist on rigorous training. This is to deter people from thinking of them as being just like a quiz in a magazine, and because the tests themselves have been developed out of many years of research. Furthermore, the most useful tests so far devised are subtle and need both intelligence and insight to interpret properly.

■ It is poor ethical practice to use tests used for one purpose as part of a selection process on a different occasion.

■ Provision for feedback. Giving individual feedback to candidates costs time and therefore money, especially if external consultants are being used. However, you should always be offered the chance to see reports and to discuss them with whoever is briefing the panel.

Assessment centres

The most common psychometric tests

■ The OPQ stands for Occupational Personality Questionnaire and is the dominant test of this kind used in the UK.

■ The 16PF stands for Sixteen Personality Factors

■ The CPI stands for California Personality Inventory.

There is also a test which is sometimes used called the Strong Interest Inventory, which links vocational interests to personality.

The aim of the tests

All ask you to fill in a questionnaire where you are asked to choose between two or more equally attractive types of answer. Essentially they are giving you many chances to answer the same question.

The great tests are more subtle than most of us will expect, especially when we first meet them. The thinking behind all of them is that the way we describe our behaviour does fall into patterns and that these patterns give clues to our underlying personality. 'Clues' is the operative word. Not even a psychometrician would claim that a questionnaire could uncover the entire complexity of human personality, though they would claim that it could give strong hints.

Tips on filling in psychometric tests

■ Don't take too long. Your first answer is usually the best. These tests are not timed, but a candidate who takes double the average length of time may raise questions in the employer's mind about the speed at which they would normally work.

■ Look out for the social desirability questions. These are the ones which many tests include to check on how truthful you are being about yourself. For instance, there may be a question about white lies. You may think (correctly) that employers will not want to employ someone who is untruthful. However, everyone tells little white lies and to claim that you never do will raise doubts about whether you have filled in the rest of the test honestly.

■ Be honest. The best tests are very subtle in their construction and it is difficult to tell what factors they are assessing. Assume that there are no right answers and be yourself. Never try to second-guess what sort of personality type the employer may be looking for. The chances are you will be wrong, and even if you are right, you will not do yourself any favours by filling in the tests in a distorted way. You will only raise suspicions in the mind of the person who will be interpreting your results.

Group tasks

Here are some examples of group tasks set in Assessment Centres:

A. The group is given three large sheets of coloured card, red, yellow and blue, two pairs of scissors, a ruler and some glue. The task is to make a toy which can be played with by the group at the end of the process. No other materials are allowed and the group has to complete its task in 40 minutes.

The underlying thinking here is:
This is a task where the resources are strictly limited and where co-operative behaviour is essential to complete the task. It is also a task where there needs to be energy and creativity. It will help is there is some leadership. So this task is not unlike the sort of thing that many people have to do, i.e.produce something with limited resources in a short space of time with other people.

What is being tested:
Your resourcefulness, creativity and your ability to co-operate with others.

B. The group is given a batch of papers which represent the papers being brought to one of the organisation's committees. They are told that they are a working party meeting just this once to make a recommendation to the main committee on one of the topics in the papers. Time is allowed to read the papers – usually about half an hour.

Assessment centres

The underlying thinking here:
The ability to absorb complex information in a short time is essential in many organisations, especially those in the public sector which have to take keen notice of accountability to the public. Most issues on which such committees ponder are complex and cannot be reduced to simple yes/no answers. It is also essential to be able to work with people whose views may be dramatically different from your own, however irritating it may seem.

What is being tested:
Your patience, resilience, willingness to see an issue from many different sides, and the ability to press for a resolution.

C. Each person in the group is given a brief which represents a different point of view. The setting is often a contentious issue which needs to be resolved in a fictional organisation. For instance, it may be that the setting is a merger between two organisations and the issue is what colour should the vans of the new company be painted. You are not required to do any role-play in this exercise, only to represent the point of view on your brief.

The underlying thinking here:
Most organisations have issues where it is difficult to reconcile strongly held views. This type of meeting simulates meetings where these issues need to be thrashed out.

What is being tested:
Your ability to negotiate, willingness to listen to others, ability to put forward a view without alienating others, persuasiveness and assertiveness.

Qualities

Usually, the person running the Centre will tell you what qualities or behaviour they are looking for. In any case this is no mystery – the qualities will relate directly to the skills that the employer has already told you (through the pre-interview paperwork) are needed.

Normally, the group task looks at how you interact with other people – are you bossy or helpful? Do you say too much or too little? Do you take the lead or wait for others? Do you listen?

Group tasks are designed on the principle that even in the stress of the assessment process, most of us will revert to type sooner or later, especially if our interest is gripped, and as nerves wear off, our true personality begins to emerge. My experience as a consultant certainly suggests that this is true.

As observer to this part of the selection process I have seen these examples among many others:

- The strongest candidate (on paper) showing undisguised irritation with other members of the group because of what he saw as their slowness to understand his point of view. The actual job involved working closely with lay committees, so patience and ability to negotiate were paramount for the successful candidate. Another candidate was appointed on this occasion.

- Another strong candidate tried through a conscious use of silky charm to bring the group round to his way of thinking. When, however, that failed because they all ignored him, he turned to the observers as if to say, "What fools they all are", and, from that point on, took no further part in the discussion, leaning well back in his chair and folding his arms. He was not appointed.

- People who have claimed to understand how to chair groups and run meetings sitting dumbly and taking very little part in the process.

- An apparently weak candidate storming past the others through the sheer skill she showed in the group discussion by subtle chairing, charm, persuasiveness and persistence. Her strong performance in the discussion forced the panel to look again at her CV and after an equally impressive interview she was offered the job.

- Two candidates having a head to head, violently conducted debate which excluded everyone else in the discussion – and no one else in the group attempting to remedy the situation. In this case, the panel did not appoint.

The striking thing about all these examples is that they were all for very senior jobs where skilled behaviour in groups was a pre-requisite of the posts, and all the candidates had claimed considerable experience in this area. Yet, alas, in four out of the five cases I quote here, the opposite seemed to be true on the evidence of the discussion.

The main point to remember here is that Assessment Centres are based on the idea that given long enough, or put under enough stress, most of us will give significant clues about our real selves. The aim is always to collect more data about you than is possible in an interview alone. So remember that everything you do is under observation. To do well in a Centre you need to cultivate the idea of the inner observer who will ask you: "Is this how I really want to come across here?"

Behaviour that helps at Assessment Centres

Learn the names of the other candidates during the day and make a point of using them during the discussion.

This is a basic courtesy and shows social skill. It will also make the other candidates listen to you – we all respond to hearing our names used.

Suggest in the first few minutes that the group spends a little time planning how you are going to have the discussion – for instance, should someone assume the formal role of facilitator/chair? How much time do you want to allocate to each aspect, should there be a period of brainstorming first?

Most groups of this sort plunge straight into the topic, without a thought for the planning aspect. A candidate who shows this kind of concern will always impress.

Enter into the discussion enthusiastically.

Enthusiasm is one of the main qualities that employers like and want; hanging back from taking part may look like shyness or diffidence.

However off the wall the other candidates' opinions seem to be, remain calm and patient.

It's only an exercise. Agitation that turns to irritation never helps in a negotiation.

Listen carefully to what the other candidates seem to be saying. Jot down main points on paper – briefly, just single words.

The majority of candidates do not do this – they quickly get into the ritual of queuing to speak – to get their own point across, regardless of what other people are saying.

Having listened, make the occasional brief summary of what other people have said, then put your views forward.

This is critically important behaviour. First it shows that you have listened – you can't summarize properly unless you have. Secondly, it reassures the other members of the group that someone is keeping track of things. Research shows consistently that good negotiators listen about twice as often as they speak.

Explain the thinking behind your views, for instance by stating your assumptions explain how you got to a particular opinion.

Making your thinking visible in this way is rare. It is virtually always persuasive and shows both assertiveness and modesty.

Remind the group of how the time is going, by saying, for example: "We've been discussing this for ten minutes and we only have twenty minutes to get to a conclusion."

This is always a helpful behaviour and again, shows you are taking responsibility for the group.

Identify common ground in the discussion – for instance: "Although we disagree on x, we all seem to agree on y, should we concentrate on that now?"

Many candidates mistakenly think that what is being assessed is their ability to trounce others by talking them down and dominating the group. This is rarely what is being assessed. On the contrary, the panel is normally looking for someone with excellent interpersonal skills who can find a path through the thickets of misunderstanding that can occur in any discussion or meeting. Identifying and drawing attention to common ground is a key behaviour here.

Make "process" observations. In other words, draw attention to the process in which the group is involved rather than just its task. For instance, say things like: "I notice that every time we touch on topic x, we all start talking across each other…"

The majority of candidates in this kind of exercise get completely absorbed in the task and forget the process, yet neglecting the process is one of the main ways in which this kind of discussion goes wrong in the "real" world. Observers seeing a candidate who is able to make process observations are normally impressed.

As you get near the end of the discussion, summarize all the main points that have been made and invite the group to draw conclusions.

This shows concern for closure – a useful leadership behaviour.

Giving a presentation

Another common technique is to ask you to make an individual presentation, sometimes to a wider group than the normal panel; for instance, when senior local authority jobs are being filled, councillors are sometimes invited to be present for this part of the selection process. The topic is normally what your strategy would be if you got the job. For example, if the job is a senior one to head up a unit, then you may be asked to outline how you would deal with staff, the unit's clients, its systems, policies, training and so on.

What is being tested here is your self-presentation:

Do you have a big presence or are you timid?
Can you hold the interest of an audience?
Are you articulate?
Do you know how to explain complex issues simply and vividly?
For any job where persuasiveness is one of the competencies, a presentation can be a useful pointer to your level of skill. Practice, coaching and feedback can improve your performance enormously.

You will normally be given advance notice of the topic and will be able to prepare the content in advance. Sometimes, you will only be told the topic on the day and will be given a limited amount of time for preparation. It would be very unusual to be given no notice and no time, though that, too, can happen occasionally.

Preparing the content

Think first about your audience. What can you assume they already know? What are their likely interests and concerns? If you were sitting in their place, what would you be interested and impressed to hear?

Now think about the points you want to make. You may be given anything from 5 to 15 minutes to make your presentation. However much time you have, bear in mind the three-point rule. This is the one that says that the human mind cannot retain any more than three points from a verbal presentation. This is a sensible rule of thumb. These three points should be the spine of what you are going to say.

Early on in your presentation, build in a reminder of your authority and credibility. For instance, if what you are going to talk about is your plan for what could (if you got the job) become your department, remind the audience of your experience and credentials in your field.

It is also useful to build in a clear verbal structure to what you are going to say. This helps you to make yourself clear, and it also helps the audience to see where you are going.

So, for instance, you could say something like: *"I'm going to talk about three things here, x, y and z topics. First, x…"* (then you talk about x).

"Now I'm going to go on to y topic…" (then you talk about y).

"And finally, I'm going to say a little about z…" (then you talk about z).

Then you summarize: *"So in this presentation I've talked about three things, x, y and z and my conclusion is that…"* (you give you conclusion).

This follows the advice of "first you tell them what you're going to tell them, then you tell them, then you tell them what you've told them."

Your first few words

It's no accident that television and radio producers devote so much time, skill and energy to working out how the first few seconds of a programme should have maximum impact. In fact, pro rata, some very simple TV programmes have more money and time spent on their opening titles than go into the programme itself. This is because people decide within a few seconds whether or not to hit that remote control button and change channels. Your audience can't do that, but they can switch off unless you grab their attention straight away.

This kind of presentation is probably not the place for a joke, unless you can tell it very well.

Some good ideas for these opening moments include:

A personal anecdote (the safest).
A startling statistic.
A quotation.
A challenging statement.

It is worth learning your opening sentence – this can carry you over the dreaded moment of actually starting to speak. It's never a good idea to read or try to learn the rest of your presentation. It inevitably comes across as stilted and conveys a lack of confidence. A presentation is not a lecture or a sermon.

If you want the confidence of writing out the whole thing first, do so, especially to time yourself, but then reduce it to a few index cards if you want the assurance of not drying up. Cards are better than large pieces of paper because they are less obtrusive. Choose large cards and write no more than few key words on each in big letters that can be read at a glance. Highlight them as well, if you think that will help.

One additional tip: punch a hole in the corner of each card then run a thread, split ring or treasury tag through them so that they stay in the right order as you turn them over. Nervousness may cause you to drop the cards – this way, they won't scatter everywhere.

Giving a presentation

As an alternative, if you know that there will be an overhead projector available and you are confident in its use, prepare overhead transparencies. These do away with the need for additional notes and also give your audience something to look at other than you. This can be useful as the thing many presenters most dread is the feeling of so many pairs of eyes gazing at them.

With OHPs, keep the content to no more than five lines in large, bold, lower-case type – for instance 20 point. Lower case is much easier to read than capitals. Even more professionally, if you are familiar with Powerpoint, and the equipment is available, you can use that, but you need to be certain that there will be no technical hitches.

Some candidates have impressed panels and appointing committees by bringing along handouts of the main points of their presentations. This is probably a good idea, though it will be worth checking out with the panel chair first. If you do decide to take this path, double check for typos, and take care with design and layout. It is unlikely that the panel will want to read anything too long – a page or two at most is likely to be the limit of their interest.

Checklist on content	Yes	Your notes
1. Have you identified what your audience's needs are?		
2. What three points are you going to make to them?		
3. How will you open your presentation?		
4. Have you reduced the content to index cards?		
5. If you want to use an OHP or Powerpoint, is the equipment available?		

Preparing yourself

However wonderful the content, you will impress even more if you convey confidence and authority in how you speak and somehow communicate that you are in touch with the audience and care how they react to you.

Long after we have forgotten the content of what even the most informative speaker has said, we may be able to remember how he or she talked.

Conveying confidence and authority

This is done in two ways: how you use your body and how you use your voice.

Always stand; never ever try to do a presentation sitting down. Standing when your audience is sitting immediately gives you a height advantage. Stand with your body in a graceful straight line with your feet planted slightly apart. Keep your shoulders down – this conveys relaxation. Don't turn one knee out or lean away with your shoulder or cock your head to one side. These (and other variants of the same thing) convey uncertainty.

Keep your hands lightly together, holding your cards, or if you can do it without notes, just leave your arms dangling loosely at your side. Don't have your hands in your pockets or clasped in front of you, nor in a "steepling" shape which may look as if you are praying for divine help.

Another good tip is to avoid waving your arms about too much or keeping them too rigidly at your sides. Either of these extremes again conveys lack of authority.

You should mentally stake out your space and occupy it while giving the presentation. This means that you should take a few steps back, forward and sideways while you are speaking. If you remain rooted to the spot, you may unwittingly convey the "tethered elephant" look, and seem as if you lack confidence. If you overdo it (unlikely in the stress of the presentation) and roam about too much, you will look like a caged lion and become distracting to your audience.

People who have fine tuned the art of this technique use the moment of moving to emphasize a point – in other words they capture the audience's attention through movement and then give their punchlines.

Your voice is very important. Check if everyone can hear you before you get going. Keep your volume up and vary the tone. Remember that it is fine to use pauses and emphasis. There is special value in pausing for two or three seconds before you start the presentation – that is at the moment when you are sure that you have everyone's attention. This shows confidence and also gives you a moment to collect your thoughts.

Keeping in touch with the audience

It can feel scary or exhilarating to be the focus of so many eyes. However, the most important point here is to remember to use the so-called lighthouse effect. This means constantly sweeping the group with eye contact. Look out for the way politicians do this to very large audiences. They slowly rake the room, giving people the impression that they are personally communicating with each and every person there. It's much easier, of course, with a smaller group. There, you really can make brief (a second or so) eye contact with each person before moving steadily on to the person sitting next to them and then back the other way.

The reason for this is that without eye contact, we lose interest in what someone is saying. Also, it gives you immediate feedback. Is the audience looking interested, bored, fidgety, entranced? Are they really listening to you? It's only by keeping the eye contact going that you will know the answer, enabling you to modify the length and content of your presentation as you go.

This simple rule of communication is the one I see broken most often. For instance, I see people whose content and voice is excellent, but who:

- Address only one side of the room.
- Persistently miss out the people sitting at the extreme edge of the group.
- Talk to the ceiling.
- Talk to the flip chart, projector or their notes.
- Talk to the tops of people's chairs.
- Talk to the spaces between people's chairs.

The other simple rule to remember is to SMILE. If you look too serious it will be off-putting. Smiling conveys that you are comfortable with what you want to say and hope that others are also enjoying it.

Warm-ups

Giving a presentation is a kind of audition. The part you are acting is yourself, and the role you are auditioning for is Best Candidate. No actor would dream of going on stage without some kind of physical warm up. The suggestions which follow come from acting training – and they do really work.

Repeat the breathing exercise described on p30. If your breathing is shallow, it will be impossible to project your voice. Correct breathing also helps you feel relaxed.

Now shake your arms for a few moments, then do the same with each leg. This warms up the muscles and helps lose tension. Flop your body downwards so that your arms just dangle loosely in front of you. Now slowly and gracefully stand up sweeping your arms above your head and around to the side in an arc, ending with your arms at your side again. Do this four or five times.

Finally, don't be afraid to show some passion. I once coached a manager who was concerned about his presentation style. When talking about relationships with suppliers in a practice presentation he was satisfactory, but his heart did not seem to be in it and he was less persuasive than he could have been. I asked him to talk to me instead about his local football team, a subject I already knew he really cared about. The difference was startling. He came alive, was funny, compelling and fluent. The challenge for him was to talk about his suppliers with this same enthusiasm.

Without passion and its close companions, sincerity and authenticity, it is hard to persuade people to listen to you. Show passion through speaking eagerly. Don't be afraid to let it show.

You are going to be doing a lot of communicating with your face, so it is important to get the facial muscles warmed up too. Three exercises help here:

- **Chewing sticky toffee: imagine you have a mouth full of really sticky toffee that you are chewing. This gets the muscles of your mouth warmed up. Do this for about a minute.**

- **Sticking your tongue vigorously in and out, in and out for about a minute. Feeling tongue tied is a handicap that many presenters bemoan. This exercise helps get around the problem by getting the tongue used to working.**

- **Massaging your face with your hands. Work around the face with both hands, massaging gently so that the whole face feels loosened up.**

'Trial by orange juice'

The Assessment Centre will often offer you the chance to meet a group of staff, usually over a buffet lunch. If it is a very senior appointment, there may be members of the board present. In making appointments in the voluntary sector, for instance, it would be common for trustees to attend at this point. There may also be more junior people present.

The purpose of the lunch is three-fold. One is to assess you and how you behave with your potential colleagues. Another is to help you decide whether or not you want to belong to the company and the third is to give you the chance to inform yourself about the organisation and its issues. It may be a very useful source of information for any points which you want to raise in your panel interview.

What to do

Remember that you are under close observation. The people who attend the lunch will be asked informally for their views of you, even if they do not have a vote in making the final selection.

■ Circulate – make sure you talk to everyone, this is not the time for social shyness. Smile, introduce yourself and ask what the other person's role is in the organisation.

■ Be prepared to give a mini-biography about yourself. People will be curious about you and how you have got to where you are and what draws you to the job. Don't go on too long – a few minutes is enough.

■ Ask questions about what the issues are as the other person sees them. Listen carefully and show that you are listening by skilful summarizing. Beware of appearing to interrogate – keep it light, it's a social occasion as well as an assessment. Don't exclude other candidates, who if they have any sense, will be doing the same.

■ Share the conversation.

■ If there are junior staff present, treat these staff graciously. They are a valuable source of information about the job and the organization but remember they will also be reporting back on you. This will apply even if they would be more junior than you if you got the job.

A better experience

A well-designed assessment centre can actually be an exhilarating and enjoyable process. It is taxing, certainly, but its very thoroughness is reassuring. It feels much fairer than a stand-alone panel interview because you have had more chance to show what you can do. It puts internal and external candidates on much more of an equal footing and is much more likely to identify the strongest candidate.

Altogether, assessment centres have one purpose and one purpose only: to increase the amount of data available to the employer about each candidate. After a properly run centre, the employer is much better informed about:

- Your intellectual capacity.
- Your aptitude for the job.
- Your personality.
- How you behave with other people.
- How you behave under stress.
- How you behave when more relaxed.
- How far what you have claimed about yourself is borne out by what you have done during the Centre.
- Whether you really want the job.

Note: if you encounter an assessment centre you should expect the following as good ethical practice:

- The purpose of every activity clearly explained.
- Clear instructions.
- The offer of proper feedback on all tests.
- Being given your test results to take away.
- No selection decision ever being based on the result of any one test.
- Due regard being paid to the way nervousness which could affect your performance.

The practice interview

You will enormously increase your chances of getting the job if you have a practice interview. This is because the practice is like a good dress-rehearsal for a play – it's the place to spot the flaws and sort them out before the performance. It gets you fluent in talking about the job and your approach to it. It helps deal with nerves because you will have the confidence that practice gives. Also, if you choose your practice – interviewer carefully, you will receive invaluable feedback on how you come across.

The ideal practice-interviewer is a sympathetic mentor – someone perhaps a little older or more experienced who is friendly, patient, cares about your career, but is not too emotionally involved with you.

This will make it easier for you to hear his or her advice. The practice interviewer does not need to know anything about the job. If you can't identify someone like this, then a friend, colleague, parent or partner can also do a good job. It is better to have even a rough-and-ready practice than none.

How to do the practice interview

- Set aside at least an hour where you can guarantee not to be interrupted.
- Brief your interviewer about the job, giving them any paperwork about it.
- Give them a copy of your application.
- Suggest some general areas for questions without giving them an actual script.
- Run the practice as if it was real – never mind if it seems embarrassing or silly, just press on.
- Ask your interviewer to take notes on how you come across.
- Devote at least twenty minutes to discussing their feedback on how you did.
- Leave time to have a second attempt at any wobbly areas revealed by the practice.

When you have run the interview, give your interviewer this check list and go through it together.

Did I	Brilliantly	OK	Area for attention
Smile?			
Convey confidence?			
Sit up, looking relaxed and alert throughout the interview?			
Remain courteous and friendly, even when probed or put under pressure?			
Keep replies to 2–3 minutes?			
Speak clearly and audibly?			
Keep my language simple and direct?			
Use direct experience to illustrate my answers?			
Maintain appropriate eye contact?			
Show enthusiasm for the tasks the job would involve?			
Show that I understand the organisation and its issues?			

Inevitably there will be some weak areas and some strong ones. If you and your helper have the time, it will pay handsomely to re-visit the weaker areas. One rehearsal is probably enough. More than one may make you seem a little over-rehearsed, and also raises the danger that you will get too fixed in your answers, so when the real interviewers ask their questions it takes you by surprise that they are not the same as the ones you practised.

4

If you don't get the job
Negotiating terms
Starting the new job

Getting feedback

Do I really want the job?

Understanding change

If you don't get the job

Most employers understand how anxious candidates are to hear the result of the interview. It is good practice to let everyone know the outcome the same day, or at the very latest the next day. This should include an explanation for any delays that may be occurring. When this doesn't happen and you are still left in limbo, you can make any of the following assumptions:

- The panel can't agree and will have to meet again.
- The panel are not enthusiastic about any of the candidates they have seen and are going to return to their original list of applicants without closing down their options on their existing shortlist.
- Another candidate has been offered the job but is holding out for better terms.
- The person delegated to tell everyone is too busy to make this a priority.
- This is a boorish group of people and can't be bothered to let you know that you were not successful.

The more time goes on without contact, the less likely it is that you are the successful candidate. However, if you still haven't heard after three days, you should ring the panel chair and ask what is happening, expressing your continuing interest in the post. Most probably you will be told that you have not got the job. If so, follow the advice in the next section.

Ask for feedback

A conscientious employer will offer this routinely, but mostly this is a duty that interviewers hate. Nobody enjoys giving bad news and many interviewers have never had any training in how to give feedback. It's just as hard to hear bad news, so it takes courage to ask, but ask you must. This is particularly important if you are getting plenty of interviews but are not getting the jobs as there may be something about your interview technique which is getting in the way.

How to do it

- By telephone – this is not a process which can be done by letter.
- Say how sorry you were not to get the job but don't make a big deal of it – the employer does not want to be made to feel guilty.
- Say: *"I'm interested in your feedback on the quality of my application. What, specifically would help me be a stronger candidate next time."* (Or, *"Is there some further experience I should specifically get?"*).
- Now ask about the interview itself. Say something like: *"Could you give me some feedback on how I came across at the interview?"* or, *"What do you think would help me be more effective another time?"*
- If you get evasive replies, such as: *"Well, you were a bit quiet"*, press the point by saying: *"Could you give me an example?"*
- Where you get vague replies, press for clarification. For instance, if the reply is something like: *"Well you seemed rather lost at that point in the interview"*, then say: *"Could you tell me what it was about what I said or did that made me seem lost?"*

When you have heard the points people make, take the time to summarize them. This has two purposes. First, it ensures that you really have heard the points that have been made. Secondly, it makes a good impression on the other person because it shows that you have got the courage to say out loud whatever critical things have been said about you or your record.

- Don't under any circumstances get defensive. Don't answer back or argue. Most especially, don't give excuses such as: *"Well I had a cold that day"*, or *"My husband was in hospital so I couldn't concentrate."* However genuine the reasons for what you feel was a poorer performance than normal, the interviewer will not want to hear it, and it will sound like an excuse. Listen carefully to make sure you've understood the feedback.
- Thank the other person enthusiastically for their time.
- Remember feedback is just feedback. It's not an instruction to change, though most feedback, even when it is clumsily given, has some truth in it from which we can learn something useful.

An exceptionally well-qualified senior manager was persistently failing at interviews. He asked me for help. My first suggestion was to encourage him to ask for feedback from panels he had recently attended. In spite of meeting considerable embarrassment, he persisted, using the techniques described on this page. The message was clear. To his utter amazement, he was told that he was conveying "arrogance". Working on his interview technique so that he conveyed more humility, ensured that he did get the next job that came up.

The follow-up letter

If you do get the job

You should still ask for feedback, and if appropriate, you should still write a thank-you letter to the panel chair.

A neglected tool

A few years ago I was chairing a panel to appoint an occupational psychologist. There were two strong candidates and one was duly offered, and accepted, the job. The other wrote a very charming letter of thanks for the opportunity to attend the interview, saying how much he had enjoyed it. As a result, although he hadn't got the job, he did get two lucrative assignments as a freelance consultant!

The follow-up letter is a severely neglected tool by virtually all job-seekers. In my experience, I can only think of a handful who bothered to do it. In every case it earned them a useful advantage.

Some examples:

- One got offered the next vacancy without having to go through another interview.
- One got offered a temporary stint as a trainee.
- One was recommended to another employer.
- One was invited back for some free career counselling.

Your ego may feel bruised because you didn't get the job. You may feel the last thing you want is any further contact with the panel. You may feel you gave a poor account of yourself.

No matter, write that letter:

- It's a further chance to demonstrate initiative.
- It gives you another shot at reminding the employer who you are.
- It's courteous and shows a proper concern for the people-side of life.

In the letter, the format should be:

Opening paragraph thanks them for the chance to be interviewed and says you were naturally disappointed not to be offered the job.

Second paragraph emphasizes your experience and maybe has something on any misleading impression you think you may have given during the interview itself.

Third paragraph conveys your continuing interest in the organization and asks them to bear you in mind for future vacancies.

Negotiating Terms

However brilliantly things appear to be going, never negotiate salary during the interview:

- You are in a weak position because you have not been offered the job.
- The tension and excitement of the interview is not the right frame of mind in which to negotiate.
- Entering a negotiation makes it seem as if you are agreeing that you want the job when actually it is too soon to say – you need time to consider such an important decision calmly.

The time to negotiate is when you have been made a definite offer and when you are clear that you want to say yes. It is unlikely that either of these conditions will be satisfied at the interview itself.

Fending off salary negotiations during the interview

A trained interviewer will not press you about you salary expectations, but an untrained one may. His or her questions may include:

- What salary are you looking for?
- What if we can't match what you're earning now?
- Would you expect a car with this job?

Your reply should be a courteous, friendly one along the lines of: *"If we both agree that I'm the best person for this job, I'd rather talk about salary then."* If the interviewer insists on a reply, then you need to have an answer prepared (see later). However, again, a safer answer is to say: *"I'm looking for something in the range of £X,000–£Y,000, depending on how the package is made up."*

Just as the interview itself is too early for the negotiation, it is also possible to leave the negotiation too late. Too late is when you have actually formally accepted the job. Much too late is when you have started the job. By then you have lost all your bargaining power.

Do I really want the job?

A surprising number of people experience a panic of indecision at this point. They do a successful interview and are offered the job. But now that the flirtation stage looks like reaching consummation, they are not so sure. Do I really want to uproot my family and go to live in Bristol? It's an interesting job all right, but the salary is less than I'd hoped for. I'm not sure I want to identify myself with that organization – and so on.

The agonizing uncertainty may be made a lot worse if your old employer, getting wind of the offer, suddenly offers you an open cheque book to stay. All of this may be made worse again if the process of going through the interview has raised doubts about how attractive the job really is.

It is useful to do a prioritizing exercise at this point, to sort out in your own mind what you really want from work.

Step 1

Fill in the dark blue boxes down the right-hand side of the chart first by answering these questions: what do you most enjoy about your present job or, if you prefer, what is most important to you in any job? Examples might be: "the chance to travel"; "the opportunity to organise my own work"; "direct contact with x or y kind of person" and so on. Just write down your answers in any order.

Step 2

In the grids on the left, compare item 1 with item 2, circling whichever is most important to you, then continue down the grid, comparing item 1 with item 3, item 2 with item 3 and so on until you have got to the end of the grid.

Step 3

Fill in the table below, looking back on your grid to see how many times you ringed each item. On the final line, write the order in which you have ranked each item.

Items for prioritising in any order

```
1

1   2
  2

1   2   3
  3     3

1   2   3   4
  4     4     4

1   2   3   4   5
  5     5     5     5

1   2   3   4   5   6
  6     6     6     6     6

1   2   3   4   5   6   7
  7     7     7     7     7     7

1   2   3   4   5   6   7   8
  8     8     8     8     8     8     8

1   2   3   4   5   6   7   8   9
  9     9     9     9     9     9     9     9

1    2    3    4    5    6    7    8    9    10
  10    10    10    10    10    10    10    10    10
```

1	2	3	4	5	6	7	8	9	10	Item number
										How many times did you ring that item?
										Final ranking

Do I really want the job?

Step 4.

Now write out your list of priorities in the final order. This should give you some robust guidance on what you want and need in any new job.

Step 5

Now stack the potential new job against the old one to provide a comparative list. Give a tick for each point that either job earns.

Your priorities	Old job	New job
1		
2		
3		
4		
5		
6		
7		
8		
9		
10		
	Total	

If you are still undecided after this exercise, ask yourself what your reservations are, and write them down or talk them through with a friend or partner. Typically, the questions still nagging away at this point are to do with unknowns in the new job. How much autonomy will I have? Do I really like this boss? Can I really do this job? Do I really want all the upheaval of the change?

Now ask yourself these questions

- On a scale of 1 to 10, how much do these reservations matter?
- What further information could help resolve them?
- Where can I obtain this information?

If your decision is "no", then let the employer know immediately so that they can re-assemble the panel and consider what to do. There may be another good candidate who is still waiting to hear. If it's "yes" then get into negotiating the salary immediately.

Preparing for the negotiation

Where the job has been advertised, a salary or salary range may have been quoted. This may still leave room for flexibility on either side, though there may be less than you think. For instance, if you are young and inexperienced in the employer's eye, he or she may try to negotiate downwards. If the employer knows that you are already earning above the ceiling quoted, he or she may expect to pay a little over the odds. In hierarchical or highly bureaucratic organizations there may be rigid rules, particularly about entry-point jobs.

The best deal is going to be the one which leaves both sides feeling satisfied – neither side feeling sore, resentful or triumphant. You have to work together and you do not want the relationship tainted from the start.

You will need to make a realistic assessment of the power balance on both sides. If your offer came quickly and the employer is clearly eager to take you on, you can probably safely assume that you have considerable bargaining power. If the reverse is true – the offer came perhaps because another candidate dropped out, or if there is a perfectly respectable internal candidate waiting in the wings, then you are on thinner ground and may have to be more modest in what you ask for.

Do I really want the job?

Your own research

First, be clear what your salary needs are. How much do you need after tax to meet your obligations and your other needs? It is worth doing some careful sums here – the total may be a lot less or a lot more than you think.

Next, find out what the going rate for the job is. What did your predecessor, if there was one, earn? What do people in similar jobs in this or other organizations earn? Look in the job ads in the papers to check out your impressions if necessary.

Finally, think carefully about the value of so-called "fringe" benefits. Many are not such a "fringe" and may be worth more than you think; for instance, some employers make generous contributions to pension schemes or will give interest-free loans for cars and season tickets. Some benefits are worth less than they seem if you have to pay tax on them. A company car can be a dubious benefit, depending on salary and likely mileage. It may pay you to use your own car instead on company business if the company has a generous mileage rate.

A fixed-term contract may pay at a higher basic rate, but if you have to fund your own pension, this may not be quite such an attractive proposition without extra salary to compensate.

It helps during the actual negotiation to find out – perhaps by asking directly – how much room for manoeuvre there is. Some employers honestly cannot go above a stated ceiling; others may have a wide range in mind.

Many employers are wary of hiring at the top of the range, especially if they feel you are still a relatively unknown quantity. Where this is the case, it is often possible to negotiate a salary review say, six months into the job, where if you are as good as you claim, you and your boss will sit down and re-negotiate your salary.

This and other details needs to be built into your contract of employment – a legal entitlement. The contract need not be a 'legalistic' document – a letter is fine, but it does need to state:

- Salary.
- Notice period on either side.
- Pension position.
- Fringe benefits.
- Hours, even if there are no set hours.
- Holiday entitlement.
- Duties.
- How performance will be judged.

Starting the job

When you start your new job, there is usually a period of disillusionment on both sides. You may feel overwhelmed with the stress of learning a new job and a new organization and getting to know a lot of new people all at once. You may experience a sudden loss of self esteem as the difficulties of the work become apparent. You may feel lonely because there is no one who feels like a friend.

You may find out that there are all sorts of niggling little details which remind you of what you have lost by moving jobs – especially if you were made redundant from your previous job and you are now earning a smaller salary. You may, for instance catch yourself thinking sentimentally about apparently trivial things like the scruffy old canteen at your former place of work. It takes effort to get to know new colleagues and to find out what the job is really about when all the excitement of the selection process is over.

On the boss's side, there is often a parallel process of disappointment. The boss discovers that you are not the perfect candidate after all because you are only human. If you take advantage of the advice offered below, you can short circuit the process of disillusionment on both sides. Your task in the first few weeks is NOT to do the new job, it's to learn how to do it – a rather different emphasis. Remember that there was a process of mutual choosing which led to the offer and acceptance of a job, and that this probably means it is the right job for you. If it isn't, then it's not a life sentence and you can move on. Much more likely, you are going through a normal process of adjustment and will learn to do – and love – the job.

It is sensible in the first few weeks of a new job to

- Turn up early.
- Show eagerness, initiative and willingness to learn.
- Book a session with your boss to explore what he or she expects from you in the first month and the five months after that.
- Introduce yourself to team members and make sure you have a meal or a drink with some of them in the first week.
- Ask team members what the do's and don'ts are of working in the organization – for instance, what customs are there which everyone knows but which no one explains unless you ask? What time do people start and finish work?

Starting the job

- Make sure that if there are other people doing the same job as you, you talk to the outstanding performers (ask who they are, everyone will know) and ask them what advice they would give you about how to succeed within the organization.
- Avail yourself of any induction programme that may be offered by the organization.
- Ask people to explain any unfamiliar jargon and anything else you don't understand.
- Get to see the core business of the organization in action: for instance, if it makes widgets, visit the place where the widgets are made. It doesn't matter if your own job is not directly connected with widget-making; you still need to know everything you can about widgets because the most important people in the organization will probably have a passion for widgets and you need to know why.
- Make sure you limit your references to how you did things at your old organization: it irritates because it suggests that you think your old way was superior.

Have I made a mistake here?

When you still feel desperately at sea with a new job, even after availing yourself of all the ideas listed above, then of course it is possible that you really have made a mistake, and so has the employer. If so, you have a number of options.

One is to stick it out for a reasonable period of time and then to start the search for a new job. This probably means committing yourself to at least eighteen months in the job. Another option is just to wait and see – you may be wrong and with time, you could settle in to the job. The third option is to get out of the job as quickly as possible. A swift exit is probably the easiest to explain to any future employer. Of course, if you do this, it leaves you without a job, and taking this path will depend on how desperate you are to have a job. You may be able to freelance while looking for another job or it may even be possible to go back to the job you have just left.

Whatever you decide to do it is essential to talk it through carefully with your boss. Sometimes in all the stress of starting a new job it is possible to get things out of proportion.

Understanding personal change

Mostly, such drastic action is rarely needed and you will settle well into the new job. However, a change of job is a major life-shift and like all such changes, the psychological adjustment can take time. In one well-known list of stressors, changing jobs comes high up for potential to cause disturbance and misery, along with the death of a loved person, moving house and getting divorced. This is in spite of the fact that the change is one that you probably initiated and wanted. If the change was forced on you through the process of redundancy or dismissal then the potential for feeling disturbed and upset is clearly much higher.

Letting go

Even a wanted change involves some loss. It will help if you have been able to "mourn" the passing of the old job; for instance if you have had a proper leaving party with speeches and presents. This makes it clear to you and to everyone else that something has really ended, just as a funeral makes it clear that a life is really over. It's probably better to cut your ties with the old job, avoiding going back to the building, and only keeping contact with the people who genuinely became your friends.

Using the transition period wisely

It is useful to have a consciously planned period of transition between the old job and the new, even if it only a few days of hanging loose and doing nothing at home between ending the old job and starting the new. Even better may be to have a short holiday, if you have the time and money. However long or short this time is, use it to refresh, to think, to take stock – for instance of your skills and where they may need to be refreshed. See friends, read, walk, see a film – whatever delights you and helps you enjoy the feeling of making the journey from one way of earning a living to another.

Committing to the new

Similarly, it helps if your new boss has been thoughtful enough to arrange a welcome party for you, or some kind of semi-formal professional and social occasion which accelerates the process of bonding with your new colleagues.

It also helps to make a point of finding one or two confidants, people who can share the funny little ways of the organization with you and help you understand its culture.

Finally, this whole process of adjustment can be speeded up considerably by throwing yourself wholeheartedly into new projects which have your signature on them and which you can shape.

In any process of personal change, it can help to remember that the actual change and the psychological adjustment may be two separate things. You can end the old job but still feel attached to it. You can start the new job on the agreed date, but the real process of adaptation is a much slower process. You can hurry it up to some extent by using the ideas on this page, but some of it will just take time. Let this happen and allow for the natural process of adjustment to take place.

Index

Further reading

What Color is Your Parachute? A Practical Manual for Job-Hunters and Career Changers: Richard Nelson Bolles (Ten Speed Press, 1997).
An American book with comprehensive material on analysing your job skills, looking for a job, preparing your CV and having the interview. Updated in a new edition every year and widely available.

The Perfect Interview: Max Eggert (Random House, 1992).
Useful summary of how to prepare for, and win, at the interview.

The Five Minute Interview: Richard H Beatty (John Wiley, 1998).
Another useful book on interview preparation, especially the long list of potential questions that may be asked by interviewers.

Jobshift – How to Prosper in a Workplace Without Jobs: William Bridges (Addison Wesley, 1994).
A book that describes the general work environment and gives good advice on how to thrive in it rather than drown.

You and Co.; Be the Boss of Your Own Career: William Bridges (Nicholas Brealey Publishing, 1997).
A racily-written "manual" for people who want to find work rather than jobs – a very useful approach for people who may want to work for themselves eventually, but also useful as a way of identifying what you want from work and life generally.

The Image Factor: Eleri Sampson (Kogan Page, 1995).
Practical book on image and why it matters, with plenty of immediately applicable advice.

Personal Power: Philippa Davies (Piatkus, 1991).
An excellent guide to boosting your confidence and improving your image. Packed with practical advice.

The Which? *Guide to Changing Careers*: S Bennett (Consumers' Association, 1998).
Good advice on all aspects of career changes, including coping with redundancy, early retirement, job-searching generally and dealing with interviews. Also has a comprehensive list of useful addresses.

Printed and bound by Chorus-France